# Analytical Morphology:
## Theory, Applications
## and Protocols

# Analytical Morphology: Theory, Applications and Protocols

Soft Cover Edition, Eaton Publishing Co.
Hard Cover Edition, Birkhäuser Boston

Jiang Gu
Institute of Molecular Morphology
Mount Laurel, NJ, USA

QH
613
.A53x
1997

**Library of Congress Cataloging-in-Publication Data**
[CATALOGING IN PROGRESS]

ISBN 1-881299-03-1 (Soft Cover Edition, Eaton Publishing Co., 154 E.
Central Street, Natick, MA 01760, USA)

ISBN 0-8176-3957-8; and 3-7643-3957-8 (Hard Cover Edition, Birkhäuser
Boston, 675 Massachusetts Ave., Cambridge, MA 02139, USA)

Printed in the United States of America

9 8 7 6 5 4 3 2 1

The front cover shows part of a summary screen from the neural network-
assisted cervical screening program PAPNET. This smear represents the first
case of cervical carcinoma *in situ* diagnosed using the program (see M. Boon
et al. in this volume).

# Analytical Morphology: Theory, Applications and Protocols

## CONTENTS

# CONTRIBUTORS

**Neera Agrawal**
Deborah Research Institute, Browns Mills, NJ, USA

**Bo Ahrén**
Department of Medicine, Malmo Univ. Hospital
University of Lund, Lund, Sweden

**Virginia M. Anderson**
Department of Pathology
SUNY Health Science Center at Brooklyn
Brooklyn, NY, USA

**S. Beck**
Leiden Cytology and Pathology Laboratory
Leiden, The Netherlands

**Victoria Belogolovokin**
Department of Pathology
Maimonides Medical Center, Brooklyn, NY, USA

**Mathilde E. Boon**
Leiden Cytology and Pathology Laboratory
Leiden, The Netherlands

**Taiying Chen**
BioGenex Laboratories, San Ramon, CA, USA

**Tak-Shun Choi**
Department of Pathology
SUNY Health Science Center at Brooklyn
Brooklyn, NY, USA

**Richard J. Cote**
Department of Pathology
University of Southern California
School of Medicine, Los Angeles, CA, USA

**Richard J. Cote**
Department of Pathology
University of Southern California
School of Medicine, Los Angeles, CA, USA

**Gorm Danscher**
The Steno Institute, Department of Neurobiology
University of Aarhus, Aarhus, Denmark

**Ursula G. Falkmer**
Department of Oncology and Pathology
Karolinska Institute and Hospital
Stockholm, Sweden

**Nancy A. Gillett**
Sierra Biomedical, Sparks, NV, USA

**Jiang Gu**
Institute of Molecular Morphology
Mount Laurel, NJ, USA

**Gerhard W. Hacker**
Medical Research Coordination Center
University of Salzburg, Salzburg, Austria

**Irene Hale**
Neuroscience Research Institute and Department
of Biological Sciences
University of California, Santa Barbara, CA, USA

**Cornelia Hauser-Kronberger**
Institute of Pathological Anatomy
Salzburg General Hospital, Salzburg, Austria

**Krishan L. Kalra**
BioGenex Laboratories, San Ramon, CA, USA

**L.P. Kok**
Institute for Theoretical Physics
Groningen, The Netherlands

**Myrthe R. Kok**
Leiden Cytology and Pathology Laboratory
Leiden, The Netherlands

**Theresa Kramer**
Department of Ophthalmology and Pathology
University of Arizona School of Medicine
Tucson, AZ

**Charles T. Ladoulis**
Department of Pathology
Maimonides Medical Center, Brooklyn, NY, USA

**Anthony S.-Y. Leong**
Division of Tissue Pathology
Institute of Medical and Veterinary Science
Adelaide, South Australia

**F. Joel Leong**
Division of Tissue Pathology
Institute of Medical and Veterinary Science
Adelaide, South Australia

**Lucy H. Lu**
PE Applied Biosystems, Foster City, CA, USA

**Sunny Luke**
Department of Pathology
Maimonides Medical Center, Brooklyn, NY, USA

**Gloria E. Meredith**
Department of Anatomy
Royal College of Surgeons in Ireland
Dublin, Ireland

**Brian Matsumoto**
Neuroscience Research and Department of
Biological Sciences
University of California, Santa Barbara, CA, USA

**Hindrik Mulder**
Department of Physiology and Neuroscience
Section for Neuroendocrine Cell Biology
University of Lund, Lund, Sweden

**Shan-Rong Shi**
Department of Pathology
University of Southern California
School of Medicine, Los Angeles, CA, USA

**Steven E. Slap**
Energy Beam Sciences, Agawam, MA, USA

**Frank Sundler**
Department of Physiology and Neuroscience
Section for Neuroendocrine Cell Biology
University of Lund, Lund, Sweden

**Clive Taylor**
Department of Pathology
University of Southern California
School of Medicine, Los Angeles, CA, USA

**Jerry A. Varkey**
Department of Pathology
Maimonides Medical Center, Brooklyn, NY, USA

**Michael M. Whittlesey**
Energy Beam Sciences
Agawam, MA, USA

# Preface

This volume is a collection of chapters contributed by leading experts in the emerging field of analytical morphology. Its contents cover a wide range of techniques for morphologic research and diagnosis and it is intended for anyone who wants to keep abreast of the rapid development in this field.

Analytical morphology is a contemporary science dealing with the analysis of shape, size and color arrangement of cell and tissue components by means of a variety of analytical maneuvers. It differs from conventional morphology and histopathology in that it employs methods beyond routine hematoxylin and eosin or histochemical staining. To a great extent, the advancement of analytical morphology is based on new advances in other disciplines, such as immunology, molecular biology, laser technology, microwave technology and computer science. Using these new methods, particular cellular components that would otherwise remain invisible, such as peptides, proteins, or nucleotide sequences, are highlighted by visible markers through chemical, physical, immunological or molecular biological reactions. These methods include immunocytochemistry, *in situ* hybridization, *in situ* polymerase chain reaction, antigen retrieval, image analysis, and the like.

Analytical morphology is the foundation of contemporary medicine. It provides concrete and visible evidence for many conceptual deductions of other life science disciplines. A survey of leading medical and biological journals reveals that more than half of the scientific and clinical articles reporting original observations were achieved by morphological means in total or in part. More than ninety percent of the articles published in recent issues of two popular pathology journals presented results obtained through advanced morphologic methods, particularly immunocytochemistry and *in situ* hybridization. This volume offers an extensive overview and practical guidance for these topics.

The chapters originated from material published recently in *Cell Vision - Journal of Analytical Morphology* and have been updated and enhanced by the authors to meet with the purpose of this volume. They are written in a comprehensive manner and generally comprise a review of a particular topic, examples of applications, protocols, discussions and comments. The authors have made significant contributions to the development or modification of the techniques presented.

I am indebted to the contributing authors for their selfless effort in reviewing these subjects and sharing their personal experiences. I want to thank those who helped me coordinate this project, in particular David R. Kersten, Tak-Shun Choi, Neera Agrawal and Gayle Englund. I am grateful to Christine McAndrews, Managing Editor of *Cell Vision*, for her assistance.

Jiang Gu, MD, PhD

# Antigen Retrieval Technique: A Novel Approach to Immunohistochemistry on Routinely Processed Tissue Sections

## Shan-Rong Shi[1], Jiang Gu[2], Krishan L. Kalra[3], Taiying Chen[3], Richard J. Cote[1] and Clive R. Taylor[1]

[1]Department of Pathology, University of Southern California School of Medicine, Los Angeles, CA; [2]Institute of Molecular Morphology, Mount Laurel, NJ; [3]BioGenex Laboratories, San Ramon, CA, USA

## SUMMARY

*The antigen retrieval (AR) technique is a non-enzymatic antigen unmasking method that is utilized prior to immunohistochemical staining of formalin-fixed, paraffin-, celloidin- or plastic-embedded tissue sections. Two variations exist; the first, based on high-temperature heating, is applicable to formalin-fixed, paraffin-embedded sections; the second is a non-heating method that involves immersing the tissue section in an AR solution, such as a NaOH-methanol solution, and is suited to celloidin-embedded tissues. As formalin is the most common fixative used in morphology, the heat-induced AR technique has emerged as a very important approach to the immunohistochemistry of routinely processed tissue sections. Since AR was developed in 1991, there have been more than two hundred articles from all over the world demonstrating the value of AR for routine immunohistochemistry. More than 200 antibodies have been tested, yielding excellent immunohistochemical staining using the AR technique on routine tissue sections in almost all instances. This article reviews the basic principles of the method and describes the application, development and standardization of the AR technique as a basis for establishing current standards of practice and as a foundation for further studies.*

## INTRODUCTION

Immunohistochemistry (IHC) has a long history, extending over half a century from 1940, when Coons developed an immunofluorescence technique to detect bacteria (25,193). However, only in the last ten years has the method found general application in surgical pathology (27,193,197). A series of technical developments of IHC have created an increasingly sensitive detection system. This system is based primarily on immunoperoxidase techniques, specifically the peroxidase anti-peroxidase (PAP), avidin-biotin-peroxidase complex (ABC) and biotin-streptavidin (BSA) methods

1

(24,29,34,193), stemming from the important contributions of Nakane, Pierce, Avrameas, Sternberger, Hsu and others (67,144). In Europe, alkaline phosphatase has found extensive use as an alternative label. The development of the hybridoma technique accelerated development of IHC through the manufacture of numerous highly specific monoclonal antibodies, many of which found early application for IHC staining (24,29,193). Initial success in cryostat sections was eventually extended to routinely processed paraffin-, celloidin- or other plastic-embedded tissue sections. Only when the IHC technique can be used for routinely formalin-fixed, paraffin-embedded tissue sections, will it finally effect a "brown revolution" in pathology (102).

The critical significance of rendering the IHC technique suitable for routinely formalin-fixed, paraffin-embedded tissue sections (routine paraffin sections) was emphasized by Taylor and colleagues, who, in 1974, showed that it was possible to demonstrate at least some antigens in routine paraffin sections (187). The collective appetites of pathologists worldwide, once whetted by these initial studies, led to serious attempts to further improve the ability to perform IHC staining on routine paraffin sections (181,187–193). Enzymatic digestion was introduced by Huang et al. (68) as a pretreatment, prior to IHC staining, to "unmask" some antigens that had been altered by formalin fixation. However, the enzymatic digestion method, while widely applied, did not improve IHC staining of the majority of antigens, as pointed out by Leong et al. (101). Another drawback of enzymatic digestion is that it proved difficult to control the optimal digestion conditions for individual tissue sections when stained with different antibodies. Difficulties in standardization provided a powerful incentive to the development of a new technique, with the requirements that it should be more powerful, more widely applicable and easier to use than enzymatic digestion. In addition, any new method should enhance immunohistochemical staining of routine paraffin sections in a reproducible and reliable manner.

The antigen retrieval (AR) technique, based on a series of biochemical studies by Fraenkel-Conrat and coworkers (39–41), was developed by Shi et al. in 1991 (154). In contrast to enzymatic digestion, the AR technique is a simple method that involves heating routine paraffin sections at high temperature (e.g., a microwave oven or other heating methods) prior to IHC staining procedures (154). An alternative method that did not utilize heating was developed for celloidin-embedded tissues (155–157). The intensity of IHC staining was increased dramatically after AR pretreatment as demonstrated by the original articles (154–157) and by numerous articles published subsequently. All published reports described excellent AR-IHC staining results, with the exception of one article (124). Only a minority of antibodies failed to show improvement following the use of the AR method (48,103). Various modifications of the AR technique have been described, the majority of which employ different buffer solutions as the AR solution in place of metal salt solutions used in the initial report (1,20,28,33,48,59,81,103,120,160,179,

194,208,212). Other studies explored different heating methods, with different temperatures and heating times (81,82,109). The purpose of this review article is to summarize the development, standardization and application of AR-IHC, both as the basis for current use and for further studies designed to optimize the method.

## AR TECHNIQUE

### Heating AR Technique

**Microwave (MW) heating method.** In the original standard method (154), after dewaxing, rehydration and blocking of endogenous peroxidase with 3% $H_2O_2$, the paraffin sections were rinsed in distilled water and then placed in plastic Coplin jars. The jars were then filled with AR solution, such as distilled water, buffer solution, metal salt solution, urea, glycine-HCl pH 3.5, etc., covered with loose-fitting screw caps, and heated in the MW oven for either 5 or 10 min. The 10-min heating time may be divided into two 5-min cycles, with an interval of 1 min between cycles to check on the fluid level in the jars. If necessary, more AR solution may be added after the first 5 min to avoid drying the tissue sections.(Note that in our experience, when more fluid was needed, it was the result of boiling over the solution and not evaporation. Replacement was therefore with fresh AR solution, not simply with distilled water as would be the case if the loss were primarily due to evaporation). After heating, the Coplin jars were removed from the oven and allowed to cool for 15 min, rinsed in distilled water twice, and then in phosphate-buffered saline (PBS) for 5 min, following which the IHC staining procedure was performed. A conventional MW oven (as used in household kitchens) with power ranging from 600 to 800 W was employed for AR heating. Some authors employ a MW oven with 800-W capacity at medium power (182). We studied the heating process when using a MW with 720-W power and found that the boiling point ($100 \pm 5°C$) for the AR solutions in the Coplin jar was reached in 140–145 s. In order to maintain consistent heating conditions, it is necessary to set the same number of jars at the same locations close to the center of the MW oven. The final temperature and the total time of heating may be established by using a small pilot study ("test battery") to find the optimal heating condition for each antibody.

The process may be modified extensively and still yield satisfactory results. For example, we studied a modified MW heating protocol as follows: the AR solution was placed in the Coplin jar in the MW oven for 3 min to boil the solution, after which the slides were immediately immersed in boiling AR solution for 15 min, while the jar was moved outside the MW oven in a cooling mode. Terada et al. (200) reported a similar approach, preheating the AR solution (PBS) in a Coplin jar in a MW oven for 5 min to a boiling point (100°C), then setting the slides in the preheated AR solution and heating by

MW for 5 min at 100°C. Finally, the slides were placed into cooled PBS (4°C). Although a few authors (43) have suggested omission of the cooling step following high-temperature heating, it is our opinion that a 15-min cooling step is necessary for the following reasons: (a) in our experience and that of others, the intensity of AR-IHC staining may be reduced for some antigens if this step is omitted; (b) tissue sections may detach from the slides if suddenly changed from high temperature to low temperature; and (c) morphologic detail may also be compromised. Recently, Tacha and Chen (186) reported a standardized calibration method for the MW heating AR technique. Slides were brought to boiling point at high power (800 W), and then changed to a lower power "simmer" setting (e.g., 250 ± 50 W) for 7–10 min. The most important concept that emerges from these studies is the idea of counting the heating time from the point at which boiling occurs in order to standardize the heating conditions as far as possible. Pertschuk et al. (140) modified MW-AR by exposing the slides to high-power MWs in a plastic pressure cooker placed in the MW oven and achieved excellent results. Recently, we also demonstrated that the MW pressure cooker method yields an intensity of staining comparable with that obtained by autoclave heating AR or elongated MW heating time (198).

**Conventional heating method.** Shi et al. (154) compared conventional heating by hot plate with MW heating for AR-IHC staining and obtained a similar intensity of staining by both methods, providing that conventional heating was prolonged. This finding has been confirmed by several recent publications (70,81,82,133). Suurmeijer and Boon (179) found no detectable difference when comparing MW with conventional heating. In Japan, Kawai et al. from Tsutsumi's group (81,82) recommended using a water bath to heat the tissue sections at 90°C for 120 min in either phosphate-buffered (pH 7.2) or citrate-buffered (pH 6.0) solution. They concluded that it may be easier to perform the AR heating procedure by using a conventional water bath, where replenishment of lost AR solution may not be necessary: the disadvantage is that the process takes much longer to achieve similar results.

**Autoclave heating method.** Shin et al. (167) described a hydrated autoclaving method to enhance immunoreactivity of tau protein in brain tissue fixed in formalin, with good results. Recently, Umemura et al. (207) compared different heating methods by using monoclonal antibody to *bcl*-2 (clone 124), concluding that hydrated autoclaving of slides in deionized water was superior to MW and water bath conventional heating. The total time required for hydrated autoclaving was 2 h (from switching on to taking out the sections). Localization of *bcl*-2 in a variety of human tissues was said to be comparable to that obtained in frozen sections. The wet autoclave AR method has also been demonstrated to provide excellent results of AR-IHC staining for antibodies such as MIB1, Ki-67, p53, etc. (4). As another alternative, a household pressure cooker has also been used for the AR technique (134), with some success and with the advantages of low cost and short operation time (134).

## Non-Heating AR Technique

In a series of biochemical studies conducted by Fraenkel-Conrat et al. in the 1940s (39–41), several approaches were employed to reverse the cross-linkage of proteins caused by formalin. Methods included both high-temperature heating and non-heating procedures, such as immersion in strong alkali or acid hydrolysis. Other authors have also tried to develop non-heating methods to unmask antigens after formalin-fixation. For example, Costa et al. (26) found that overnight pretreatment with 6 M guanidine or urea was beneficial for specimens fixed in formalin 24 h, when staining with monoclonal antibody to human transthyretin (TTR); Kitamoto et al. (89) used formic acid pretreatment to enhance immunostaining of cerebral and systemic amyloidosis. Also, pretreatment with 2 N HCl has been advocated as a means of enhancing IHC staining for nuclear antigens such as PCNA, albeit that the outcome is still controversial (218).

Based on the studies of Fraenkel-Conrat and coworkers (39–41), Shi et al. (155) developed a non-heating AR technique by using NaOH-methanol as the AR-solution for IHC staining on routinely formalin-fixed, acid-decalcified, celloidin-embedded human temporal bone sections. The method is simple and effective for retrieval of a variety of antigens in celloidin-embedded tissue sections. The NaOH-methanol AR-solution is prepared by adding 50–100 g of NaOH to 500 mL methanol in a brown-colored bottle and mixing vigorously. The solution can be stored at room temperature for 1–2 weeks (also available commercially from BioGenex, San Ramon, CA, USA). The clear, saturated solution is diluted 1:3 by addition to methanol prior to use (see Appendix 1 for the protocol).

A combination of MW heating and NaOH-methanol non-heating methods has been developed recently (159) in an attempt to retrieve additional antigens at increased staining intensity. In this approach, celloidin sections are immersed in AR solution (e.g., 5% urea or metal solutions) and then boiled in a MW oven for 5 to 15 min. Sections are then immersed in 1:3 diluted NaOH-methanol solution for 30 to 50 min. The non-heating AR technique has been used not only for human temporal bone collections but also for neuroanatomical studies of human by Dr. Jean Moore et al. (125). Similarly, Linthicum et al. (107,203) reported a series of AR-IHC studies concerning constituents of the endolymphatic sac and tubules in human temporal bone by using non-heating methods of AR-IHC. More recently, Su and Yu have developed an effective AR solution that does not require a high heating temperature (K.L. Kalra and T. Chen, BioGenex; personal communication). This solution contains 0.01% sodium periodate as an oxidizer. The intensity of immunostaining of monoclonal antibodies to CD29 and dystrophin II may be enhanced by pretreating paraffin sections at 37°C for 30 min with this solution (Figure 1, M–O, see color plate in Addendum, p. A-1). Exploration of a broader range of antibodies is continuing.

## Major Factors that Influence the Effectiveness of Heat-Induced AR

**Heating conditions.** The duration and intensity of heating are the most important factors. As mentioned above, the heating AR-IHC method is based on biochemical studies of Fraenkel-Conrat and coworkers, who documented that the chemical reactions that occur between protein and formalin may be reversed, at least in part, by high-temperature heating or strong alkaline hydrolysis (39–41). We demonstrated that the use of conventional heating at 100°C may achieve similar results to those obtained by MW heating, and also that distilled water could be used as the AR solution, albeit with less effect (154). Subsequently, several publications have reported similar results by using conventional heating (70,81,82). Malmstrom et al. (111) performed AR-IHC on PCNA (PC10 and 19F4) by using distilled water as the AR solution on routine paraffin sections of urinary bladder carcinoma and obtained good results.

The chemical reactions occurring during the formalin-fixation process remain obscure. The AR technique may improve our understanding of how formaldehyde produces cross-linkages and shed light on the exact mechanism of cross-linked proteins. Mason and O'Leary (112) demonstrated that the process of cross-linking does not result in discernible alteration of protein secondary structure, at least as determined by calorimetric and infrared spectroscopic investigation. They found that significant denaturation of unfixed purified proteins occurred at temperature ranges from 70° to 90°C, whereas similar temperatures had virtually no adverse effect on formalin-fixed proteins (i.e., formalin-fixed proteins are more heat-stable). Thus, the AR heating technique, employing high-temperature heating of tissue sections fixed in formalin, appears to take advantage of the fact that the cross-linkage of protein produced by formalin fixation may protect formalin-modified epitopes from denaturation during the heating phase. Although the mechanism of action of the AR technique is not clear, it appears unlikely, based upon the above observation that "protein denaturation alone is the mechanism" as advocated by Cattoretti et al. (20). Suurmeijer and Boon (178) have proposed several possible mechanisms of AR, such as: (a) breaking the formalin-induced cross-links between epitopes and unrelated proteins; (b) extraction of diffusible blocking proteins; (c) precipitation of proteins; and (d) rehydration of the tissue section allowing better penetration of antibody and increasing accessibility of epitopes. The breaking of cross-linkages hypothesis has been challenged by recent studies concerning AR heating pretreatment on methacarn-fixed tissue sections (48).

In our original article, we found significantly different intensities of immunostaining between formalin-fixed and alcohol-fixed tissue sections after AR treatment (154), an observation in accordance with Suurmeijer and Boon's data (179). Igarashi et al. (70) conducted a careful study using four heating methods as described below and obtained very good retrieval by heating alcohol-fixed tissues refixed with formalin, suggesting that the existence of cross-linkages is a prerequisite for any effective enhancement of staining by the AR method.

We believe that the major point for understanding the mechanism of AR-IHC may be based on the heat- or chemical-induced modification of the three-dimensional structure of protein, reversing the condition from a formalin-modified protein structure back towards its original structure. In other words, the mechanism of AR-IHC appears to involve a "re-modification" of the structure of fixed proteins through a series of conformational changes, including the possible breaking of formalin-induced cross-linkages, with the whole process being driven by thermal energy from the heat source (6,85, 126,141,151,210,217,219). The term "denaturation" is not appropriate to express this process, as denaturation has already occurred during formalin-fixation of the tissue; in a sense, AR "unfixes" the tissue, returning proteins to the unfixed state. The term "renaturation" may even be applicable, in terms of restoration of structure, if not function. Because the antigen-antibody recognition is solely based on reciprocity of protein structures, further studies concerning the mechanism of AR-IHC should be focused on studying the alterations of protein structure that take place during fixation and "unfixation" or retrieval.

The fact that high temperature is a very important factor in AR-IHC has been demonstrated by a variety of studies. Lucassen et al. (109) studied AR-IHC of monoclonal antibody to MG-160 antigen, a sialoglycoprotein of the medial cisternae of the Golgi apparatus, by using different heating temperatures, and emphasized the importance of the temperature during MW-AR treatment. Kawai et al. (82) found that heating at 90°C for 10 min was more effective than heating at 60°C for 120 min; however, overnight heating at 60°C gave satisfactory AR effect of PCNA and p53 IHC staining. Igarashi et al. (70) studied the heating AR technique by using 56 antibodies with several heating methods: hydrated autoclaving (121°C, 2 atm for 20 min), MW oven (100°C, 10 min), conventional heating in distilled water (60°C overnight), and in 20% $ZnSO_4$ at (90°C for 10 min). They found that higher temperature heating (MW and autoclave) was superior for most antibodies tested, although a few antibodies (HHF35 and CGA7, smooth muscle actin) gave better results at 60°C overnight. Our experience also indicates an inverse correlation between heating temperature and heating time as a formula: AR (optimal AR-IHC result) $\alpha$ T (temperature of AR heating) $\times$ t (time of AR-heating treatment). Recently, we conducted an experiment to study further the correlation between the heating temperature and the heating time, under a variety of different heating conditions. Our conclusion was based on careful comparison of different heating temperatures ranging from 60°–100°C with heating times ranging from 5 min to 10 h by using MW, conventional heating (water bath and hot plate), pressure cooker and autoclave. A similarly strong intensity of AR-IHC staining was generated by the following heating

7

conditions (T × t): $100°C \times 20$ min, $90°C \times 30$ min, $80°C \times 50$ min, and $70°C \times 10$ h (162,198). Shibuya et al. (165) also determined that the intensity of AR-IHC may be dependent on both the heating temperature and time. Furthermore, Munakata and Hendricks (128) demonstrated that an extended heating time may be required to obtain an optimal AR-IHC result for tissue sections "over-fixed" in formalin. Suurmeijer and Boon (178) found that repeating the boiling cycles was more effective than extending the boiling time of single cycle when using the AR technique; three cycles of 5 min was superior to using MW continuously for 15 min. Emanuels et al. (35) reported a modified MW heating program with 3 min heating followed by 5 min cooling, and then repeating the cycle 1–6 times in order to avoid drying the sections. Our experience of extending heating times up to 30 min and beyond, in 5-min increments, indicates that two cycles of 5 min in a MW oven at high-power setting are suitable for most commonly used antibodies. Although absolute maximal or optimal staining may be achieved using 20 min for some antibodies, the difference in quality is not sufficient to warrant routine use of the 20-min heat exposure. For any individual antibody, an optimal AR-protocol may be established by using a test battery method, as described below.

**pH of the AR solution.** The pH value of the AR solution is also an important factor. Recently, we tested the hypothesis that pH of the AR solution may influence the quality of immunostaining by using seven different AR buffer solutions at a series of different pH values ranging from 1 to 10. We evaluated the staining of monoclonal antibodies to cytoplasmic antigens (AE1, HMB45, NSE), nuclear antigens (MIB1, PCNA, ER), and cell surface antigens (MT1, L26, EMA) on routinely formalin-fixed, paraffin-embedded sections under different pH conditions with MW-AR heating for 10 min. The pH value of the AR buffer solution was carefully measured before, immediately after, and 15 min after the AR procedure. From this study, we drew the following conclusions:

I. There were three types of patterns reflecting the influence of pH. First, several antigens showed no significant variation utilizing AR solutions with pH values ranging from 1.0 to 10.0 (L26, PCNA, AE1, EMA and NSE); second, other antigens (MIB1, ER) showed a dramatic decrease in the intensity of the AR-IHC at middle range pH values (pH 3.0–6.0), but strong AR-IHC results above and below these critical zones; and third, still other antigens (MT1, HMB45) showed negative or very weak focally positive immunostaining with a low pH (1.0–2.0), but excellent results in the high pH range (Figure 1, A–L).

II. Among the seven buffer solutions at any given pH value, the intensity of AR-IHC staining was very similar, with the single exception of Tris-HCl buffer, which tended to produce better results at a higher pH, compared with other buffers.

III. Optimization of the AR system should include optimization of the pH of the AR solution.

IV. A high pH AR solution, such as Tris-HCl or sodium acetate buffer at pH 8.0–9.0, may be suitable for most antigens.

V. Low pH AR solutions are most useful for nuclear antigens such as retinoblastoma protein (RB), estrogen receptor protein (ER) and androgen receptor.

VI. Focal weak false positive nuclear staining may be found when using a low pH AR solution; the use of negative control slides is important to exclude this possibility (161,196).

Evers and Uylings (38) also found that the AR-IHC is pH- and temperature-dependent. They tested two antibodies, MAP-2 and SMI-32, and indicated that the optimal pH values were pH 4.5 for MAP-2 and pH 2.5 for SMI-32. In addition, they demonstrated that the use of 4% $AlCl_3$ as the AR solution for SMI-32 could achieve a similar result to that obtained by using citrate buffer of pH 2.5, from which they concluded that it is not important what kind of solution is used as long as the pH is at an appropriate level.

**Molarity of the AR solution.** Suurmeijer and Boon (178) tested the effect of molarity by using aluminum chloride solution with concentrations ranging from 0.5%, 1%, 2% and 4% for AR-IHC staining of vimentin on routine paraffin sections. They found that the optimal staining for vimentin in 10 different tissue components was achieved with 4% aluminum chloride. Suurmeijer (180) compared AR results with the use of 0.01 M and 0.3 M aluminum chloride, and found that different antigens showed significantly different responses to changes in molarity.

**Chemical composition of the AR solution.** In our initial studies of the AR technique, we used a metal salt as the AR solution and obtained good results for several antibodies tested. The use of metal salt solutions was based on earlier studies concerning zinc formalin fixation of tissue sections (77). We speculated that the metal salt solution might effectively influence the structure of protein by playing a role in refixation of the retrieved antigens. The increase of immunostaining intensity resulting from use of metal salt AR solutions has subsequently been demonstrated by several studies (49,52,54,81,109,120,139,169,173,177,178,194). However, a major drawback of metal salts, particularly lead salts, is the potentially toxic effect as mentioned by Suurmeijer and Boon (178). It is, therefore, better to avoid using lead or other toxic metal salts if other kinds of solution may serve with equal effect in AR procedure. It should be noted also that some authors reported poor results of AR-IHC staining with the use of metal salt solutions (124); we can only speculate that alterations in pH or other variables may account for these discrepancies.

From our more recent studies regarding AR-IHC under the influence of pH (161), we found that, while the pH is critical, the chemical composition of the AR solution may also play a role as a cofactor in AR-IHC. For example, for antibodies to RB, including a polyclonal antibody (RB-WL-1) and a monoclonal antibody (PMG3-245; BioGenex), we found that a better staining result was obtained using acetate buffer solution of pH 1.0–2.0 with MW heating at 100°C for 10 min, when comparing it with citrate buffer at the same pH value. Imam et al. (71) reported that the use of glycine-HCl solution at pH 3.6 as the AR solution yielded stronger immunostaining with antibodies to some antigens, such as androgen receptor, estrogen receptor, Ki-67, MIB1, EMA, MT-1, actin, etc., when comparing with that obtained by using citrate buffer, pH 6.0, as the AR solution, raising the possibility that the chemical composition of glycine-HCl may play a role in AR treatment. Another example of the possible influence of chemical component in the AR solution was provided by Hazelbag et al. (60), who demonstrated that a simple detergent solution could yield the same result of AR immunostaining as obtained by using the commercial AR solution known as "TUF" for a variety of antibodies to keratin. The potential functions of chemical components in the AR solution include: (a) secondary fixation after "unfixation" by high-temperature heating; (b) stabilization of antigens during heating or strong alkaline hydrolysis; (c) maintenance of optimal molarity; and (d) unknown cofactors in reconfiguring the unfixed protein thereby recovering antigenicity. Analysis of these separate factors may allow development of new AR solutions applicable to those cell surface antigens that do not respond well to current retrieval methods. For example, we recently tested the influence of different chemicals in AR and found that a mixed AR solution of Tris-HCl, pH 9.5, with 5% urea may yield stronger intensity (163). Similarly, Kwaspen et al. (94) reported the use of 1% periodic acid as AR solution for MW-AR-IHC to study keratin expression on archival tissue sections providing strong positive staining.

**Method of Heating: Choices**

The next important issue concerns selection of the heating method. As noted above, several different methods have been employed, each of which has its own idiosyncrasies. Different commercial MW ovens may provide different heating conditions, which may not be directly comparable even when using the same heating protocol. Therefore, it is necessary to standardize the heating conditions considering both temperature, time and placement of slides in the oven. If a MW oven can display the exact temperature, this value (e.g., 100°C) may be used to determine the start time. A simpler way to standardize the heating time is by beginning to count the time only after the solution has reached boiling. Either way, the heating conditions can be calculated more exactly from the point of boiling, which may allow better comparison between different laboratories. Microwave ovens with temperature readouts are available; their power output is regulated by a temperature feedback mechanism and a timer. For example, the oven provided by Energy Beam Sciences (Agawam, MA, USA) is capable of monitoring both temperature and time, and is recommended for AR-IHC as well as for fixation and accelerated immunostaining.

An easier way may be the use of a MW oven plus pressure cooker, which has an alarm or manometer/thermometer to indicate the level of heating temperature. It may provide advantages of standardization, ease of use, and even heating of many slides in one test.

## Application of AR

Table 1 summarizes published data concerning AR-IHC for a variety of antibodies. Although use of AR continues to grow, it has already shown broad utility in surgical pathology and in morphologic research. The following highlights some of the general principles derived from review of the literature.

**Retrieval of antigens showing negative or very weakly positive staining even with use of other unmasking procedures.** Among this group of antibodies are Ki-67, MIB1, ER-1D5, *bcl*-2, androgen receptor, CD5, CD8, CD35, CD38, CD25, CD45R, CD34, CD68, CD44, CD55, VCAM-1, CGA7 (smooth muscle-specific actin), IB5 (MHC class II), IgD, delta-TCS, JC1 (proliferation marker) (20,28,48) and CD19 (Leu-12) (103). The AR technique has provided new opportunities for retrospective studies and for the use of the antibodies in pathological diagnosis of routinely processed tissue. In addition, Kawai et al. (81) advanced the prospect that insoluble secretory proteins which are deposited in tissue and improved by enzyme digestion for immunostaining may also be enhanced by the AR-IHC technique. It should be noted, however, that the AR technique may not produce any enhancement of those proteins that can be immunostained very well without any pretreatment, and are resistant to the adverse effects of fixation. Likewise, ideally fixed tissue may not benefit.

**Reduced IHC detection thresholds (increased sensitivity) for a wide range of monoclonal and polyclonal antibodies.** The use of AR appears to enhance sensitivity of IHC in routine paraffin sections, in some instances providing a detection threshold similar to that observed in frozen sections. Background staining is also reduced in paraffin sections after AR treatment, for unknown reasons. For example, while a dramatic enhancement of the staining of Reed-Sternberg cells for CD30 was achieved following AR-IHC, the nonspecific staining of plasma cells disappeared simultaneously (22). After investigating this interesting finding, Charalambous et al. (22) strongly recommended the MW heating AR-IHC method as superior to digestion for IHC staining of CD30 for Reed-Sternberg cells. In other studies, immunoreactivity of some antigens after AR treatment is consistent with the immunoreactivity of alcohol-fixed or frozen tissues, as indicated by Igarashi et al. (70), Gown et al. (48), Leong et al. (103), Cuevas et al. (28) and Tenaud et al. (199). As a result, the reliability and utility of IHC in pathology have been greatly improved (27,193,195,197). In particular, there is a significant reduction in false-negative immunostaining results, with corresponding increase in the accuracy of diagnosis (122).

This finding was illustrated in our initial studies in which formalin-fixed

**Table 1. Antibodies Tested for AR-IHC with Good Results**

| Antibody | Clone No. | Source | Heating Method | AR Solution | References |
|---|---|---|---|---|---|
| ACTH | P | BG,D | MW | P,Z,W,C | 48 154 |
| Actin, smooth muscle | 1A4,CGA7 | BG*,S,D, A.Gown | MW | C,U | 20 48 |
| AFP | A-013-01 | BG | MW | P,Z,W,C | 103 154 |
| Albumin | P | BG | MW | P,Z,W | 154 |
| Alpha H chain | P | D | MW | C | 48 |
| Alpha hCG | 02-310-94 | BG | MW | P,Z,W | 154 |
| Androgen receptor | F39.4.1, P:AR70,AR52 | M,N,BG* | MW | C,U | 48 73 75 108 158 |
| B + Macrophage | LN5 | BG | MW | C,U | 20 |
| B cell | MB1,MB2, DBA44 | BG,BT,D | MW | P,Z,W,C,U | 20 154 |
| B cell CD45Ra? | Ki-B3 | B,Kiel | MW | C,U | 20 28 |
| B-cell receptor | HM57 | MS | MW | C,U | 20 |
| bcl-2 | 124,BCL-2 | D,K,BG* K.Gatter,CB | MW,WB,A | C,U,W,Z,PB | 4 20 21 28 48 65 87 103 207 214 |
| Blood Group A | 81 FR2.2 | BG | MW | P,Z,W | 154 |
| Blood Group B | 81/11 | BG | MW | P,Z,W | 154 |
| C-erb-B2 | CB11,P, P: Lot#061 | BG,T,D,NI | MW | P,Z,W,C,U | 20 48 62 70 75 103 154 |
| c-kit | P | O | MW | C,U | 20 |
| c-trk | P | O | MW | C,U | 20 |
| CA125 | M | CIS,BG* | MW | W | 70 |
| Carbonic anhydrase II | P | ? | MW | C | 118 |
| Cathepsin B | P | BG | MW | P,Z,W | 154 |
| CD15 | MMA,Leu-M1,M1 | BD,D | MW | C | 22 28 48 |
| CD2 | CDT1,CDT2 | CRB | MW | C,U | 20 |
| CD2 | MT910,F92 -3A11,TS2/18.1.1 | D | MW | C,U | 20 |
| CD2 | CLB-T11.1, MT110, T11/3PT2H9 | R.VanLier | MW | C,U | 20 |
| CD20 | L26 | D,BG* | MW | C,U | 20 28 48 65 |
| CD23 | BU38 | Binding Site | MW | C | 28 |
| CD24 | OKB2 | OR | MW | C,U | 20 |
| CD25 | TU69 | A.Ziegler | MW | C | 28 |
| CD29VLAB1 | K20 | IT | MW | C,U | 20 |
| CD3 | P | D,BG* | MW | C,U | 20 28 103 |
| CD30 | Ber-H2 | D,BG* | MW | C,U | 20 22 28 48 |
| CD31 | JC/70A | D,BG* | MW | C,U | 20 48 103 |
| CD34 | My10,QBEnd 10 | BD,N,OX,BG* | MW | C,U | 20 28 48 103 |
| CD35 | E11 | K.Gatter,BG* | MW | C | 28 |
| CD38 | A10,AT13/5 | MF,TG,Tutt | MW | C,U | 20 28 |

**Table 1. (cont.)**

| Antibody | Clone No. | Source | Heating Method | AR Solution | References |
|---|---|---|---|---|---|
| CD4 T cell | OPD4 | D | MW | C | 103 |
| CD43 | WR14 | W | MW | C | 28 |
| CD44 | F10442,HK23 | 4thIW | MW | C,U | 20 |
| CD44 | BRIC235&222 | D.Anstee | MW | C | 28 |
| CD44 | 2H1,IF1,IE8, Hermes1&3 | S.Jalkenen | MW | C | 28 |
| CD45 | PD7/26,2B11, KC56 | D,C | MW | C,U | 20 48 |
| CD45 | 9.4,HLe-1,EO1, T29/33,BMAC2 | 2ndIW | MW | C,U | 20 |
| CD45R | WR16 | W | MW | C | 28 |
| CD45RA | 4KB5,F8.11.13 | D,ST | MW | C,U | 20 48 |
| CD45RO | UCHL-1,OPD4 | D,TG | MW | C,U | 20 48 65 103 178 |
| CD5 | CD5 | K.Gatter | MW | C | 28 |
| CD54 ICAM | 84H10 | IT | MW | C,U | 20 |
| CD55 | BRIC 128 | D.Anstee | MW | C | 28 |
| CD57 | HNK-1 (Leu-7) | BD,BG* | MW | C,U | 20 48 |
| CD66 NCA | CLB gran/10 | 2ndIW | MW | C,U | 20 |
| CD68 | KP1, | D,B,2rdIW | MW | C,U | 20 103 116 118 |
| CD68 | EBM11 | K.Gatter | MW | C | 28 |
| CD68 | PGM1 | B.Falini | MW | C | 28 |
| CD71 | HB21 | ATCC | MW | C | 28 |
| CD74 | LN2 | BT,BG* | MW | C,U | 20 |
| CD8 | C8/144B,CD8,P | MS,K.Gatter | MW | C,U,W,M | 20 28 113 |
| CD9 | FMC56,ALB6 | 3rdIW | MW | C,U | 20 |
| CDw60 | M-T32 | 3rdIW | MW | C,U | 20 |
| CDw75 | LN1 | BT,BG* | MW | C | 28 |
| CEA | SP651,P,CE J065,A5B7 | BG,BM,D | MW | P,Z,W,C | 48 154 |
| Chromogranin | LK3H10 | BG | MW | P,Z,W | 154 |
| Chromogranin A | LK2H10,PHE5 | BD,A.Gown | MW | C | 48 |
| Chromogranin A | NM-08-308 | CRB | MW | C,U | 20 |
| CMV | P | BG | MW | P,Z,W | 154 |
| Collagen type IV | Collagen IV | D,BG* | MW | C | 28 |
| alph,beta-Crystallin | P | J.Love | MW | C | 118 |
| Cyclin D1 | HD64,P | W.Yang | MW | C | 221 |
| Cytokeratin | AE1,AE3,AE8 | BG,BM,H | MW | P,Z,W,A,C,U | 20 48 154 179 |
| Cytokeratin 8,18 & 19 | Cam5.2 | BD | MW | C,P,A,U | 20 48 103 178 |
| Cytokeratin | 34BE12 | D,OR, A.Gown | MW | C,U | 20 48 103 |
| Cytokeratin | 35BH11 | D | MW | C | 103 |

**Table 1. (cont.)**

| Antibody | Clone No. | Source | Heating Method | AR Solution | References |
|---|---|---|---|---|---|
| Cytokeratin | 145818 | DPC | MW | P,A | 179 |
| Cytokeratin 8,18,19 | 5D3 | BG | MW | P,A,Z,W | 154 179 |
| Cytokeratin | NA | BG | MW | P,A | 179 |
| Cytokeratin 13 | KS-1A3,AE8 | S,BG* | MW | C,U | 20 |
| Cytokeratin 13 & 16 | Ks8.12 | S | MW | C,U | 20 |
| Cytokeratin 14 | CKB1 | S | MW | C,U | 20 |
| Cytokeratin 18 | CK5,Cy90, KS-B17.2 | S | MW | C,U | 20 |
| Cytokeratin 19 | Ks19.1,Ks4.62 | ICN,S | MW | C,U | 20 103 |
| Cytokeratin 7 | CK7 | BG | MW | P,Z,W | 154 |
| Cytokeratin peptide 7 | LDS68 | S | MW | C,U | 20 |
| Cytokeratin,pan | KC2 cocktail | H.Battifora | MW | C | 48 |
| Cytokeratins | KL-1 | IT | MW | C,U | 20 |
| Cytokeratins pan | F12-19,K8.13 | BG,S | MW | P,Z,W,C,U | 20 154 |
| Desmin | D33,DE-R-11, DE-B-5 | D,BM | MW | C,U | 20 48 103 178 |
| E-cadherin | HECD 1 | S.Hirohashi | MW | C | 28 211 |
| EBV LMP | CS1-4 | D | MW | C,U | 20 |
| EMA | E29 | D,BG* | MW | C | 48 103 |
| Endothelium | EN4 | SB,TG | MW | C,U | 20 |
| ER | 1D5,H222 | D,AM,AB, BG* | MW | C,G | 37 47 48 63 78 83 103 104 152 |
| ER | 1C5,1D5 | T.AlSaati | MW | C | 1 |
| ERRP | D5 | BG | MW | P,Z,W | 154 |
| Factor VIII | P | BG | MW | P,Z,W | 154 |
| Ferritin | P | D | MW | C | 118 |
| Fibronectin | FN-3E2 | S | MW | C,U | 20 |
| FSH | P | BG | MW | C | 48 |
| Gamma H Chain | P | D | MW | C | 48 |
| Gastrin | P | D | MW | C | 48 |
| GCDFP-15 | D6 | SI | MW | C | 48 |
| GFAP | P,GA5,6F2, RPN1106 | BG,D,AS,BM | MW | P,Z,W,C,U | 20 48 118 154 178 220 |
| GFAP | GFP8 | A.Gown | MW | C | 48 |
| Glucagon | P | BG | MW | C | 48 |
| Glycophorin | JC159 | D | MW | C,U | 20 |
| Glycoprotein hormone alpha-subunit | P | UCB | MW | C | 31 |
| Growth hormone | P | BG | MW | C | 48 |
| HCG | P | D | MW | C | 48 |

Table 1. (cont.)

| Antibody | Clone No. | Source | Heating Method | AR Solution | References |
|---|---|---|---|---|---|
| Heat shock protein72/73 | W27 | O | MW | C,U | 20 |
| Hepatitis C | Tordjii-22 | AM | MW | C | 48 |
| Herpes simplex I | P | D | MW | C,U | 20 |
| Herpes simplex II | P | D | MW | C,U | 20 |
| HLA-DR | LN3 | BT | MW | C,U | 20 118 |
| HMFG-2 | 3.14.A3 | OX | MW | C | 48 |
| Human r TCR | deltaTCS | BD | MW | C | 28 |
| IgD | IADB6,P, | BG,D | MW | P,Z,W,C,U | 20 28 154 |
| IgM,H & L Chains | P,R169 | D,BG* | MW | P,Mn,U | 120 |
| Insulin | P | D | MW | C | 48 |
| Kappa L Chain | KP53,P | BG,D | MW | P,Z,W,C | 48 154 |
| Ki-1 CD30 | Ber-H2 | D,BG* | MW | C | 103 |
| Ki-67 | MIB1&3,Ki67, P,Ki-S5 | AM,D,BG,IT | MW,CH,A | C,U,Al,G | 4 5 19 20 23 28 42 45 48 61 64 76 86 90 92 103 128 131 148 160 170 171 178 183 184 194 202 |
| L26 | M | D,BG* | MW | C | 178 |
| Lambda L chain | HP6054,P | BG,D | MW | P,Z,W,C | 48 154 |
| Laminin | Laminin | S | MW | C | 28 |
| LCA | M | D,BG* | MW | W | 70 |
| Leu-12 CD19 | 4G7 | BD | MW | C | 103 |
| Leu-7 CD57 | HNK-1 | BD,BG* | MW | C | 103 |
| Leu-M1CD15 | MMA | BD | MW | C | 103 |
| LH | P | BG | MW | C | 48 |
| LN1 CDw75 | NA | BD | MW | C | 103 |
| LN2 CD74 | NA | BD | MW | C | 103 |
| Macrophage | LN5 | BG | MW | P,Z,W | 154 |
| MAP-2 | M4403 | S | MW | C(pH4.5) | 38 |
| Measles virus | P | M.Godec | MW | C | 118 |
| Melanoma | HMB45 | D,OR,BG* | MW | C,U | 20 103 |
| MHC Class II | IB5,WR18 | ICRF,W | MW | C | 28 |
| MT1 CD43 | MT1 | BD | MW | C | 103 |
| Mu H chain | P | D | MW | C | 48 |
| Multi-drug resistance | C219 | CIS | MW | C,U | 20 |
| Myeloid,CD15 | Tu9 | BG | MW | P,Z,W | 154 |
| Neurofilaments 160-200kD | MNF | E | MW | C,U | 20 |
| Neurofilaments 160kD | RPN1104 | AS | MW | C,U | 20 |
| Neurofilaments 200kD | RPN1103 | AS | MW | C,U | 20 |

15

Table 1. (cont.)

| Antibody | Clone No. | Source | Heating Method | AR Solution | References |
|---|---|---|---|---|---|
| Neurofilaments 68kD | NR4,RPN1105 | AS,BM,BG* | MW | C,U | 20 220 |
| NF,non-phosphor-ylated | SMI-32 | SM | MW | C(pH2.5) | 38 |
| Neurofilaments | 2F11 | BG,D | MW | P,Z,W,C | 48 95 154 |
| Neuron-specific enolase | MIG-N3 | SB,BG* | MW | C,U | 20 |
| NSE | P | BG* | MW | Pb, Zn & DW | 154 |
| NSE | MIG-N3, BBS/NC/V1-H14 | BG,D,SB | MW | P,Z,W,C | 48 103 154 178 |
| P acid P | P | D | MW | C | 48 |
| P-cadherin | NCC-CAD | S.Hirohashi | MW | C | 28 |
| p30/32-M1C2 | O13 | SI | MW | C | 48 |
| p53 | PAb1801,1802, 1803,240, 248, DO7,DO1,BP53 -11-1 BP53-12-1, CM-1 | OS,D,BG,N,O J.Bartek, D.Lane L.Crawford Tanner | MW,WB,A | C,Z,P,G,U PBS,TUF, W | 3 4 12 14 17 20 28 32 48 51 57 58 69 70 72 82 91 96 99 100 103 105 114 117 127 129 132 136 137 139 142 143 145 147 172 194 200 206 208 213 215 222 |
| PAI-1 | 380 or 3785 | AD | MW | C | 13 |
| PAP | P | D | MW | C | 48 |
| PTHrP | 9H7 | | MW | C | 74 |
| PCNA | 19A2,PC10,19F4 | AB,D,N,BM,C, BG* | MW,WB,A | C,U,P,W,G PBS,Z,TUF C,U | 20 30 49 55 56 70 82 88 103 111 123 130 138 150 166 169 173 202 206 |
| PDGF | P | O,GZ | MW | | 201 220 |
| PDGF Receptor | P:-7(alpha), -3(beta) | A.Erikson | MW | C | 201 |
| PGP9.5 | P | U | MW | C | 98 118 |
| PgR-ICA | Rat MAb | AB | MW | C,U | 20 |
| PLAP | P | D | MW | C | 103 |
| Pneumocystis | 3F6 | D | MW | C | 103 |
| PR | MPRI,MPR3, PgR-ICA,PgR1A6, KD68,10A9 | TB,AB,N,BG IT | MW | C,G,P,W | 46 48 104 185 |
| Prolactin | P | BG | MW | C | 48 |
| Proliferation marker | JC1 | K.Gatter | MW | C | 28 |
| PSA | 8 | BG | MW | P,Z,W,C | 48 154 |
| RB | 3H9,PMG3-245, P:WL-1 | UBI,P,T, H.-J.Xu,BG* | MW,CH | C,U,AC | 20 43 48 |

Table 1. (cont.)

| Antibody | Clone No. | Source | Heating Method | AR Solution | References |
|---|---|---|---|---|---|
| Ribonucleotide reductase | AD203 | IR,TG | MW | C,U | 20 |
| S-100 | P | D | MW | C,U | 20 48 178 |
| Serotonin | P | BG | MW | P,Z,W | 154 |
| Sialoglycoprotein | MG-160 | P.Lucasson | MW | Z | 109 |
| Somatostatin | P | D | MW | C | 48 |
| Substance P | CA-08-355 | CRB | MW | C,U | 20 |
| Surfactant | PE10 | NA | MW | C | 103 |
| Synaptophysin | P,SY38, | D,BM,BG* | MW | C,U | 20 48 118 174 |
| T cell | MT1,MT2 | BG | MW | P,Z,W | 154 |
| TCR2 alpha | Alpha F1 | TS | MW | C,U | 20 |
| TdT | 6A6.09,P | BG,TG | MW | C,U | 20 |
| Thyroglobulin | P | BG,D | MW | P,Z,W,C | 48 154 |
| TIA-1 | Cat.#6604593 | C | MW | C | 149 |
| TSH | P | D | MW | C | 48 |
| VCAM-1 | 1.4C3 | D.Haskard | MW | C | 28 |
| Villin | BDID2C3 | AM | MW | C | 48 |
| Vimentin | V9,P,RPN1102, Fil-3 | BG,D,E,AS, EP,DPC | MW,A | P,Z,W,C,Al U | 2 4 20 28 48 70 103 154 179 |
| VIP | P,CA-08-345 | BG,D,CRB | MW | P,Z,W,C,U | 20 48 154 |
| Vitamin D receptor | P | R. Kumar | MW | PB+Tween | 93 |
| von W's factor | F8/86 | D | MW | C | 48 |
| VS38(Pasma cell) | VS38 | Turley | MW | C | 205 |

**Abbreviations:** P = polyclonal antibody; PAI-1 = plasminogen activator inhibitor-1; PTHrP = parathyroid hormone related protein. **Source of antibody:** AB = Abbott; AD = American Diagnostica; AM = AMAC; AS = Amersham; B = Behring; BC = Behring Calbiochem; BD = Becton Dickinson; BG = BioGenex; BM = Boehringer Mannheim; BT = Biotest; C = Coulter; CB = Cambridge Biotechnology; D = Dako; E = Eurodiagnostic; EP = Euro-Path; GZ = Genzyme; H = Hybritech; IR = InRo Biomedtek; IT = Immunotech; IW = International Workshop on Human Leucocyte Differentiation, 2nd, 3rd & 4th; M = Monosan; MF = Malavasi F; MS = Mason; N = Novocastra Labs; NI = Nichirei; O = Oncogene Science Inc; OR = Ortho; OX = Oxoid; P = Pharmingen; S = Sigma; SB = Sanbio; SI = Signet; SM = Sternberger Monoclonals; ST = Serotec; T = Triton; TB = Transbio Sarl; TG = Technogenetics; TS = T Cell Sciences; U = UltraClone; UCB = UCB-Bioproducts; W = Wessex. **Heating Method:** MW = microwave oven; WB = water bath; A = autoclave; CH = conventional heating; P = pressure cooking. **AR-solution:** A = aluminum chloride; AC = acetic buffer pH1; C = citrate buffer pH 6; M = metal; Mn = manganese chloride; P = lead thiocyanate; PB = phosphate buffer; U = urea 6 M or 5%; W = distilled water; Z = zinc sulphate or chloride; * = personal communication from T. Chen and K.L. Kalra.

tissue from a poorly differentiated adenocarcinoma could only be demonstrated as keratin-positive after using the AR-IHC technique (154). Yachinis and Trojanowski (220) demonstrated the enhanced detection of neuronal and glial antigens in archival tissues by using AR-IHC for studies of childhood brain tumors. Similarly, Battifora (11) reported a case of lobular carcinoma of the breast metastatic to the stomach, which was initially diagnosed as primary gastric carcinoma, and was subsequently immunostained positively for GCDFP-15 protein and ER-1D5 by using the AR-IHC technique. Recently, other authors (81,82,103) have applied AR-IHC for use in routine surgical pathology with markedly improved results. A further sequel is that primary antibodies may be further diluted, while at the same time giving much better staining. For example, Haerslev and Jacobson (56) diluted the PC10 (monoclonal antibody to PCNA) up to 1:600 for routine paraffin sections by using the AR-IHC technique, contrasted to a dilution of 1:10 if applied to tissue sections without AR treatment. In our experience, some antibodies may be diluted to more than 1000-fold following effective antigen retrieval.

One important issue should be raised concerning the increased sensitivity obtained after AR treatment, particularly if semiquantitative studies are to be performed. As a result of the increased sensitivity, the number of positively stained cells and intensity of staining are dramatically increased. For example, in one study, p53-positivity was 57% for routine paraffin sections of cancer tissue with regular IHC staining, but increased to 93% with the AR-IHC staining procedure (70). Also, the use of ER-1D5 (anti-estrogen receptor antibody) with AR on routine paraffin sections has achieved much improved staining, both in intensity and number of positive cells, equaling or surpassing not only the previously used monoclonal antibody ER-ICA (antibody H222; Abbott Labs, Abbott Park, IL, USA) on paraffin sections, but also the results obtained in frozen sections (9). Our studies indicate that the intensity and the number of positive cells for ER-1D5 with AR-IHC are maximized by using the AR solution at low pH (161). One explanation for the stronger intensity of ER-1D5 on paraffin sections after AR-IHC, as compared to frozen sections, may be that some receptor molecules or epitopes are lost in the frozen sections by autolysis, diffusion or the fixation employed. The major question raised here is—which of these techniques gives the right answer? Battifora (9,10) published two editorial articles analyzing this issue with respect to AR-IHC of ER and p53. Studies have shown that the increased sensitivity achieved using a ER-1D5 antibody in paraffin sections after AR-IHC provides a better correlation with overall survival and with disease-free survival than does the DCC cytosol assay or ER-ICA with antibody H222 (9). Thus, in the context of predicting prognosis and therapy, the AR-IHC method for ER using ERID5 is the preferred method. Goulding et al. (47) demonstrated the significance of assessment of ER status by using ER-1D5-AR-IHC based on a careful comparative study between ER-1D5 and ER-H222 and found that ER-1D5-AR-IHC method gave accurate results on routine paraffin sections, which was correlated with response to tamoxifen therapy with a sensitivity of 90%.

Battifora asked the question: has the time arrived to abandon the cytosol- and frozen-section-based receptor assays altogether in favor of the paraffin-based ICA? (9). We would answer in the affirmative, providing that an effective AR procedure is employed.

**Examples of application of AR-IHC in surgical pathology.** Applications are manifold. However, review of the literature reveals two main new areas of interest: proliferation related cell markers (e.g., PCNA, Ki67) and tumor suppressor genes (e.g., p53, RB).

**(A) Proliferation Related Markers.** Prototypic studies have demonstrated a significant correlation between the growth fraction of cancer cells and the histopathological grade of malignancy (45,50,86,92). Following the development of the monoclonal antibody Ki-67 (44), there have been a number of articles regarding the use of this antibody. However, antibody to Ki-67 was only applicable in early studies to frozen sections, severely limiting utilization in surgical pathology. Other approaches were therefore sought for the demonstration of proliferation markers in routine paraffin sections. Greenwell et al. (49) documented the use of a variation of AR for monoclonal antibody to PCNA (antibody 19A2) in 1991, with successful demonstration of antigens in animal tissues fixed in formalin over 24 months. In 1992, Cattoretti and Gerdes et al. (19,45) used the AR-IHC for MIB1, a newly developed monoclonal antibody to the Ki-67 antigen. MIB1 was developed by using bacterially expressed protein of Ki-67 as immunogen to manufacture antibodies that can recognize epitopes encoded by the 66 base pairs (bp) of the Ki-67 DNA sequence (86,92). The authors obtained excellent staining results using a retrieval process on routine paraffin sections, even after exposure to formalin for up to 72 h or storage for more than 60 years. Cattoretti et al. (19) summarized four advantages of using the AR-IHC technique for antibodies MIB1 or MIB3: (a) better recognition of cellular details, allowing more certain identification of positive cells; (b) high sensitivity, facilitating distinction between positive and negative cells; (c) feasibility of semiquantitative retrospective studies on archival material; and (d) utility of the method in many laboratories worldwide. These conclusions have received ample confirmation. More recently we have also been able to achieve staining of paraffin sections with the original Ki-67 antibody following AR, although the results remain inferior to those obtained with MIB1.

To evaluate the utility of MIB1-positive nuclear staining, it is important to compare the AR-IHC of MIB1 on routine paraffin sections with staining obtained in frozen sections. Barbareschi et al. (5) compared AR-IHC staining in routine paraffin sections using antibodies Ki-67 and MIB1 (both against Ki-67 antigen) on frozen sections, in a series of 40 breast carcinomas. The Ki-67 labeling index (LI) on frozen sections was $12.9 \pm 8.9$ (median 12). For MIB1 the LI was $21.2 \pm 11.9$ (median 19.5) on frozen sections and $24 \pm 15.2$ (median 21.5) on routine paraffin sections. While results of staining of Ki-67 LI and MIB1 LI on frozen and paraffin sections are strictly correlated, the cutoff values to define high and low proliferative activity are different, as reflected by the differing mean and median values. Onda et al. (135) reported a close correlation between the MIB1 proliferating cell index obtained by

AR-IHC, and the *in vivo* BUdR LI in serial sections of gliomas, recommending that MIB1 immunostaining is a useful technique for analyzing the proliferative potential. Another comparative study of AR-IHC of cell-proliferation antibodies, including monoclonal and polyclonal antibodies to Ki-67, MIB1, Ki-S5 and PC10, was carried out by Kelleher et al. (84). They found that the paraffin section AR-IHC staining results of antibodies to Ki-67 epitopes were similar to those obtained with monoclonal antibody Ki-67 on frozen sections. In 100 cases of different non-Hodgkin's lymphomas, Kreipe et al. (90) revealed almost identical results for antibody Ki-S5 on routinely fixed tissue sections (post-AR) and fresh tissues. Skalova et al. (170) studied 30 cases of acinic cell carcinoma of salivary glands by using AR-IHC with MIB1, and concluded that 5% of MIB1-positive cells represents a clinically significant threshold: none of 17 patients from total 30 cases with less than 5% MIB1 LI developed recurrences during a follow-up period of 30 years.

Siitonen et al. (169) compared the AR-IHC technique with other antigen unmasking methods, including enzyme digestion and HCl hydrolysis, using monoclonal antibodies to PCNA (19A2 and PC10) on routine paraffin sections of 109 axillary-node-negative breast cancer cases. They reported that the best immunostaining results were obtained by the AR-IHC technique using saturated lead thiocyanate with the 19A2 antibody. There was a significant correlation among the PCNA-19A2 staining, the DNA flow cytometric S-phase fraction, the mitotic count, and DNA aneuploidy. ER-positivity correlated inversely. The follow-up data showed that axillary-node-negative patients with a high PCNA-19A2 score (cutoff point 8%) had a significantly worse prognosis. They advocated the use of AR-IHC of PCNA-19A2 as a useful alternative to DNA flow cytometry for the analysis of cell proliferation activity on routine paraffin sections of breast carcinoma. In a series of 83 astrocytic tumors, Haapasalo et al. (55) demonstrated that the proliferation index, derived by AR-IHC using PCNA-19A2, showed a close association with the cellularity; they suggested that the combination of PCNA-AR-IHC and morphometry may give prognostic information that is independent of the histopathological grade.

That the use of a metal salt as the AR solution may be effective for antibody PCNA-19A2 antibody was also demonstrated by Taylor et al. (194), and Shin et al. (166). Considering the potentially toxic effect of lead, viable alternatives include 4% aluminum chloride solution or 1% zinc sulfate as introduced by Suurmeijer and Boon (178) and Shin et al. (166). Kong and Ringer (88) developed a scanning microfluorometric analysis of PCNA for archival paraffin sections of rat liver by using AR technique and crystal violet quenching procedure to reduce the autofluorescence in routine paraffin sections, and demonstrated the differences of PCNA in hyperplastic and neoplastic rat liver.

Spires et al. (173) studied 40 archival needle biopsies of prostate adenocarcinoma by using monoclonal antibody to PCNA-PC10 with AR-IHC technique and found significant differences between increasing grade and increasing proliferation index among prostate carcinomas of Gleason grades (DGG) 3, 4 and 5. They advocated the use of PCNA-PC10 as an adjunctive technique to histologic grading, particularly in needle biopsies where sample size is limited. Boon et al. (15,209) conducted a similar study of proliferating cells in cervical smears and bladder washings by using MIB1-AR-IHC.

In analyzing these data, one critical issue should be emphasized, namely that standardization of the optimal AR-IHC condition for PCNA and MIB1 is a prerequisite for any semiquantitative study. There are three major factors to consider: the nature of the AR solution, the heating conditions and pH value of the AR solution; all should be optimized and standardized. For MIB1 antibody, we have demonstrated that the strongest intensity of AR-IHC staining on routine paraffin sections may be obtained by using a low pH AR solution (Figure 1). According to Munakata and Hendricks (128), no significant difference in the total percentage of positive area is observed between tissue microwaved for 14 or 21 min, regardless of fixation time; there is reason to believe, however, that shorter or longer times outside of this range may affect the result. As noted previously, the original Ki-67 clone can also be used for immunostaining on routine paraffin sections by using pH 3.5 glycine-HCl buffer or urea solution as AR solutions (160). Using the microwave AR technique, Gu et al. (53) demonstrated PCNA and Ki-67 immunoreactivities in the nucleus of mature muscle cells in porcine heart undergoing coronary arterial angioplasty. They speculated that the muscle cells might display a potential for proliferation. The authors did, however, caution that false positivity could not be entirely excluded.

**(B) "Tumor suppressor" genes.** p53 protein has been studied extensively as a prognostic marker in pathology. Wild type p53 plays a role in regulation of cell cycle, inducing arrest in the G1 phase (36). p53 mutations are perhaps the most common molecular aberrations found in human tumors to date (16). The mutant protein is no longer functional, but having a longer half-life, it accumulates in cells and provides a convenient marker of the mutational event by IHC. Esrig et al. (36) studied p53 nuclear protein accumulation by IHC using monoclonal antibodies 1801 and BP53-12 on 4- to 12-h formalin-fixed tissues of 73 bladder carcinomas. They discovered that a cutoff point of less than 10% of p53-positive cells correlated with data from single-stranded conformational polymorphism analysis (SSCP). It appears that the study of p53 expression by IHC must be semiquantitative for an accurate interpretation of prognosis. However, the level of p53-positivity (mutation) varies greatly in different neoplasms. Individual tumor tissues also are heterogenous, e.g., ranging from 13% to 82% for ductal breast carcinomas (69). Variations may also be caused by the condition of tissues (frozen or paraffin), inequality in fixation or processing, whether tissues are freshly processed or stored, the quality of the antibody and reagents used, and the criteria employed in the interpretation of the results, as emphasized by Taylor (195).

While the literature reveals considerable variation, it also prompts the question as to whether it may be possible to retrieve masked p53-epitopes in formalin-fixed tissues in such a way as to reach a standardized result based on AR-IHC. Although p53-IHC staining can be carried out on paraffin-embedded tissue sections without pretreatment (7,8,14,36), by using tissues fixed in formalin for only a short time or by modified fixation methods, the immunoreactivity is influenced significantly by the fixation as indicated by Igarashi et al. (70) and by storage of cut slides (80,146). Therefore, it is important to standardize p53 staining and scoring when doing retrospective studies using archival routine paraffin sections, based on an optimal AR-IHC technique. The use of an effective standardized AR method levels the playing field! Binks et al. (14) compared the p53-immunostaining results in an analysis of routine paraffin sections of 54 lung cancers with and without AR. They found much higher p53-positivity in the AR-IHC group. Fifteen of 28 tumors scoring less than 5 without AR pretreatment were scored 5–200 when using MW-AR-IHC. Tenaud et al. (199) compared the results of AR-IHC by using 5 AR solutions with 8 antibodies to p53 on 21 lung carcinomas for which frozen, formalin-fixed and Bouin's fixative-fixed tissues samples were available. They concluded that p53 expression can be detected on routine paraffin sections with AR-IHC, with a reliability similar to that obtained using frozen tissue for a panel of at least 3 antibodies, with epitopes on the N-terminal or midpart of the molecule. A comparative study of p53 gene and protein alterations in 83 colorectal carcinomas based on AR-IHC using PAb 1801 was recently carried out by Bertorelle et al. (12), indicating a significant correlation between p53-AR-IHC staining and p53 gene mutations, by both PCR-SSCP and LOH (loss of heterozygosity). However, they also found that 16 out of 47 (34%) tumors that had no detectable p53 gene alterations were p53-positive by AR-IHC, a finding similar to other publications (36).

Another issue raised by using the AR-IHC of p53 is that some normal tissues and benign tumors show p53-positivity. Prior to adoption of AR, most articles reported that p53-positive immunostaining was found in malignant or premalignant lesions, but not in normal or benign tissues (7,8). Recently, using AR-IHC as a more sensitive technique for immunolocalization of p53 in routine paraffin sections, p53-positivity has been demonstrated in some normal and benign tissues, including mesothelial cells (213), paracortical area and germinal centers of lymph nodes, cortex of thymus (142), and in 63% of adenomas of thyroid (222). Pezzella et al. (142) reported cytoplasmic-perinuclear p53-positive immunostaining using the monoclonal antibody to p53 (PAb248) in a variety of normal tissues. Oohashi et al. (136) found that p53-positive cells could be demonstrated in all 6 cases of acute cholecystitis showing a sporadic pattern, that was different from carcinoma of the gallbladder with a more diffuse pattern. Linden et al. (105) demonstrated that p53 immunostaining within the proliferating components of actinic keratosis, esophagitis, proliferative phase endometrium and reactive tonsils, usually

<25% of cells, restricted to proliferating cell compartments. Lavieille et al. (99) studied 107 cases of preneoplastic lesions and squamous cell carcinoma for p53 localization by AR-IHC and found them to be p53-positive in 9% of normal-appearing but carcinogen-exposed mucosal specimens, 37% of hyperplasias, 68% of dysplasia, and 75% of in situ carcinomas. However, the significance of p53-detecting staining in premalignant lesions was supported by Krishnedath et al. (91) on Barrett's oesophagus, which is an established model of multi-step changes of malignancies, from epithelial metaplastic dysplasia to carcinoma. They demonstrated that p53 is a progression marker with high expression in high-grade dysplasia (89%). Thus, while p53 does appear to show preferential staining of tumor tissue, some quantitation is required for which it is necessary to standardize the AR method, to document patterns of reactivity of various normal and neoplastic tissues in routine paraffin sections.

## OTHER APPLICATIONS

AR-IHC has been used successfully for plastic-embedded tissue sections with excellent results (115,178). According to a study by Hopwood et al. (66), the ultrastructure of tissues is well preserved after high-temperature heating. This leads to the possibility of using the AR-IHC for immunoelectron microscopy (IEM), as indicated by the recent success of AR IEM by Stirling and Graff (176) and Wilson et al. (216). Recent studies of AR-IHC by Kaczorowski et al. (79) and Sherriff et al. (153) demonstrated that after MW treatment, the morphology was much better preserved than with enzyme digestion or formic acid treatment. We reached the same conclusion by comparing AR-IHC and enzyme digestion in detecting micrometastases by immunostaining for keratin (unpublished data).

Lan et al. (97) reported the use of AR heating method for multiple immunoenzyme staining of both routine paraffin and frozen tissue sections, in order to block antibody cross-reactivity. For complete blocking of cross-reactivity of the second antibody (link) and the label [PAP or alkaline phosphatase anti-alkaline phosphatase (APAAP)], the two 5-min MW heating treatments are used between each round of the immunohistochemical staining procedure. They demonstrated that this is an excellent method for accomplishing multiple immunolabeling without cross-reactivity. Triple immunohistochemical staining for bromodeoxyuridine and catecholamine biosynthetic enzymes by using MW AR was demonstrated by Tischler (204).

The high-temperature heating AR technique has also been used for enhancement of in situ hybridization (ISH) (174) with considerable success both in our laboratory and in the hands of Dr. Page Erickson, who used an automated stainer for this purpose (BioTech Solutions, Santa Barbara, CA, USA; personal communication). More recently, Sibony et al. (168) documented a detailed protocol of mRNA in situ hybridization by using MW heating method and obtained excellent results. The enhancement could be estimated at 60% to 120%, by computer-assisted quantification of the signal, and the morphology was very well maintained.The MW high-temperature heating method has also been adopted for enhancement of the TdT-mediated nick end-labeling technique (TUNEL) by a short MW heating exposure (1 min) (175), and silver staining of the nucleolar organizer (Ag-NOR) (119). However, Lucassen et al. (110) found that the MW heating was not suitable for *in situ* end-labeling on post-fixed, perfused brain, immersion-fixed rat brain or human postmortem brain, all of which showed strong nonspecific labeling.

## DEVELOPMENT AND STANDARDIZATION

The above studies demonstrate the value and limitations of AR. Although the AR-IHC is widely used in pathology and other fields of morphology, there are a number of unsolved and overlapping problems. These include (1) determination of the optimal AR solution and pH value for use in heating the tissue sections; (2) establishing the optimal heating conditions for a variety of antibodies tested; (3) evaluation of the results obtained by using AR-IHC based on comparative studies between frozen and routine paraffin sections, and particularly comparison of clinical behavior with results of AR-IHC; and (4) standardization and automation of the AR-immunohistochemical procedure.

As noted previously, the pH value of the solution is an important factor; the chemical composition is less so. Most, if not all, AR solutions currently used for AR-IHC, including some commercial AR solutions such as TUF (208), may be replaced by optimal buffer solutions of predetermined pH value. Chemical composition and molarity are also factors to be considered in developing any new AR solution. Ideally, an optimal AR solution should be sought for all antibodies, with predefined heating conditions.

To date there is no such universal AR solution. Commercial solutions are available from BioGenex and others. Citrate buffer pH 6.0 is perhaps most widely used, although as noted, the pH 6.0 may not be optimal for many antibodies, and Tris-HCl or Tris-HCl/5% urea may show slightly superior results for some antibodies (161,163,194). In addition, Merz et al. (121) have recently reported an ImmunoMax method that is a combination of AR with a series of enhancement procedures for the detection system, including a biotinylated tyramine step in order to make IHC staining applicable for some antibodies, such as IgD, IgM and CD7. These developments, while providing improved performance, further complicate the practical issue of achieving standardization or comparability among different reports in different laboratories.

### Practical Approach to Developing an AR Protocol

With the variety of published procedures and a huge range of antibodies

**Table 2. A Test Battery for Optimal Protocol of AR-IHC**

| AR Technique | pH 1.0 | pH 6.0 | pH 10.0 |
|---|---|---|---|
| MW 100°C 4 × 5 min | Slide #1 | Slide #4 | Slide #7 |
| MW 100°C 2 × 5 min | Slide #2 | Slide #5 | Slide #8 |
| MW 90°C 5 min | Slide #3 | Slide #6 | Slide #9 |

**Note.** In total, 10 tissue sections are used for this pilot test. One slide without AR is used as control. An optimal protocol of AR-IHC for certain antibodies may be developed on the basis of the comparison among the 10 slides tested. Tris-HCl buffer solution is recommended for this test.

available, a critical problem for every laboratory is how to develop a suitable protocol for AR-IHC for each antibody. We recommend the use of a test battery, allowing rapid comparison of selected major variables in order to define the optimal protocol.

**Test battery of AR-IHC.** The basic principle of the test battery is of a simple pilot study to evaluate the major factors that influence the effectiveness of AR, predominantly the heating conditions (temperature × time) and the pH value of the AR solution. An initial test battery is based on three pH values (pH 1.0, 6.0 and 10.0), combined with MW heating at 90°C for 5 min, 100°C for 10 min and 100°C for 20 min as summarized in Table 2 (10 slides total including control). Tris buffer is chosen as the AR solution for the test battery based upon our experience of overall broad utility. Elsewhere we have described the testing of 14 antibodies using this test battery, and have reported satisfactory results in terms of developing an optimal protocol for each antibody (162). For example, by this approach, we established an optimal protocol of MW heating at 100°C for 10 min in low pH buffer (pH 1.0–2.0) for polyclonal (RB-WL-1) and a monoclonal antibody to RB G3-245 provided by PharMingen (San Diego, CA, USA) and BioGenex. In another variation of the test battery, we compared high-temperature heating by autoclave (120°C), with elongated heating by MW time (1 h at 100°C), and the MW with pressure cooker approach, obtaining similar excellent staining of RB (164,198). Following identification of the preferred pH range and preferred heating condition, other buffers may be substituted. This type of simple pilot study (test battery) provides a convenient starting point in the establishment of any new IHC stain in the laboratory and is preferable to the often-used default alternative of random trial and error.

## PROTOCOLS OF AR TECHNIQUE

### 1. Requirements for routinely formalin-fixed, paraffin-embedded tissue sections prior to AR treatment.

(1) Mount the tissue sections on slides coated by using either poly-L-lysine or 3-aminopropyltriethoxysilane (APES), or charged slides provided by

Fisher Scientific (Pittsburgh, PA, USA). Dry the slides at 60°C for at least 1 h to adhere the tissue sections to the slides. An overnight incubation of the slides produces optimal adhesion.

(2) Deparaffinize by using routine procedure: Histoclear or xyline, rehydration by graded alcohol, blocking the endogeneous peroxidase by $H_2O_2$-methanol solution.

(3) Rinse the slides in distilled water 3 changes for 15 min, followed by AR treatment.

## 2. Microwave (MW) heating method (154)

(1) Slides are placed in plastic Coplin jars containing AR solution, such as distilled water, buffer solution, metal salt solution, urea, pH 3.5 glycine-HCl, etc.

(2) Jars are covered with loose-fitting screw caps and heated in the microwave oven to boiling point (100°C) for either 5 or 10 min. The 10-min heating time is divided into two 5-min cycles with an interval of 1 min between cycles to check on the fluid level in the jars. If necessary, more AR solution is added after the first 5 min to avoid drying the tissue sections.

(3) After heating, the Coplin jars are removed from the oven and allowed to cool for 15 min.

(4) Slides are then rinsed in distilled water twice and in PBS for 5 min, followed by IHC staining.

## 3. H2550 Laboratory MW Processor (Energy Beam Sciences).

(1) This MW processor with temperature readouts is available. Its power output is regulated by a temperature feedback mechanism and a timer, and it is capable of monitoring both temperature and heating time.

(2) Fix a Coplin jar on a thick capboard or plastic plate and set it in the center portion of the turntable, filling the jar with distilled water or tap water. Set the temperature probe into the Coplin jar through a hole in the cap to measure the temperature.

(3) Set all test jars around the central probe jar as closely as possible.

(4) Turn on the MW processor. Set the temperature and the time at temperature as required.

(5) Turn on the turntable, making sure that the jars are not moved by the probe and the table.

(6) Start the MW heating; the timer is automatically controlled.

(7) For heating at 100°C, the heating time should be divided into 5-min cycles as mentioned above.

(8) This MW processor is particularly useful for test battery (see below).

## 4. Calibration technique for MW oven (BioGenex).

(1) Microwave at high power (800 W) for 2–3 min until solution comes to

a rapid boil; then turn off the oven. Note the exact time it takes for the solution to boil.

(2) Set oven power on 3 or 4 (30%–40% power or 250+50 W) or on defrost (low power). Heat for 7–10 min (If using a 500-W oven, set power level at 5 or 6 or 50%–60%, i.e., 250 W). Adjust the setting so oven cycles on and off every 20–30 s and the solution boils about 5–10 s each cycle.

(3) The following formula can be used to determine the power setting: S = 250/P × 10, where S is the MW power setting for AR and P is the output power of the individual MW oven. For example, if a MW oven output power is 800 W, then the power setting for AR (S) is: S = 250/800 × 10 = 3.1. Therefore, it should be set at 3 and heated for 7–10 min (186).

## 5. Autoclave heating method (4)

(1) Slides are set in Coplin jars or other kinds of containers filled with AR solutions. The covers are fixed in place with a special tape designed for the autoclave.

(2) Set the jars with slides in the center portion of the autoclave.

(3) Tightly close the door of the autoclave as required by the instructions.

(4) Set the temperature at 120°C for 10 min.

(5) Allow a cool down period of 20–30 min, after heating.

(6) For IHC staining, follow the same procedure as for MW heating.

## 6. Pressure cooking method (134)

(1) A domestic cooker with an operating pressure of 103 kPa/15 psi is one-third filled with AR solution and heated by hotplate to boil.

(2) Slides suspended in metal slide racks are placed quickly into the AR solution, and the pressure lid is tightly replaced.

(3) Timer starts when the pressure indicator valve reaches the maximum (around 4 min). The optimal period of pressurized boiling is 1–2 min.

(4) After heating, the cooker is depressurized and cooled under running water, then the lid is removed. Cold tap water is added to replace the hot AR solution. A 15–20 min cooling time may be required as in the MW heating method.

## 7. MW + plastic pressure cooker (140,198)

(1) Three plastic staining jars, containing as many as 25 slides, filled with AR solution are placed into a plastic pressure cooker (Nordicware, MN). About 600 mL of distilled water is added in the pressure cooker to reach one-half of the volume, making sure that all three jars stand in the water in a stable position.

(2) The plastic pressure cooker containing three staining jars with tissue slides as described above is placed in the MW oven (A Sharp carousel MW oven, Model R-4A46 with multiple sequence cooking which can switch from

one power level setting to another automatically; Sharp, Mahwah, NJ, USA). The oven is set at maximum power (900 W, 2450 MHz) for 15 min to boil the water, followed by a 40% power setting for another 15 min (simmer) to maintain boiling.

(3) Cool down for 15 to 20 min, followed by the procedure as for the regular MW AR method.

### 8. Steam heating AR (198)

(1) We recommend using the "Steam HIER" (Heat-Induced Epitope Retrieval) available from BioTek, which allows the convenient use of the BioTek autostainer.

(2) The slides are set in the TechMate slide holder in the regular face-to-face orientation maintaining the capillary gap.

(3) The steamer is filled with distilled water to the top line and heated to boiling point. The dial is than turned to 30 min for the next step.

(4) Add 10 mL to each of the 10-well trays in the grey tile provided by BioTek, and place the slide holder on the grey tile with that the tips of the slide pairs in the AR solution locate in the 10-well trays. Then, place the grey tile with slide holder into the center of the steam chamber.

(5) Place the whole steam chamber base on the steamer while boiling vigorously for 20 min.

(6) After cooling the heated slides for 15 min, the remaining procedure for IHC is as the MW heating method.

### 9. Test battery to develop an optimal protocol of AR for certain antibody tested (162,164)

(1) Ten slides numbered 1 to 10 are used for three pH values of AR solution and three different heating conditions as follows:

| AR Solution | pH 1 | pH 6 | pH 10 |
|---|---|---|---|
| Super-high temperature | #1 | #4 | #7 |
| High temperature | #2 | #5 | #8 |
| Mid-high temperature | #3 | #6 | #9 |

A tenth slide stands as a non-AR control.

(2) Super-high temperature: Place slides in the microwave oven at 100°C for 4 × 5 min, or heat in an autoclave at 120°C for 10 min. High temperature: Microwave at 100°C for 10 min as in the original protocol. Mid-high temperature: Place in a temperature-controlled MW oven or water bath at 90°C for 10 min. If possible, a H2550 MW processor may be used to maintain the heating conditions accurately, as listed above.

(3) The use of an autostainer is recommended to perform IHC staining.

(4) Evaluate the intensity for all 10 slides in order to identify the optimal protocol for the antibody under test.

(5) Additional studies, including different AR solutions, or different heating methods may be used if necessary.

## 10. Non-heating AR method for routinely formalin-fixed, acid-decalcified, celloidin-embedded tissue section (155, 157)

(1) Preparation of AR solution. Saturated sodium hydroxide (NaOH) methanol solution is made of 50–100 g of NaOH mixed with 500 mL methanol in a brown-colored bottle by vigorous shaking. After standing for 1–2 weeks, the supernant can be used diluted 1:3 in methanol. (This NaOH-methanol solution, DECAL™, is commercially available from BioGenex, Cat. # HK089-5K).

(2) The celloidin-embedded tissue sections are washed in distilled water for 10 min and mounted on either 0.1% poly-L-lysine (Sigma Chemical, St. Louis, MO, USA) or aqueous mounting media (BioGenex, #HK009-5K) coated slides by pressing the section down with filter paper and trimming the tissue along the edges of the slides.

(3) Place a few drops of 0.1% poly-L-lysine on the slide to cover the whole section and dry briefly in an oven. Pay attention to controlling the optimal time of drying the sections in the oven. Do not dry excessively.

(4) Immerse the slides in freshly prepared AR solution (DECAL, diluted in 1:3) for 30 min. Rinse the slides in 100% methanol for 2× 15 min, 70% methanol 2 × 15 min, PBS 2 × 15 min, 0.3% Triton® X-100 for 10 min, and PBS 15 min. The slides are then ready for immunohistochemical staining.

(5) For thicker (50-μm) celloidin tissue sections, a floating method may be used to avoid mounting the sections on the glass slides.

The DECAL AR solution has also been used for enhancement of immunostaining on acid-decalcified, formalin-fixed paraffin sections for antibodies such as keratin AE1, AE3, vimentin, GFAP, desmin, muscle-specific actin, S-100, NSE, kappa, lambda, bcl-2, L-26, and T cell (UCHL) (Harkins, L.E. and W.E. Grizzle. 1995. Advanced techniques in immunohistochemistry, Workshop #35, National Society for Histotechnology Annual Symposium 199 5, Buffalo, New York, October 7-13).

(6) A combined AR heating and non-heating (DECAL) method may be used to improve some antibodies that give poor results using non-heating method alone. In this combined method, the heating induced AR is performed before immersing the slides in the DECAL solution.

## 11. AR multiple immunoenzyme staining (97)

For PLP- or formalin-fixed frozen tissue section:

(1) Pre-incubate with 10% fetal calf serum and 10% normal goat serum for 10 min.

(2) Sections are drained, then incubated with the primary antibody, for the required time, followed by PBS wash.

(3) Incubate with secondary antibody, followed by another PBS wash.

(4) Incubate with label (such as APAAP) for 30 min and develop with the Fast Blue BB base (Sigma) substrate in the presence of 2 mM levamisole (for APAAP).

(5) Sections are then microwaved in citrate buffer, pH 6.0, for $2 \times 5$ min as routine AR treatment in order to block completely all antibodies bound to the tissue.

(6) Sections are then washed in PBS, prior to adding a second different primary antibody and repeating steps 1 to 4. If using peroxidase label, a blocking step of incubation in 0.3% $H_2O_2$ in methanol for 20 min may be added after step 1.

(7) If a third or fourth primary antibody is required for multiple immuno-labeling, step 5 (MW heating procedure) should be repeated between each of the immunostaining procedures. For paraffin sections: The same procedure (i.e., step 5, MW heating) is used between each immunostaining sequence, except that: (1) MW heating is used prior to the first primary antibody staining as in AR (i.e., before step 1); (2) As some surface antigens may not tolerate the $2 \times 5$ min MW heating, this method may be limited to nuclear and cytoplasmic antigens.

## 12. AR immunoelectron microscopy (176)

(1) Routinely processed epoxy resin-embedded tissue block is cut, flattened with chloroform vapor and collected on formvar-coated nickel grids.

(2) The grids are etched with either a saturated aqueous solution of sodium metaperiodate for 1 h or 10% fresh saturated solution of sodium ethoxide (prepared overnight) diluted with anhydrous ethanol for 2 min.

(3) Heat the grids with a pre-heated by MW oven to boiling point; sections are heated at 95°-100°C by a hotplate for 10 min. (citrate buffer pH 6.0 is recommended as the AR solution).

(4) After heating, the grids are allowed to cool for 15 min before immnostaining.

## 13. MW heating for enhancement of mRNA *in situ* hybridization (168)

(1) After deparaffinization as mentioned above, rinse in 0.85% NaCl for 5 min.

(2) The slides are immersed in a sodium citrate buffer (0.01 M) at pH 6.0. Sodium citrate buffer is prepared by diluting 1:10 in autoclaved water consisting of a mixture of two autoclaved stock solutions: one part of a 0.1 M citric acid solution and four parts of a 0.1 M trisodium citrate solution. Slides are immersed in 600 mL of this buffer and are heated at full power in the MW oven for 7 min to reach the boiling point and another 5 min with a short interruption to check the liquid level. If necessary, more distilled water is added to

restore the initial level of buffer solution as in the MW AR method. After heating, the sections are allowed to remain in the same solution for 20 min to cool.

(3) Follow with the ISH procedure.

## 14. MW heating for the TUNEL method (175)

(1) For paraffin sections for enhacement of TUNEL.

(2) Tissue sections that are immersed in 10 mM citrate buffer pH 6.0 and heated by microwave for 1 min at 750 W show enhanced sensitivity with the TUNEL method.

### ACKNOWLEDGMENTS

We thank Dr. Ashraf Imam for his help in the cooperative study of AR technique. Ms. Lillian Young and Christina Yang for their excellent help in technical assistance, Linda Montellano and Lisa Sanchez for kind help in processing this manuscript. This paper was supported by the University Pathology Associates of U.S.C., by the American Cancer Society (ACS IN-21-31) and by the Firestein-Gertz Cancer Research Fund.

### REFERENCES

1. **Al Saati, T., S. Clamens, E. Cohen-Knafo, J.C. Faye, H. Prats, J.M. Coindre, J. Wafflart, P. Caveriviere, F. Bayard and G. Delsol.** 1993. Production of monoclonal antibodies to human estrogen-receptor protein (ER) using recombinant ER (RER). Int. J. Cancer *55*:651-654.
2. **Arber, D.A., P.L. Kandalaft, P. Mehta and H. Battifora.** 1993. Vimentin-negative epithelioid sarcoma. The value of an immunohistochemical panel that includes CD34. Am. J. Surg. Pathol. *17*:302-307.
3. **Baas, I.O., J.-W. R. Mulder, G.J.A. Offerhaus, B. Vogelstein and S.R. Hamilton.** 1994. An evaluation of six antibodies for immunohistochemistry of mutant p53 gene product in archival colorectal neoplasms. J. Pathol. *172*:5-12.
4. **Bankfalvi, A., H. Navabi, B. Bier, W. Bocker, B. Jasani and K.W. Schmid.** 1994. Wet autoclave pretreatment for antigen retrieval in diagnostic immunohistochemistry. J. Pathol. *174*:223-228.
5. **Barbareschi, M., S, Girlando, F.M. Mauri , S. Forti, C. Eccher, F.A. Mauri, R.Togni, P.D. Palma and C. Doglioni.** 1994. Quantitative growth fraction evaluation with MIB1 and Ki67 antibodies in breast carcinomas. Am. J. Clin. Pathol. *102*:171-175.
6. **Barlow, D.J., M.S. Edwards and J.M. Thornton.** 1986. Continuous and discontinuous protein antigenic determinants. Nature *322*:747-748.
7. **Bartek, J., J. Bartkova, B. Vojtesek, Z. Staskova, A. Rejthar, J. Kovarik and D.P. Lane.** 1990. Patterns of expression of the p53 tumour suppressor in human breast tissues and tumours in situ and in vitro. Int. J. Cancer *46*:839-844.
8. **Bartek, J., J. Bartkova, B. Vojtesek, Z. Staskova, J. Lukas, A. Rejthar, J. Kovarik, C.A. Midgley, J.V. Gannon and D.P. Lane.** 1991. Aberrant expression of the p53 oncoprotein is a common feature of a wide spectrum of human malignancies. Oncogene *6*:1699-1703.
9. **Battifora, H.** 1994. Immunocytochemistry of hormone receptors in routinely processed tissues. Appl. Immunohistochem. *2*:143-145.
10. **Battifora, H.** 1994. p53 Immunohistochemistry: a word of caution. Hum. Pathol. *25*:435-437.
11. **Battifora, H.** 1994. Metastatic breast carcinoma to the stomach simulating linitis plastica. Appl. Immunohistochem. *2*:225-228.
12. **Bertorelle, R., G. Esposito, A.D. Mistro, C. Belluco, D. Nitti, M. Lise and L. Chieco-Bianchi.** 1995. Association of p53 gene and protein alterations with metastases in colorectal cancer. Am. J. Surg. Pathol. *19*:463-471.
13. **Bianchi, E., R.L. Cohen, A. Dai, A.T. Thor, M.A. Shuman and H.S. Smith.** 1995. Immunohistochemical localization of the plasminogen activator inhibitor-1 in breast cancer. Int. J. Cancer 60:597-

603.

14. **Binks, S., C.A. Clelland, P. Gadsdon and J. Ronan.** 1994. Enhancement of p53 oncogene product expression in lung cancers using microwave antigen retrieval. J. Pathol. Suppl. *173*:178.

15. **Boon, M.E., E.D. Kleinschmidt-Guy and N.E. Quwerkerk.** 1994. PAPNET for analysis of proliferating (MIB-1 positive) cell populations in cervical smears. Eur. J. Morphol. *32*:78-85.

16. **Bosl, G.J., W.R. Fair, H.W. Herr, D.F. Bajorin, G. Dalbagni, A.S. Sarkis, V.E. Reuter, C. Cordon-Cardo, J. Sheinfeld and H.I. Scher.** 1994. Bladder cancer: advances in biology and treatment. Crit. Rev. Oncol. Hematol. *16*:33-70.

17. **Cagle, P.T., R.W. Brown and R.M. Lebovitz.** 1994. p53 Immunostaining in the differentiation of reactive processes from malignancy in pleural biopsy specimens. Hum. Pathol. *25*:443-448.

18. **Cattoretti, G.,F. Rilke, S. Andreola, L. D'Amato and D. Delia.** 1988. p53 Expression in breast cancer. Int. J. Cancer *41*:178-183.

19. **Cattoretti, G., M.H.G. Becker, G. Key, M. Duchrow, C. Schluter, J. Galle and J. Gerdes.** 1992. Monoclonal antibodies against recombinant parts of the Ki-67 antigen (MIB 1 and MIB 3) detect proliferating cells in microwave-processed formalin-fixed paraffin sections. J. Pathol. *168*:357-363.

20. **Cattoretti, G., S. Pileri,C. Parravicini, M.H.G. Becker, S. Poggi, C. Bifulco, G. Key, L. D'Amato, E. Sabattini, E. Feudale, F. Reynolds, J. Gerdes and F. Rilke.** 1993. Antigen unmasking on formalin-fixed, paraffin-embedded tissue sections. J. Pathol. *171*:83-98.

21. **Chan, W.K., M.M. Mole, D.A. Levison, R.Y. Ball, Q.-L. Lu, K. Patel and A.M. Hanby.** 1995. Nuclear and cytoplasmic bcl-2 expression in endometrial hyperplasia and adenocarcinoma. J. Pathol. *177*:241-246.

22. **Charalambous,C., N. Singh and P.G. Issacson.** 1993. Immunohistochemical analysis of Hodgkin's disease using microwave heating. J. Clin. Pathol. *46*:1085-1088.

23. **Cho, K.J., A.K. El-Naggar, N.G. Ordonez, M.A. Luna, J. Austin and J.G. Batsakis.** 1995. Epithelial-myoepithelial carcinoma of salivary glands. A clinicopathologic, DNA flow cytometric, and immunohistochemical study of Ki-67 and HER-2/neu oncogene. Am. J. Clin. Pathol. *103*:432-437.

24. **Colvin, R.B., A.K.Bhan and R.T.McCluskey (Eds.).** 1988. Diagnostic Immunopathology. Raven Press, New York.

25. **Coons, A.H., H.J. Creech and R.N. Jones.** 1941. Immunological properties of an antibody containing a fluorescent group. Proc. Soc. Exp. Biol. Med. *47*:200-202.

26. **Costa, P.P., B. Jacobsson, V.P. Collins and P. Biberfeld.** 1986. Unmasking antigen determinants in amyloid. J. Histochem. Cytochem. *34*:1683-1685.

27. **Cote, R.J. and C.R. Taylor.** 1996. Immunohistochemistry and related marking techniques, p. 136-175. *In* I. Damjanov and J. Linder (Eds.), Anderson's Pathology, 10th ed. CV Mosby, Philadelphia.

28. **Cuevas, E.C., A.C. Bateman, B.S. Wilking, P.A. Johnson, J.H. Williams, A.H.S. Lee, D.B. Jones and D.H. Wright.** 1994. Microwave antigen retrieval in immunocytochemistry: a study of 80 antibodies. J. Clin. Pathol. 47:448-452.

29. **DeLellis, R.A. (Ed.).** 1988. Advances in Immunohistochemistry. Raven Press, New York.

30. **de Luque, M.M.D. and E.H. Luque.** 1995. Effect of microwave pretreatment on proliferating cell nuclear antigen (PCNA) immunolocalization in paraffin sections. J. Histotechnol. *18*:11-16.

31. **Desai, B., J.M. Burrin, C.A. Nott, J.F. Geddes, E.J. Lamb, S.J.B. Aylwin, D.F. Wood, C. Thakkar and J.P. Monson.** 1995. Glycoprotein hormone alpha-subunit production and plurihormonality in human corticotroph tumours—and in vitro and immunohistochemical study. Eur. J. Endocrinol. *133*:25-32.

32. **DiGiuseppe, J.A., R.H. Hruban, S.N. Goodman, M. Polak, F.M. van den Berg, D.C. Allison, J.L. Cameron and J.A. Offerhaus.** 1994. Over-expression of p53 protein in adenocarcinoma of the pancreas. Am. J. Clin. Pathol. *101*:684-688.

33. **Dookham, D.B., A.J. Kovatich and M. Miettinen.** 1993. Non-enzymatic antigen retrieval in immunohistochemistry. Comparison between different antigen retrieval modalities and proteolytic digestion. Appl. Immunohistochem. *1*:149-155.

34. **Elias, J.M., M. Margiotta and D. Gaborc.** 1989. Sensitivity and detection efficiency of the peroxidase antiperoxidase (PAP), avidin-biotin peroxidase complex (ABC), and peroxidase-labeled avidin-biotin (LAB) methods. Am. J. Clin. Pathol. *92*:62-67.

35. **Emanuels, A., H. Hollema, A. Suurmeyer and J. Koudstaal.** 1994. A modified method for antigen retrieval MIB-1 staining of vulvar carcinoma. Eur. J. Morphol. Suppl. *32*:335-337.

36. **Esrig, D., C.H. Spruck III, P.W. Nichols, B. Chaiwun, K. Steven, S. Groshen, S.-C. Chen, D.G. Skinner, P.A. Jones and R.J. Cote.** 1993. p53 Nuclear protein accumulation correlates with mutations in the p53 gene, tumor grade, and stage in bladder cancer. Am. J. Pathol. *143*:1389-1397.

37. **Esteban, J.M., C. Ahn, H. Battifora and B. Felder.** 1994. Quantitative immunohistochemical assay for hormonal receptors: technical aspects and biological significance. J. Cell. Biochem. Suppl.

*19*:138-145.
38.**Evers, P. and H.B.M. Uylings.** 1994. Microwave-stimulated antigen retrieval is pH and temperature dependent. J. Histochem. Cytochem. *42*:1555-1563.
39.**Fraenkel-Conrat, H., B.A. Brandon and H.S. Olcott.** 1947. The reaction of formaldehyde with proteins. IV. Participation of indole groups. Gramicidin. J. Biol. Chem. *168*:99-118.
40.**Fraenkel-Conrat, H. and H.S. Olcott.** 1948. Reaction of formaldehyde with proteins. VI. Cross-linking of amino groups with phenol, imidazole, or indole groups. J. Biol. Chem. *174*:827-843.
41.**Fraenkel-Conrat, H. and H.S. Olcott.** 1948. The reaction of formaldehyde with proteins. V. Cross-linking between amino and primary amide or guanidyl groups. J. Am. Chem. Soc. *70*:2673-2684.
42.**Gee, J.M.W., A. Douglas-Jones, P. Hepburn, A.K. Sharma, R.A. McClelland, I.O. Ellis and R.I. Nicholson.** 1995. A cautionary note regarding the application of Ki-67 antibodies to paraffin-embedded breast cancers. J. Pathol. *177*:285-293.
43.**Geradts, J., S.-X. Hu, C.E. Lincoln, W.F. Benedict and H.-J. Xu.** 1994. Aberrant RB gene expression in routinely processed, archival tumor tissue determined by three different anti-RB antibodies. Int. J. Cancer *58*:161-167.
44.**Gerdes, J., U. Schwab, H. Lemke and H. Stein.** 1983. Production of a mouse monoclonal antibody reactive with a human nuclear antigen associated with cell proliferation. Int. J. Cancer *31*:13-20.
45.**Gerdes, J., M.H.G. Becker, G. Key and G. Cattoretti.** 1992. Immunohistological detection of tumour growth fraction (Ki-67) in formalin-fixed and routinely processed tissues. J. Pathol. *168*:85-87.
46.**Goldberg, D.E., J. Stuart and F.C. Koerner.** 1994. Progesterone receptor detection in paraffin sections of human breast cancers by an immunoperoxidase technique incorporating microwave heating. Mod. Pathol. 7:401-406.
47.**Goulding, H., S. Pinder, P. Cannon, D. Pearson, R. Nicholson, D. Snead, J. Bell, C.W.E. Elston, J.F. Robertson, R.W. Blamey and I.O. Ellis.** 1995. A new immunohistochemical antibody for the assessment of estrogen receptor status on routine formalin-fixed tissue samples. Hum. Pathol. *26*:291-294.
48.**Gown, A.M., N. de Wever and H. Battifora.** 1993. Microwave-based antigenic unmasking. A revolutionary new technique for routine immunohistochemistry. Appl. Immunohistochem. *1*:256-266.
49.**Greenwell, A., J.F. Foley and R.R. Maronpot.** 1991. An enhancement method for immunohistochemical staining of proliferating cell nuclear antigen in archival rodent tissues. Cancer Lett. *59*:251-256.
50.**Greenwell, A., J.F. Foley and R.R. Maronpot.** 1993. Detecting proliferating cell nuclear antigen in archival rodent tissues. Environ. Health Perspect. Suppl. *101*:207-210.
51.**Grizzle, W.E., R.B. Myers, M.M. Arnold and S. Srivastava.** 1994. Evaluation of biomarkers in breast and prostate cancer. J. Cell. Biochem. Suppl. *19*:259-266.
52.**Gu, J.** 1994. Microwave in immunocytochemistry, p. 67-80. *In* J. Gu and G.W. Hacker (Eds.), Modern Methods in Analytical Morphology. Plenum Press, New York.
53.**Gu, J., M. Forte, C. Xenachis, N. Tarazona, J. Windsor and E.C. Santoian.** 1994. Immunohistochemical demonstration of PCNA and Ki 67 positivities in stimulated myochariocytes indicates dividing potential for cardiac muscle cells in adult heart. Cell Vision *1*:91-92.
54.**Gu, J., M. Forte, H. Hance, N. Carson, C. Xenachis and R. Rufner.** 1994. Microwave fixation, antigen retrieval and accelerated immunocytochemistry. Cell Vision *1*:76-77.
55.**Haapasalo, H.K., P.K. Sallinen, P.T. Helen, I.S. Rantala, H.J. Helin and J.J. Isola.** 1993. Comparison of three quantitation methods for PCNA immunostaining: applicability and relation to survival in 83 astrocytic neoplasms. J. Pathol. *171*:207-214.
56.**Haerslev, T. and G.K. Jacobsen.** 1994. Microwave processing for immunohistochemical demonstration of proliferating cell nuclear antigen (PCNA) in formalin-fixed and paraffin-embedded tissue. APMIS *102*:395-400.
57.**Haerslev, T. And G.K. Jacobsen.** 1995. An immunohistochemical study of p53 with correlations to histopathological parameters, c-erbB-2, proliferating cell nuclear antigen, and prognosis. Hum. Pathol. *26*:295-301.
58.**Hamana,T., K. Kawai, A. Serizawa,Y. Tsutsumi and K. Watanabe.** 1994. Immunohistochemical demonstration of p53 protein in colorectal adenomas and adenocarcinomas. Reliable application of the heat-induced antigen retrieval method to formalin-fixed, paraffin-embedded material. Pathol. Int. *44*:765-770.
59.**Happerfield, L.C., E.A. Dublin and L.G. Bobrow.** 1993. Microwave-stimulated antigen retrieval: A new way of overcoming problems caused by fixation? J. Pathol. Suppl. *170*:382.
60.**Hazelbag, H.M., L.J.C.M.v.d. Broek, E.B.L. van Dorst, G.J.A. Offerhaus, G.J. Fleuren and P.C.W. Hogendoorn.** 1995. Immunostaining of chain-specific keratins on formalin-fixed, paraffin-embedded tissues: a comparison of various antigen retrieval systems using microwave heating and

proteolytic pre-treatments. J. Histochem. Cytochem. *43*:429-437.

61. **Hellquist, H.B., H. Hellquist, L.Vejlens and C.E. Lindholm.** 1994. Epithelioid leiomyoma of the larynx. Histopathology *24*:155-159.

62. **Hellquist, H.B., M.G. Karlsson and C. Nilsson.** 1994. Salivary duct carcinoma—a highly aggressive salivary gland tumour with overexpression of c-erbB-2. J. Pathol. *172*:35-44.

63. **Hendricks, J.B. and E.J.Wilkinson.** 1993. Comparison of two antibodies for evaluation of estrogen receptors in paraffin-embedded tumors. Mod. Pathol. *6*:765-770.

64. **Hendricks, J.B. and E.J. Wilkinson.** 1994. Quality control considerations for Ki-67 detection and quantitation in paraffin-embedded tissue. J. Cell. Biochem. Suppl. *19*:105-110.

65. **Hernandez, A.M., B.H.N. Nathwani, D. Nguyen, D. Shibata, W. Chuan, P. Nichols and C.R. Taylor.** 1995. Nodal benign and malignant monocytoid B cells with and without follicular lymphomas: a comparative study of follicular colonization, light chain restriction, bcl-2, and t(14;18) in 39 cases. Hum. Pathol. *26*:625-632.

66. **Hopwood, D., G. Milne and J. Penston.** 1990. A comparison of microwaves and heat alone in the preparation of tissue for electron microscopy. Histochem. J. *22*:358-364.

67. **Hsu, S.-M., L. Raine and H. Fanger.** 1981. Use of avidin-biotin-peroxidase complex (ABC) in immunoperoxidase techniques: a comparison between ABC and unlabeled antibody (PAP) procedures. J. Histochem. Cytochem. *29*:577-580.

68. **Huang, S.-N.** 1975. Immunohistochemical demonstration of hepatitis B core and surface antigens in paraffin sections. Lab. Invest. *33*:88-95.

69. **Hurlimann, J., P. Chaubert and J. Benhattar.** 1994. p53 Gene alterations and p53 protein accumulation in infiltrating ductal breast carcinomas: correlation between immunohistochemical and molecular biology techniques. Mod. Pathol. *7*:423-428.

70. **Igarashi.H., H. Sugimura, K. Maruyama, Y. Kitayama, I. Ohta, M. Suzuki, M. Tanaka, Y. Dobashi and I. Kino.** 1994. Alteration of immunoreactivity by hydrated autoclaving, microwave treatment, and simple heating of paraffin-embedded tissue sections. APMIS *102*:295-307.

71. **Imam, S.A., L. Young, B. Chaiwun and C.R. Taylor.** 1995. Comparison of 2 microwave based antigen retrieval solutions in unmasking epitopes in formalin-fixed tissue for immunostaining. Anticancer Res. *15*:1153-1158.

72. **Inoue, M., M. Fujita, T. Enomoto, H. Morimoto, T. Monden, T. Shimano and O. Tanizawa.** 1994. Immunohistochemical analysis of p53 in gynecologic tumors. Am. J. Clin. Pathol. *102*:665-670.

73. **Iwamura, M., P.-A. Abrahamsson, C.M. Benning, A.T.K. Cockett and P.A. di Sant'Agnese.** 1994. Androgen receptor immunostaining and its tissue distribution in formalin-fixed, paraffin-embedded sections after microwave treatment. J. Histochem. Cytochem. *42*:783-788.

74. **Iwamura, M. S. Gershagen, O. Lapets, R. Moynes, P.-A. Abrahamsson, A.T.K.Cockett, L.J. Deftos and P.A. di Sant'Agnese.** 1995. Immunohistochemical localization of parathyroid hormone-related protein in prostatic intraepithelial neoplasia. Hum. Pathol. *26*:797-801.

75. **Janssen, P.J., A.O. Brinkmann, W.J. Boersma and T.H. van der Kwast.** 1994. Immunohistochemical detection of the androgen receptor with monoclonal antibody F39.4 in routinely processed, paraffin-embedded human tissues after microwave pre-treatment. J. Histochem. Cytochem. *42*:1169-1175.

76. **Jensen, V. and M. Ladekarl.** 1995. Immunohistochemical quantitation of oestrogen receptors and proliferative activity in oestrogen receptor positive breast cancer. J. Clin. Pathol. *48*:429-432.

77. **Jones, M.D., P.M. Banks and B.L. Caron.** 1981. Transition metal salts as adjuncts to formalin for tissue fixation. Lab. Invest. *44*:32.

78. **Jotti, G.S., S.R.D. Johnston, J. Salter, S. Detre and M. Dowsett.** 1994. Comparison of new immunohistochemical assay for oestrogen receptor in paraffin wax embedded breast carcinoma tissue with quantitative enzyme immunoassay. J. Clin. Pathol. *47*:900-905.

79. **Kaczorowski, S., M. Kaczorowska and B. Christensson.** 1994. Expression of EBV encoded latent membrane protein 1 (LMP-1) and bcl-2 protein in childhood and adult Hodgkin's disease: application of microwave irradiation for antigen retrieval. Leuk. Lymphoma *13*:273-283.

80. **Kato, J, S. Sakamaki and Y. Niitsu.** 1995. More on p53 antigen loss in stored paraffin slides. N. Engl. J. Med. *333*:1507-1507.

81. **Kawai, K., S.Umemura and Y.Tsutsumi.** 1994. Antigen retrieval by heating treatment. Saibo (Japan) *26*:152-157.

82. **Kawai, K., A. Serizawa, T. Hamana and Y. Tsutsumi.** 1994. Heat-induced antigen retrieval of proliferating cell nuclear antigen and p53 protein in formalin-fixed, paraffin-embedded sections. Pathol. Int. *44*:759-764

83. **Kell, D.L., O.W. Kamel and R.V. Rouse.** 1993. Immunohistochemical analysis of breast carcinoma estrogen and progesterone receptors in paraffin-embedded tissue. Correlation of clones ER1D5 and 1A6 with a cytosol-based hormone receptor assay. Appl. Immunohistochem. *1*:275-281.

84. Kelleher, L., H.M. Magee and P.A. Dervan. 1994. Evaluation of cell-proliferation antibodies reactive in paraffin sections. Appl. Immunohistochem. 2:164-170.

85. Kelly, D.P., M.K. Dewar, R.B. Johns, W.-L. Shao and J.F. Yates. 1977. Cross-linking of amino acids by formaldehyde. Preparation and $^{13}$C NMR spectra of model compounds, p. 641-647. *In* M. Friedman (Ed.), Protein Crosslinking. Symposium on Protein Crosslinking, San Francisco, 1976. Plenum Press, New York.

86. Key, G., M.H.G. Becker, B. Baron, M. Duchrow, C. Schluter, H.-D. Flad and J. Gerdes. 1993. New Ki-67-equivalent murine monoclonal antibodies (MIB 1-3) generated against bacterially expressed parts of the Ki-67 cDNA containing three 62 base pair repetitive elements encoding for the Ki-67 epitope. 1993. Lab. Invest. *68*:629-636.

87. Khan, G., R.K. Gupta, P.J. Coates and G. Slavin. 1993. Epstein-Barr virus infection and bcl-2 proto-oncogene expression. Separate events in the pathogenesis of Hodgkin's disease? Am. J. Pathol. *143*:1270-1274.

88. Kong, J. And D.P. Ringer. 1995. Scanning microfluorometric analysis of proliferating cell nuclear antigen in formalin-fixed sections of hyperplastic and neoplastic rat liver. Cytometry *20*:86-93.

89. Kitamoto, T., K. Ogomori, J. Tateishi and S.B. Prusiner. 1987. Formic acid pretreatment enhances immunostaining of cerebral and systemic amyloids. Lab. Invest. *57*:230-236.

90. Kreipe, H, H.-H. Wacker, H.J. Heidebrecht, K. Haas, M. Hauberg, M. Tiemann and R. Parwaresch. 1993. Determination of the growth fraction in non-Hodgkin's lymphomas by monoclonal antibody Ki-S5 directed against a formalin-resistant epitope of the Ki-67 antigen. Am. J. Pathol. *142*:1689-1694.

91. Krishnadath, K.K., H.W. Tilanus, M. Van Blankenstein, F.T. Bosman and A.H. Mulder. 1995. Accumulation of p53 protein in normal, dysplastic, and neoplastic barrett's oesophagus. J. Pathol. *175*:175-180.

92. Kubbutat, M.H.G., G. Key, M. Duchrow, C. Schluter, H.-D. Flad and J. Gerdes. 1994. Epitope analysis of antibodies recognizing the cell proliferation associated nuclear antigen previously defined by the antibody Ki-67 (Ki-67 protein). J. Clin. Pathol. *47*:524-528.

93. Kumar, R., J. Schaefer, J.P. Grande and P.C. Roche. 1994. Immunolocalization of calcitriol receptor, 24-hydroxylase cytochrome P-450, and calbindin D28k in human kidney. Am. J. Physiol. *266*:477-485.

94. Kwaspen, F., F. Smedts, J. Blom, A. Peonk, M.-J. Kok, M. Van Dijk and F. Ramaekers. 1995. Periodic acid as a nonenzymatic enhancement technique for the detection of cytokeratin immunoreactivity in routinely processed carcinomas. Appl. Immunohistochem. *3*:54-63.

95. Lafer, D.J., S. Masood, L. Lu and G. Fromont-Hankard. 1995. Immunohistochemical analysis of neuropeptide distribution in Hirschsprung's disease utilizing formalin-fixed, paraffin-blocked tissue. Cell Vision 2:373-381.

96. Lambkin, H.A., C.M. Mothersill and P. Kelelhan. 1994. Variations in immunohistochemical detection of p53 protein overexpression in cervical carcinomas with different antibodies and methods of detection. J. Pathol. *172*:13-18.

97. Lan, H.Y., W. Mu, D.J. Nikolic-Paterson and R.C. Atkins. 1995. A novel, simple, reliable, and sensitive method for multiple immunoenzyme staining: use of microwave oven heating to block antibody cross reactivity and retrieve antigens. J. Histochem. Cytochem. *43*:97-102.

98. Langlois, N.E.I., G. King, R. Herriot and W.D. Thompson. 1994. Non-enzymatic retrieval of antigen permits staining of follicle centre cells by the rabbit polyclonal antibody to protein gene product 9.5. J. Pathol. *173*:249-253.

99. Lavieille, J.P., E. Brambilla, C. Riva-Lavieille, E. Reyt, R. Charachon and C. Brambilla. 1995. Immunohistochemical detection of p53 protein in preneoplastic lesions and squamous cell carcinoma of the head and neck. Acta Otolaryngol. (Stockh.) *115*:334-339.

100. Ledet, S.C., R.W. Brown and P.T. Cagle. 1995. P53 immunostaining in the differentiation of inflammatory pseudotumor from sarcoma involving the lung. Mod. Pathol. *8*:282-286.

101. Leong, A.S.-Y., J. Milios and C.G. Duncis. 1988. Antigen preservation in microwave-irradiated tissues: a comparison with formaldehyde fixation. J. Pathol. *156*:275-282.

102. Leong, A.S.-Y. 1992. Commentary. Diagnostic immunohistochemistry—problems and solutions. Pathology 24:1-4.

103. Leong, A.S.-Y. and J. Milios. 1993. An assessment of the efficacy of the microwave antigen-retrieval procedure on a range of tissue antigens. Appl. Immunohistochem. *1*:267-274.

104. Leong, A.S.-Y. and J. Milios. 1993. Comparison of antibodies to estrogen and progesterone receptors and the influence of microwave-antigen retrieval. Appl. Immunohistochem. *1*:282-288.

105. Linden, M.D., S.D. Nathanson and R.J. Zarbo. 1994. Evaluation of anti-p53 antibody staining. Quality control and technical considerations. Appl. Immunohistochem. 2:218-224.

35

106.**Linden, M.D., S.D. Nathanson and R.J. Zarbo.** 1995. Evaluation of anti-p53 antibody staining immunoreactivity in benign tumors and nonneoplastic tissues. Appl. Immunohistochem. *3*:232-238.

107.**Linthicum, F.H., Q. Tian and M. Milicic.** 1995. Constituents of the endolymphatic tubules as demonstrated by three-dimensional morphometry. Acta Otolaryngol. (Stockh.) *115*:246-250.

108.**Loda, M., F. Fogt, F.S. French, M. Posner, B. Cukor, H.T. Aretz and N. Alsaigh.** 1994. Androgen receptor immunohistochemistry on paraffin-embedded tissue. Mod. Pathol. *7*:388-391.

109.**Lucassen, P.J., R. Ravid, N.K.Gonatas and D.F.Swaab.** 1993. Activation of the human supraoptic and paraventricular nucleus neurons with aging and in Alzheimer's disease as judged from increasing size of the Golgi apparatus. Brain Res. *632*:105-113.

110.**Lucassen, P.J., W.C.J. Chung, J.P. Vermeulen, M.V.L. Campagne, J.H.V. Dierendonck and D.F. Swaab.** 1995. Microwave-enhanced in situ end-labeling of fragmented DNA: parametric studies in relation to postmortem delay and fixation of rat and human brain. J. Histochem. Cytochem. *43*:1163-1171.

111.**Malmstrom, P.-U., K. Wester, J. Vasko and C. Busch.** 1992. Expression of proliferative cell nuclear antigen (PCNA) in urinary bladder carcinoma. Evaluation of antigen retrieval methods. APMIS *100*:988-992.

112.**Mason, J.T. and T.J. O'Leary.** 1991. Effects of formaldehyde fixation on protein secondary structure: a calorimetric and infrared spectroscopic investigation. J. Histochem. Cytochem. *39*:225-229.

113.**Mason, D.Y., J.L. Cordell, P. Gaulard, A.G.D. Tse and M.H. Brown.** 1992. Immunohistological detection of human cytotoxic/suppressor T cells using antibodies to a CD8 peptide sequence. J. Clin. Pathol. *45*:1084-1088.

114.**Matias-Guiu, X., M. Cuatrecasas, E. Musulen and J. Prat.** 1994. p53 expression in anaplastic carcinomas arising from thyroid papillary carcinomas. J. Clin. Pathol. *47*:337-339.

115.**McCluggage, W.G., S. Roddy, C. Whiteside, P. Maxwell and H. Bharucha.** 1994. Immunohistochemistry on plastic-embedded bone marrow trephine biopsies following microwave pretreatment in citrate buffer. J. Pathol. Suppl. *173*:202.

116.**McHugh, M. and M. Miettinen.** 1994. KP1 (CD68). Its limited specificity for histiocytic tumors. Appl. Immunohistochem. 2:186-190.

117.**McKee, P.H., C. Hobbs and P.A. Hall.** 1993. Antigen retrieval by microwave irradiation lowers immunohistological detection thresholds. Histopathology *23*:377-379.

118.**McQuaid, S., R. McConnell, J. McMahon and B. Herron.** 1995. Microwave antigen retrieval for immunocytochemistry on formalin-fixed, paraffin-embedded post-mortem CNS tissue. J. Pathol. *176*:207-216.

119.**Medina, F.-J., A. Cerdido and R. Marco.** 1995. Microwave irradiation improvements in the silver staining of the nucleolar organizer (Ag-NOR) technique. Histochemistry *103*:403-413.

120.**Merz, H., O. Rickers, S. Schrimel, K. Orscheschek and A.C. Feller.** 1993. Constant detection of surface and cytoplasmic immunoglobulin heavy and light chain expression in formalin-fixed and paraffin-embedded material. J. Pathol. *170*:257-264.

121.**Merz, H., R. Malisius, S. Mannweiler, R. Zhou, W. Hartmann, K. Orscheschek, P. Moubayed and A.C. Feller.** 1995. ImmunoMax. A maximized immunohistochemical method for the retrieval and enhancement of hidden antigens. Lab. Invest. *73*:149-156.

122.**Miettinen, M.** 1993. Immunohistochemistry in tumour diagnosis. Ann. Med. *25*:221-233.

123.**Mintze, K., N. Macon, K.E. Gould and G.E. Sandusky.** 1995. Optimization of proliferating cell nuclear antigen (PCNA) immunohistochemical staining: a comparison of methods using three commercial antibodies, various fixation times, and antigen retrieval solution. J. Histotechnol. *18*:25-30.

124.**Momose, H., P. Mehta and H. Battifora.** 1993. Antigen retrieval by microwave irradiation in lead thiocyanate. Appl. Immunohistochem. *1*:77-82.

125.**Moore, J.K.,Y.-L. Guan and S.-R. Shi.** Axonal sprouting in the fetal brainstem auditory pathway, identified by neurofilament immunohistochemistry. Anat. Embryol. (In press).

126.**Morgan, J.M., H. Navabi, K.W. Schimid and B. Jasani.**1994. Possible role of tissue-bound calcium ions in citrate-mediated high-temperature antigen retrieval. J. Pathol. *174*:301-307.

127.**Mulder, J.W.R., I.O. Bass, M.M. Polak, S.N. Goodman and G.J.A. Offerhaus.** 1995. Evaluation of p53 protein expression as a marker for long-term prognosis in colorectal carcinoma. Br. J. Cancer *71*:1257-1262.

128.**Munakata, S. and J.B. Hendricks.** 1993. Effect of fixation time and microwave oven heating time on retrieval of the Ki-67 antigen from paraffin-embedded tissue. J. Histochem. Cytochem. *41*:1241-1246.

129.**Myers, R.B., D. Oelschlager, S. Srivastava and W.E. Grizzle.** 1994. Accumulation of the p53 protein occurs more frequently in metastatic than in localized prostatic adenocarcinomas. Prostate *25*:243-248.

130. **Nadasdy, T., Z. Laszik, K.E. Blick, D.L. Johnson and F.G. Silva.** 1994. Tubular atrophy in the end-stage kidney: a lectin and immunohistochemical study. Hum. Pathol. *25*:22-28.

131. **Nagao, T., F. Kondo, T. Sato, Y. Nagato and Y. Kondo.** 1995. Immunohistochemical detection of aberrant p53 expression in hepatocellular carcinoma: correlation with cell proliferation activity indices, including mitotic index and MIB-1 immunostaining. Hum. Pathol. *26*:326-333.

132. **Nielsen, A.L. and H.C.J. Nyholm.** 1994. p53 protein and c-erbB-2 protein (p185) expression in endometrial adenocarcinoma of endometrioid type. An immunohistochemical examination on paraffin sections. Am. J. Clin. Pathol. *102*:76-79.

133. **Norton, A.J.** 1993. Microwave oven heating for antigen unmasking in routinely processed tissue sections. J. Pathol. *171*:79-80.

134. **Norton, A.J.,S. Jordan and P. Yeomans.** 1994. Brief, high-temperature heat denaturation (pressure cooking): a simple and effective method of antigen retrieval for routinely processed tissues. J. Pathol. *173*:371-379.

135. **Onda, K., R.L. Davis, M. Shibuya, C.B. Wilson and T. Hoshino.** 1994. Correlation between the bromodeoxyuridine labeling index and the MIB-1 and Ki-67 proliferating cell indices in cerebral gliomas. Cancer *74*:1921-1926.

136. **Oohashi, Y., H. Watanabe, Y. Ajioka and K. Hatakeyama.** 1995. P53 immunostaining distinguishes malignant from benign lesions of the gall-bladder. Pathol. Int. *45*:58-65.

137. **Orazi, A., G. Cattoretti, N.A. Heerema, G. Sozzi, K. John and R.S. Neiman.** 1993. Frequent p53 overexpression in therapy related myelodysplastic syndromes and acute myeloid leukemias: an immunohistochemical study of bone marrow biopsies. Mod. Pathol. *6*:521-525.

138. **Ortego, L.S., W.E. Hawkins, W.W. Walker, R.M. Krol and W.H. Benson.** 1994. Detection of proliferating cell nuclear antigen in tissues of three small fish species. Biotech. Histochem. *69*:317-323.

139. **Pavelic, Z.P., L.G. Portugal, M.J. Gootee, P.J. Stambrook, C. Smith, R.E Mugge, L. Pavelic, K. Wilson, C.R. Buncher, Y.-Q. Li, J.S. McDonald and J.L. Gluckman.** 1993. Retrieval of p53 protein in paraffin-embedded head and neck tumor tissues. Arch. Otolaryngol. Head Neck Surg. *119*:1206-1209.

140. **Pertschuk, L.P., Y.-D. Kim, C.A. Axiotis, A.S. Braverman, A.C. Carter, K.B. Eisenberg and L.V. Braithwaite.** 1994. Estrogen receptor immunocytochemistry: the promise and perils. J. Cell. Biochem. Suppl. *19*:134-137.

141. **Perutz, M.F. (Ed.).** 1992. Protein Structure, New Approaches to Disease and Therapy. W.H. Freeman and Company, New York.

142. **Pezzella, F., K. Micklem, H. Turley, K. Pulford, M. Jones, S. Kocialkowsk, D. Delia, A. Aiello, R. Bicknell, K. Smith, A.L. Harris, K.C. Gatter and D.Y. Mason.** 1994. Antibody for detecting p53 protein by immunohistochemistry in normal tissues. J. Clin. Pathol. *47*:592-596.

143. **Piffko, J., A. Bankfalvi, D. Ofner, U. Joos, W. Bocker and K.W. Schmid.** 1995. Immunohistochemical detection of p53 protein in archival tissues from squamous cell carcinomas of the oral cavity using wet autoclave antigen retrieval. J. Pathol. *176*:69-75.

144. **Pinkus, G.S.** 1982. Diagnostic immunocytochemistry of paraffin-embedded tissues. Hum. Pathol. *13*:411-415.

145. **Pons, C., I. Costa, B. Von Schilling, X. Matias-Guiu and J. Prat.** 1995. Antigen retrieval by wet autoclaving for p53 immunostaining. Appl. Immunohistochem. *3*:265-267.

146. **Prioleau, J. and S.J. Schmitt.** 1995. Antigen loss in stored paraffin slides. N. Engl. J. Med. *332*:1521-1522.

147. **Resnick, J.M., D. Cherwitz, D. Knapp, D. Uhlman and G.A. Niehans.** 1995. A microwave method that enhances detection of aberrant p53 expression in formalin-fixed, paraffin-embedded tissues. Arch. Pathol. Lab. Med. *119*:360-366.

148. **Risio, M.** 1994. Methodological aspects of using immunohistochemical cell proliferation biomarkers in colorectal carcinoma chemoprevention. J. Cell. Biochem. Suppl. *19*:61-67.

149. **Sale, G.E., M. Beauchamp and D. Myerson.** 1994. Immunohistologic staining of cytotoxic T and NK cells in formalin-fixed paraffin-embedded tissue using microwave TIA-1 antigen retrieval. Transplantation *57*:287-289.

150. **Sarli, G., C. Benazzi, R. Preziosi and P.S. Marcato.** 1994. Proliferative activity assessed by anti-PCNA and Ki67 monoclonal antibodies in canine testicular tumours. J. Comp. Pathol. *110*:357-368.

151. **Scheraga, H.A.** 1971. Theoretical and experimental studies of conformations of polypeptides. Chem. Rev. *71*:195-217.

152. **Scott, F.R., L. More and A.P. Dhillon.** 1995. Hepatobiliary cystadenoma with mesenchymal stroma: expression of oestrogen receptors in formalin-fixed tissue. Histopathology *26*:555-558.

153. **Sherriff, F.E., L.R. Bridges and P. Jackson.** 1994. Microwave antigen retrieval of beta-amyloid precursor protein immunoreactivity. Neuroreport *5*:1085-1088.

37

154.**Shi, S.-R., M.E. Key and K.L. Kalra.** 1991. Antigen retrieval in formalin-fixed, paraffin-embedded tissues: An enhancement method for immunohistochemical staining based on microwave oven heating of tissue sections. J. Histochem. Cytochem. *39*:741-748.

155.**Shi, S.-R., C. Cote, K.L. Kalra, C.R. Taylor and A.K. Tandon.** 1992. A technique for retrieving antigens in formalin-fixed, routinely acid-decalcified, celloidin-embedded human temporal bone sections for immunohistochemistry. J. Histochem. Cytochem. *40*:787-792.

156.**Shi, S.-R., A.K. Tandon, C. Cote and K.L. Kalra.** 1992. S-100 protein in human inner ear: use of a novel immunohistochemical technique on routinely processed, celloidin-embedded human temporal bone sections. Laryngoscope *102*:734-738.

157.**Shi, S.-R., A.K. Tandon, R.R.M. Haussmann, K.L. Kalra and C.R. Taylor.** 1993. Immunohistochemical study of intermediate filament proteins on routinely processed, celloidin-embedded human temporal bone sections by using a new technique for antigen retrieval. Acta Otolaryngol. Stockh. *113*:48-54.

158.**Shi, S.-R., B. Chaiwun, L. Young, R.J. Cote and C.R. Taylor.** 1993. Antigen retrieval technique utilizing citrate buffer or urea solution for immunohistochemical demonstration of androgen receptor in formalin-fixed paraffin sections. J. Histochem. Cytochem. *41*:1599-1604.

159.**Shi, S.-R. and Q.Tian.** 1993. Development of an antigen retrieval technique for immunohistochemistry on archival celloidin-embedded sections. J. Histochem. Cytochem. *41*:1121.

160.**Shi, S.-R., B. Chaiwun, L. Young, A. Imam, R.J. Cote and C.R. Taylor.** 1994. Antigen retrieval using pH 3.5 glycine-HCl buffer or urea solution for immunohistochemical localization of Ki-67. Biotech. Histochem. *69*:213-215.

161.**Shi, S.-R., A. Imam, L. Young, R.J. Cote and C.R. Taylor.** 1995. Antigen retrieval immunohistochemistry under the influence of pH using monoclonal antibodies. J. Histochem. Cytochem. *43*:193-201.

162.**Shi, S.-R., R.J. Cote, L. Young, C. Yang, C. Chen, G.D. Grossfeld, D.A. Ginsberg, F.L. Hall and C.R. Taylor.** 1995. Development of optimal protocols for antigen retrieval immunohistochemistry based on the effects of variation in temperature and pH: use of a 'test battery', p. 828-829. *In* G.W. Bailey, M.H. Ellisman, R.A. Hennigar and N.J. Zaluzec (Eds.), JMSA Proceedings Microscopy and Microanalysis 1995. Jones & Begell, New York.

163.**Shi, S.-R., R.J. Cote, L.Young, S.A. Imam and C.R. Taylor.** 1995. Tris-HCl buffer pH 9.5 mixed with 5% urea used for antigen retrieval immunohistochemistry. *In* G.W. Bailey, M.H. Ellisman, R.A. Hennigar and N.J. Zaluzic (Eds.), JMSA Proceedings Microscopy and Microanalysis 1995. Jones & Begell, New York.

164.**Shi, S.-R., R.J. Cote, C. Yang, C. Chen, H.-J. Xu, W.F. Benedict and C.R. Taylor.** 1996. Development of an optimal protocol for antigen retrieval: a 'test battery' approach exemplified with reference to the staining of retinoblastoma protein (pRB) in formalin-fixed paraffin sections. J. Pathol. *179*:347-352.

165.**Shibuya, M., H. Utsunomiya and R.Y. Osamura.** 1993. Immunohistochemical determination of the proliferating cells with monoclonal antibody MIB-1 on paraffin-embedded section—using antigen retrieval method. Byori-to-Rinsho (Japan) *11*:373-377.

166.**Shin, H.J.C., D.M. Shin and J.Y. Ro.** 1994. Optimization of proliferating cell nuclear antigen immunohistochemical staining by microwave heating in zinc sulfate solution. Mod. Pathol. 7:242-248.

167.**Shin, R.-W., T. Iwaki, T. Kitamoto and J. Tateishi.** 1991. Hydrated autoclave pretreatment enhances TAU immunoreactivity in formalin-fixed normal and Alzheimer's disease brain tissues. Lab. Invest. *64*:693-702.

168.**Sibony, M., F. Commo, P. Callard and J.-M. Gasc.** 1995. Enhancement of mRNA in situ hybridization signal by microwave heating. Lab. Invest. *73*:586-591.

169.**Siitonen, S.M., O.-P. Kallioniemi and J.J. Isola.** 1993. Proliferating cell nuclear antigen immunohistochemistry using monoclonal antibody 19A2 and a new antigen retrieval technique has prognostic impact in archival paraffin-embedded node-negative breast cancer. Am. J. Pathol. *142*:1081-1089.

170.**Skalova, A., I. Leivo, K. von Boguslawsky and E. Saksela.** 1994. Cell proliferation correlates with prognosis in acinic cell carcinomas of salivary gland origin. Immunohistochemical study of 30 cases using the MIB 1 antibody in formalin-fixed paraffin sections. J. Pathol. *173*:13-21.

171.**Smith, T.W. and C.F. Lippa.** 1995. Ki-67 immunoreactivity in Alzheimer's disease and other neurodegenerative disorders. J. Neuropathol. Exp. Neurol. *54*:297-303.

172.**Sparrow, L.E., R. Soong, H.J. Dawkins, B.J. Iacopetta and P.J. Heenan.** 1995. P53 gene mutation and expression in nevi and melanomas. Melanoma Res. *5*:93-100.

173.**Spires, S.E., C.D. Jennings, E.R. Banks, D.P. Wood, D.D. Davey and M.L. Cibull.** 1994. Proliferating cell nuclear antigen in prostatic adenocarcinoma: correlation with established prognostic indicators. Urology *43*:660-666.

174.**Spaulding, D.C.** 1994. Applications of combined ISH and IHC methods for simultaneous detection of nucleic acids and proteins in pathological specimens. J. Histochem. Cytochem. *42*:1011.

175. **Strater, J., A.R. Gunthert, S. Bruderlein and P. Moller.** 1995. Microwave irradiation of paraffin-embedded tissue sensitizes the TUNEL method for in situ detection of apoptotic cells. Histochemistry *103*:157-160.

176. **Stirling, J.W. and P.S. Graff.** 1995. Antigen unmasking for immunoelectron microscopy: labeling is improved by treating with sodium ethoxide or sodium metaperiodate, then heating on retrieval medium. J. Histochem. Cytochem. *43*:115-123.

177. **Suurmeijer, A.J.H.** 1992. Microwave-stimulated antigen retrieval. A new method facilitating immunohistochemistry of formalin-fixed, paraffin embedded tissue. Histochem. J. *24*:597.

178. **Suurmeijer, A.J.H. and M.E. Boon.** 1993. Notes on the application of microwaves for antigen retrieval in paraffin and plastic tissue sections. Eur. J. Morphol. *31*:144-150.

179. **Suurmeijer, A.J.H. and M.E. Boon.** 1993. Optimizing keratin and vimentin retrieval in formalin-fixed, paraffin-embedded tissue with the use of heat and metal salts. Appl. Immunohistochem. *1*:143-148.

180. **Suurmeijer, A.J.H.** 1994. Optimizing immunohistochemistry in diagnostic tumor pathology with antigen retrieval. Eur. J. Morphol. Suppl. *32*:325-330.

181. **Swanson, P.E.** 1993. Methodologic standardization in immunohistochemistry. A doorway opens. Appl. Immunohistochem. *1*:229-231.

182. **Swanson, P.E.** 1994. Microwave antigen retrieval in citrate buffer. Lab. Med. *25*:520-522.

183. **Szekeres, G., J. Audouin and A. Le Tourneau.** 1994. Is immunolocalization of antigens in paraffin sections dependent on method of antigen retrieval? Appl. Immunohistochem. *2*:137-140.

184. **Szekeres, G., A. Le Tourneau, J. Benfares, J. Audouin and J. Diebold.** 1995. Effect of ribonuclease A and deoxyribonuclease I on immunostaining of Ki-67 in fixed-embedded sections. Pathol. Res. Pract. *191*:52-56.

185. **Szekeres, G., Y. Lutz, A.L. Tourneau and M. Delaage.** 1994. Steroid hormone receptor immunostaining on paraffin sections with microwave heating and trypsin digestion. J. Histotechnol. *17*:321-324.

186. **Tacha, D.E. and T. Chen.** 1994. Modified antigen retrieval procedure: calibration technique for microwave ovens. J. Histotechnol. *17*:365-366.

187. **Taylor, C.R. and J. Burns.** 1974. The demonstration of plasma cells and other immunoglobulin containing cells in formalin-fixed, paraffin-embedded tissues using peroxidase labelled antibody. J. Clin. Pathol. *27*:14-20.

188. **Taylor, C.R.** 1976. An immunohistological study of follicular lymphoma, reticulum cell sarcoma and Hodgkin's disease. Eur. J. Cancer *12*:61-75.

189. **Taylor, C.R.** 1978. Immunocytochemical methods in the study of lymphoma and related conditions. J. Histochem. Cytochem. *26*:495-512.

190. **Taylor, C.R.** 1978. Immunoperoxidase techniques: theoretical and practical aspects. Arch. Pathol. Lab. Med. *102*:113-121.

191. **Taylor, C.R.** 1979. Immunohistologic studies of lymphomas: new methodology yields new information and poses new problems. J. Histochem. Cytochem. *27*:1189-1191.

192. **Taylor, C.R.** 1980. Immunohistologic studies of lymphoma: past, present and future. J. Histochem. Cytochem. *28*:777-787.

193. **Taylor, C.R. and R.J.Cote (Eds.).** 1994. Immunomicroscopy: A Diagnostic Tool for the Surgical Pathologist, 2nd ed. W.B. Saunders, Philadelphia.

194. **Taylor, C.R., S.-R. Shi, B. Chaiwun, L. Young, S.A. Imam and R.J. Cote.** 1994. Strategies for improving the immunohistochemical staining of various intranuclear prognostic markers in formalin-paraffin sections: androgen receptor, estrogen receptor, progesterone receptor, p53 protein, proliferating cell nuclear antigen, and Ki-67 antigen revealed by antigen retrieval technique. Hum. Pathol. *25*:263-270.

195. **Taylor, C.R.** 1994. An exaltation of experts: concerted efforts in the standardization of immunohistochemistry. Hum. Pathol. *25*:2-11.

196. **Taylor, C.R., S.-R. Shi, B, Chaiwun, L. Young, S.A. Imam and R.J. Cote.** 1994. Standardization and reproducibility in diagnostic immunohistochemistry. Hum. Pathol. *25*:1107-1109.

197. **Taylor, C.R.** 1994. The current role of immunohistochemistry in diagnostic pathology. Adv. Pathol. Lab. Med. *7*:59-105.

198. **Taylor, C.R., C. Chen, S.-R. Shi, L.Young, C.Yang and R.J.Cote.** 1995. A comparative study of antigen retrieval heating methods. CAP Today *9*:16-22.

199. **Tenaud, C., A. Negoescu, F. Labat-Moleur, Y. Legros, T. Soussi and E. Brambilla.** 1994. p53 immunolabeling in archival paraffin-embedded tissues: optimal protocol based on microwave heating for eight antibodies on lung carcinomas. Mod. Pathol. *7*:853-859.

200. **Terada, T., K. Shimizu, R. Izumi and Y. Nakanuma.** 1994. p53 expression in formalin-fixed, paraf-

fin-embedded archival specimens of intrahepatic cholangiocarcinoma: retrieval of p53 antigenicity by microwave oven heating of tissue sections. Mod. Pathol. 7:249-252.

201. Tesch, G.H., M. Wei, Y.-Y. Ng, R.C. Atkins and H.Y. Lan. 1995. Enhancement of immunodetection of cytokines and cytokine receptors in tissue sections using microwave treatment. Cell Vision 2:435-439.

202. Teter, K.P., D.C. Holloway and G.E. Sandusky. 1995. Assessment of PCNA (19A2) and Ki-67 (MIB1) cell proliferation markers in formalin fixed tissues. J. Histotechnol. 18:99-104.

203. Tian, Q., H. Rask-Andersen and F.H. Linthicum. 1994. Identification of substances in the endolymphatic sac. Acta Otolaryngol. Stockh. 114:632-636.

204. Tischler, A.S. 1995. Triple immunohistochemical staining for bromodeoxyuridine and catecholamine biosynthetic enzymes using microwave antigen retrieval. J. Histochem. Cytochem. 43:1-4.

205. Turley, H., M. Jones, W. Erber, K. Mayne, M. de Waele and K. Gatter. 1994. VS38: a new monoclonal antibody for detecting plasma cell differentiation in routine sections. J. Clin. Pathol. 47:418-422.

206. Ulbright, T.M., A. Orazi, W. deRiese, C. deRiese, J.E. Messemer, R.S. Foster, J.P. Donohue and J.N. Eble. 1994. The correlation of p53 protein expression with proliferative activity and occult metastases in clinical stage I non-seminomatous germ cell tumors of the testis. Mod. Pathol. 7:64-68.

207. Umemura, S, K. Kawai, R.Y. Osamura and Y. Tsutsumi. 1995. Antigen retrieval for bcl-2 protein in formalin-fixed, paraffin-embedded sections. Pathol. Int. 45:103-107.

208. van den Berg, F.M., I.O. Baas, M.M. Polak and G.J.A. Offerhaus. 1993. Detection of p53 overexpression in routinely paraffin-embedded tissue of human carcinomas using a novel target unmasking fluid. Am. J. Pathol. 142:381-385.

209. van der Poel, H.G. and M.E. Boon. 1994. Microwave-antigen retrieval for proliferation analysis (MIB-1) and quanticyt karyometry of bladder washings. Eur. J. Morphol. 32:71-78.

210. Vasquez, M., G. Nemethy and H.A. Scheraga. 1994. Conformational energy calculations on polypeptides and proteins. Chem. Rev. 94:2183-2239.

211. Vessey, C.J., J. Wilding, N. Folarin, S. Hirano, M. Takeichi, P. Soutter, G.W.H. Stamp and M. Pignatelli. 1995. Altered expression and function of E-cadherin in cervical intraepithelial neoplasia and invasive squamous cell carcinoma. J. Pathol. 176:151-159.

212. Von Wasielewski, R., M. Werner, M. Nolte, L. Wilkens and A. Georgii. 1994. Effects of antigen retrieval by microwave heating in formalin-fixed tissue sections on a broad panel of antibodies. Histochemistry 102:165-172.

213. Walts, A.E., J.W. Said and H.P. Koeffler. 1994. Is immunoreactivity for p53 useful in distinguishing benign from malignant effusions? Localization of p53 gene product in benign mesothelial and adenocarcinoma cells. Mod. Pathol. 7:462-468.

214. Westin, P., P. Stattin, J.-E. Damber and A. Bergh. 1995. Castration therapy rapidly induces apoptosis in a minority and decreases cell proliferation in a majority of human prostatic tumors. Am. J. Pathol. 146:1368-1375.

215. Westra, W.H., J.A. Offerhaus, S.N. Goodman, R.J.C. Slebos, M. Polak, I.O. Baas, S. Rodenhuis and R.H. Hruban. 1993. Overexpression of the p53 tumor suppressor gene product in primary lung adenocarcinomas is associated with cigarette smoking. Am. J. Surg. Pathol. 17:213-220.

216. Wilson, D.F., D.-J. Jiang, A.M. Pierce and O.W. Wiebkin. 1995. Antigen retrieval in plastic-embedded ultrathin tissue sections using microwave heating techniques. J. Dent. Res. 74:58-58.

217. Wilson, J.E. 1991. The use of monoclonal antibodies and limited proteolysis in elucidation of structure-function relationships in proteins, p. 207-250. In C.H. Suelter (Ed.), Methods of Biochemical Analysis. John Wiley & Sons, New York.

218. Wolf, H.K. and K.L. Dittrich. 1992. Detection of proliferating cell nuclear antigen in diagnostic histopathology. J. Histochem. Cytochem. 40:1269-1273.

219. Wong, S.S. (Ed.). 1991. Chemistry of Protein Conjugation and Cross-Linking. CRC Press, Boca Raton.

220. Yachnis, A.T. and J.Q. Trojanowski. 1994. Studies of childhood brain tumors using immunohistochemistry and microwave technology: methodological considerations. J. Neurosci. Methods 55:191-200.

221. Yang, W.I., L.R. Zukerberg, T. Motokura, A. Arnold and N.L. Harris. 1994. Cyclin D1 (Bcl-1, PRAD1) protein expression in low-grade B-cell lymphomas and reactive hyperplasia. Am. J. Pathol. 145:86-96.

222. Yanni, A., M. Harrison and P. Dervan. 1994. p53 protein expression in thyroid tumours. J. Pathol. Suppl. 173:204.

# CHAPTER 2

# Immunogold-Silver Staining — Autometallography: Recent Developments and Protocols

**Gerhard W. Hacker[1,2], Gorm Danscher[2] and Cornelia Hauser-Kronberger[3]**

[1]Medical Research Coordination Center, University of Salzburg, Salzburg, Austria; [2]The Steno Institute, Department of Neurobiology, University of Aarhus, Aarhus, Denmark; [3]Institute of Pathological Anatomy, Salzburg General Hospital, Salzburg, Austria

## SUMMARY

*In the early eighties, a series of papers were published to introduce a reliable and easy-to-handle technique for light microscopical and ultrastructural autometallographic studies. Specific methods were developed to demonstrate endogenous zinc pools of zinc ions in synaptic and secretory vesicles; exogenous mercury, silver bound as-sulphide or selenide crystals, in lysosomes; and tissue-bound gold ions resulting from medicamented aurothiocompounds. It was demonstrated that gold ions chemically bound in the lysosomes had to be reduced to metallic gold before they could be silver-amplified. This understanding made it feasible to apply autometallographic amplification for magnifying colloidal-gold particles in thick, semithin and ultrathin sections. As a result, the highly sensitive* in situ *colloidal gold-labeled detection of peptides, proteins and amines by immunocytochemistry (immunogold-silver staining [IGSS]), carbohydrates by lectin histochemistry, and DNA and RNA by* in situ *hybridization,* in situ *PCR and* in situ *3SR techniques were born. Here, we present an overview of the literature on IGSS techniques used for light and electron microscopic studies and also give guidelines for using IGSS based on our own experience.*

## INTRODUCTION

Classical silver stains are widely used in histology and histopathology to visualize different tissue components and cell types (23–25). Silver enhancement of the invisible pictures of an undeveloped film by a photographic developer containing silver ions is not considered to be a silver staining but rather a silver amplification. Liesegang introduced the technique for silver amplification in tissue sections (58,59). As only crystal lattices consisting of gold atoms or molecules composed of silver, mercury or zinc ions and the anions sulphide or selenide have been demonstrated in tissues by this particular photographic developer, the term autometallography (AMG) was coined to identify applications different from those in the photographic laboratory (14,15,17–19).

In 1981 it was discovered that gold chemically bound to human and animal tissues could not ignite an AMG development before it had been reduced to metallic gold atoms (11). Two years later, the AMG setup was applied to amplify colloidal gold used to label enzymes (13,17–19) and immunoglobulins (44). The latter was termed the immunogold-silver staining (IGSS) technique by Holgate et al. (44).

A series of applications of IGSS and various modifications of the technique have been introduced (e.g., 21,27,29,32,34–37,47,49–53,55,68,69,79, 82,90). AMG has also been used for lectin histochemistry (47,72,77), for sensitive nucleic acid detection by *in situ* hybridization (33,36,60,89), *in situ* polymerase chain reaction (PCR) (91–93) and for the *in situ* self-sustained sequence replication-based amplification (*in situ* 3SR) reaction for low copy mRNA detection as reported in *Cell Vision* (94). Excellent reviews on existing IGSS techniques and other AMG applications were given by Hayat and various contributing authors (41–43).

## AUTOMETALLOGRAPHY

The AMG developer contains both a silver salt (i.e., a silver ion donator) and a reducing agent, most often hydroquinone. Originally, silver nitrate was used as the silver ion donator (58). Later, it was replaced by silver lactate in order to increase the specificity and sensitivity of the same reaction (10–12). Silver acetate was introduced as a less light-sensitive alternative to silver lactate (31,77). In our experience, AMG developers perform best at a low pH of about 3.8. Often a protecting agent is added to reduce the speed and further increase the specificity of the reaction (17).

The AMG technique is highly specific and extremely sensitive. It allows silver amplification of catalytic crystals smaller than 0.5 nm. Crystals of gold and silver and crystals where the lattices contain both a heavy metal (Ag, Au, Zn, Hg, Cu) and sulfide or selenide ions can be amplified. Such AMG crystals or crystal lattices have the ability to convey electrons from reducing molecules adhering to the surface of the particle to likewise adhering silver ions. Once started, the AMG process will build up a shell of metallic silver around the crystals and reveal their exact position in the tissue. AMG, therefore, is most suited to tracing gold particles used as labels of, for example, immunoglobulin, streptavidin, or enzymes at light (LM) (42,43) and electron microscopic (EM) (27,35,81) levels.

## IGSS FOR IMMUNOCYTOCHEMISTRY

The first IGSS technique introduced was an indirect method (44). Direct, bridge, streptavidin-biotin, protein A-gold-AMG and various other combinations have been described (e.g., 21,56,67–71,84). The IGSS application of AMG resulted in greatly increased sensitivity and detection efficiency compared with other immunocytochemical techniques (30,44,71,79).

Both LM and EM applications of IGSS possess a number of advantages over other immunostaining methods. In thick paraffin and semithin resin sections, IGSS may give positive immunostaining where other methods fail. The technique is particularly suited for detection of antigenic substances present in very small quantities (Figure 1) (30,79). For ultrastructural studies, small gold particles are recommended because they give high labeling densities (27,55). The silver-amplified colloidal gold particles can be clearly seen even in low-power magnification in the LM or EM (35,75). This fact may also facilitate semiquantitative analysis of EM preparations (27,75).

Hazardous and potentially carcinogenic reagents are avoided by using AMG. In LM applications, immunoreactions can be easily identified due to the intense signal, facilitating the screening of sections at low magnifications. This is advantageous for immunohistopathologists. The gold-cored AMG silver particles allow the highest LM and EM resolutions (Figures 1–3). Conventional counterstaining can be used to improve the assessment of morphological changes and is carried out by hematoxylin and eosin, Nuclear Fast Red or Neutral Red on cryostat or paraffin sections. Azure II, Methylene Blue and Basic Fuchsin or Toluidine Blue can be used on semithin resin sections (31,79).

In IGSS and other AMG techniques, the detection step is often based on gold particles (either colloidal gold or clustered gold-Nanogold) adsorbed to macromolecules (antibodies, protein A or streptavidin). The high detecting sensitivity and efficiency of IGSS is only reached when an adequate AMG silver intensification is used (81). It should result in sections with dark-gray or black specifically stained spots containing only tiny quantities of the labeled molecules (Figures 1–3, see color plate in Addendum, p. A-2 and A-3).

## PRECEPTS FOR IGSS AND OTHER AMG$_{Au}$ TECHNIQUES

All of the commonly used tissue fixatives have been found to be workable. None of them was found to interfere with AMG development. IGSS in LM performs well with 4% neutral phosphate-buffered formaldehyde-, Stefanini's/Zamboni's- (80) or Bouin's- fixed paraffin sections.

While paraffin sections seldom create problems for IGSS, the following precautions should be observed when using cryostat sections. Frozen sections should be dried onto glass slides coated with poly-L-lysine (PLL) (45) or, preferably, aminopropyl-triethoxysilane (APES) (61) for more than 1 h. Sometimes it also helps if the sections are coated with gelatin by dipping in a 0.5% gelatin solution and drying (10,18). They are then subjected to a dehydration/rehydration process by going through increasing concentrations of alcohols to xylene and then back to water (37). Frozen sections from tissues that are not or only mildly fixed before freezing, can be postfixed after the dehydration/rehydration treatment using 10% buffered formalin or Stefanini's solution. This step can also be carried out before the dehydration/rehydration step. It further improves staining and reduces background reactions (37).

Semithin resin sections should be pretreated with sodium ethoxide or methoxide to soften the resin (31). Pre-embedding AMG amplification can be carried out using gold particles of 1–5 nm in diameter on Stefanini's-fixed cryostat sections. These sections are subjected to $AMG_{Au}$ and further processed for EM examination after being block-stained with osmium tetroxide and uranyl acetate (7,54,55,92). In post-embedding AMG ultrastructural studies, fixation with buffered glutaraldehyde and paraformaldehyde solution or Stefanini's/Zamboni's solution is suggested. Osmium tetroxide fixation should never take place before AMG development, but can be performed after silver amplification together with uranyl block staining (10–12,18). It should be stressed that EM post-embedding immunocytochemistry sometimes demands low-temperature polymerization. For some antigens, it should not be higher than 40°C.

For LM applications, the indirect IGSS methods appear to be superior to the direct methods or protein A-gold. Streptavidin-gold-silver techniques in combination with Nanogold™ give even better results (see below). Recently, new staining protocols for indirect IGSS have been published (34,35,37). Most IGSS protocols use oxidation of the sections with Lugol's iodine followed by decolorization with sodium thiosulfate. This step has empirically been found to increase AMG staining efficiency in most LM applications. If iodine treatment is omitted, the AMG amplification process needs to be prolonged. For AMG staining of EM thin sections on grids, Lugol's iodine can be used but is not mandatory.

Concerning washing buffers, most protocols recommend Tris-buffered saline (TBS) or phosphate-buffered saline (PBS) with a pH of around 7.2–7.6. High salt concentrations and the addition of Triton® X-100 or Tween® 80 to the buffer used before applying the primary antibody may improve the staining (79). Addition of 0.1% gelatin (e.g., cold water fish gelatin) to the washing buffer is a very effective way to reduce nonspecific reaction (32,34, 35).

Polyclonal rabbit or guinea pig antisera and monoclonal mouse or rat antibodies have been applied to detect various substances at the LM and EM levels (e.g., 8,9,20,29–37,44,49–53,57, 62,79,85). Primary antibody incubation for 60–90 min at room temperature is convenient and sufficient, and this time span can be drastically reduced by microwave irradiation (3,28,86). Prolonged incubation and/or double application of the same antibody may further increase the detection efficiency and may permit higher antibody dilution (5,26,30). Optimal dilution of antibodies should be established individually.

Antigen detectability in formalin-fixed paraffin sections of most substances can be considerably improved by applying antigen retrieval techniques (73) for IGSS (Hacker et al., unpublished observations). For this purpose, deparaffinized sections are boiled several times in 0.01 M HCl sodium citrate buffer, pH 6.0. Further improvement for some antigens can be obtained by boiling in Target Unmasking Fluid (TUF®; Kreatech, Amsterdam, NL). The antigen retrieval process is carried out by immersing the sections in excessive amounts of buffer or TUF and boiling for 15–45 min by heating in a microwave oven ($5 \times 5$ min, with intermittent breaks of 3 min each) set to high power (e.g., 750 W), or in an autoclaving sterilizer (set to 120°C, 15 psi, 5–15 min) (76) or other types of pressure cookers (1). After the boiling period, the glass slides should not be disturbed until they have reached room temperature. This prevents sections from detaching. For antigen unmasking techniques, it is crucial to use a high-quality section adhesive, such as aminopropyl-triethoxysilane (61) (Sigma Chemical, Deisenhofen, FRG). We have observed considerable improvement by using these techniques in demonstrating many substances, including cytokeratins; S-100 proteins; PGP-9.5; glial fibrillary acidic protein; vimentin; neurofilament proteins; desmin; lymphoid markers such as CD-3, CD- 43, CD-20 (L-26), CD-45R0 (UCHL-1); proliferation markers such as MIB-1 (Ki-67) or PCNA; histiocytic markers; endothelial markers, such as BMA-120 or CD-31, most regulatory peptides and receptors (Hacker et al., unpublished observations).

The choice of the initial gold particle size is crucial for obtaining optimum results because it determines labeling density and level of penetration. If diameters of about 1 to 5 nm are used, stronger labeling results because more gold particles adhere to each antigenic site to be labeled, and these tiny gold particles allow the tagged molecule to penetrate deep into the sections (27, 55). AMG silver amplification makes these initially tiny gold particles visible even at low magnifications for both LM and EM (2,7,35,74,78,87). In our hands, gold particles of 5 nm in diameter give the best results compared to larger and smaller (1 nm) particles. "Ultrasmall" immunogold reagents (1 nm and below) appear to be less stable. This phenomenon was also reported by de Valk et al. (84) and by Gu et al. (27) and was confirmed by Danscher et al. (17).

"Nanogold" is a new AMG label that is often superior to traditional colloidal-gold probes where the gold particles are coated with large, bulky proteins for stabilization (17). Nanogold probes use 1.4-nm diameter gold particles covalently bound to F(ab) fragments. Especially in combination with streptavidin, Nanogold gives a dramatic increase in staining quality and sensitivity. If this setup is combined with CARD (catalyzed reporter deposition), even single molecule sensitivity may be achieved (Hacker, unpublished observations). Penetration properties and AMG are better for Nanogold probes than colloidal-gold probes (38,88). A very interesting approach to enhancing the accuracy of immunocytochemistry is the anti-horseradish peroxidase antibody-gold complexes for cytochemistry and *in situ* hybridization introduced by Roth et al. (69,70).

It is advisable to test different immunogold reagents available on the market, as their quality and prices differ considerably. We presently recommend use of 5-nm EM-grade colloidal-gold probes obtained from BioCell (Cardiff, UK), Amersham (Amersham, UK) or Aurion (Wageningen, NL) for both LM and EM applications, or better, Streptavidin-Nanogold (Stony Brook, NY, USA). In our hands, reagents from those companies gave high labeling densities and showed acceptable penetration properties. We advise use of mixtures of several optimally diluted immunogold batches with different colloidal-gold sizes for LM or only one gold size for EM from different suppliers, mixed in equal amounts. Dilutions of immunogold reagents should be optimized by titration and are usually between 1/25 and 1/200. TBS (pH 7.6 to 8.2) is used as diluent for the gold reagent and should contain 0.8% bovine serum albumin (BSA) and 0.1% gelatin. This procedure helps prevent aggregation of gold particles and results in suppression of nonspecific staining. A postfixation step using 2% glutaraldehyde in PBS (pH 7.2) should be applied after the washes in order to remove unbound immunogold reagent. This postfixation also reinforces binding of the gold-labeled molecules to their binding sites in the tissue and is most convenient if the AMG developer operates at low pH.

Before AMG gold-silver amplification, LM and EM sections should be washed carefully in double-distilled water. The purity of water is crucial; therefore, only very clean glassware and plastic or teflonized forceps should be used. To ensure purity, all tools and vials can be cleaned for 30 min in 10% Farmer's solution (one part sodium thiosulphate and nine parts 10% potassium ferricyanide) (17).

LM sections should be placed vertically in a glass container, e.g., in a Schiefferdecker staining dish (Schott, Mainz, Germany), and covered by 80 mL of AMG developer. If a silver lactate developer is applied, the container should be covered from light with a cardboard box (10–12,18). Silver acetate developers are less light sensitive than silver lactate or nitrate and do not require covering for LM applications of IGSS (31,37,53,77). Both silver lactate and silver acetate AMG developers permit monitoring of the staining intensity visually. For silver amplification of immunogold-labeled thin sections on EM grids, both silver lactate and silver acetate AMG developers should be covered with a dark box. Development of EM grids can be carried out by floating them face-down on top of drops of freshly prepared developer on dental wax. To avoid "crunchy preparations", grids must not be dipped completely in the AMG solution. Development of ultrathin sections should be carried out for 3 to 15 min at room temperature. The duration of development depends on the size of the colloidal-gold particles and the AMG developer used (35,75,81). To ensure optimal results for EM, slow developers are preferred (16,31,81).

The detection efficiency of IGSS is closely related to the type of AMG developer used. A number of AMG developers have been introduced since Danscher's first description of the acid silver lactate developer (10,12). Key points of the AMG technique include 1) prevention of nonspecific silver precipitation, 2) reproducibility and 3) homogeneity (81). We do not find it necessary to use commercially available developers. Every laboratory technician should be able to make up an AMG developer, and the results obtained in most cases are far better than those obtained with the expensive commercial products. The original silver lactate and silver acetate developers, for example, have proved to be extremely efficient and sensitive for AMG detection of colloidal-gold particles (81).

## DETECTION OF DNA AND mRNA WITH IMMUNOGOLD-AMG (IGS-AMG)

*In situ* hybridization (ISH) methods have been frequently used for molecular biological analysis of nucleic acid (DNA and mRNA) sequences at LM and EM levels. The techniques can be applied to detect viral genomes in infected cells, to investigate biosynthesis of peptides and/or proteins or to study genetic disease. Biotin or digoxigenin are used in non-isotopic ISH in order to avoid hazardous radioactivity. These reporter molecules satisfy the safety requirements of most pathological laboratories. Also, nonradioactive labels are cheaper, easier to handle and give a higher resolution than the radioactive probes (6,33,34,36,48,60,89). Biotin, digoxigenin or other labels, such as fluorescein-isothiocyanate (FITC) used for nucleic acid detection, can be easily visualized by using direct or indirect IGSS methods (33,36,91–94).

Optimized protocols for *in situ* DNA hybridization with biotin-labeled probes and IGSS techniques have been described (33,34,36). *In situ* hybridization methods using IGSS detection are very specific and sensitive (Figure 4, see color plate in addendum, p. A-3). Most recently, applications of AMG for PCR *in situ* hybridization (PISH) and *in situ* 3SR have been presented (91–94). These methods have the highest sensitivity as they allow detection of a single copy of DNA or mRNA at the cellular level with AMG. With pre-embedding PISH or 3SR on formalin- or paraformaldehyde-fixed cells followed by a "pop-off" procedure, osmium tetroxide postfixation and resin embedding, investigation of AMG-amplified DNA staining in the EM becomes possible (54,92; and W.H. Muss, Salzburg, Austria, personal communication). In our hands, streptavidin-Nanogold (Nanoprobes, Stony Brook, NY, USA) gives an extremely sensitive detection of biotinylated DNA or RNA hybrids. It can be combined with CARD techniques using biotinylated tyramide and allows single or very low copy sensitivity (Hacker, Zehbe and Tubbs, unpublished observations) for both DNA and RNA and will potentially replace both *in situ* PCR and *in situ* 3SR (Figure 4).

## NON-MICROSCOPICAL IGSS (IGS-AMG$_{Au}$)

AMG techniques can also be successfully applied to non-microscopical, biochemical and molecular biological techniques. A review on their use in the latter was given recently (40). Electrophoretic separation of proteins in polyacrylamide gel is a standard procedure for characterization of proteins and protein mixtures (22,83), and the recently introduced electrotransfer of the protein bonds generated in the gel onto porous membranes has unveiled new aspects for applications of refined detection procedures because the possibility of binding antibodies and other substances to proteins on the membrane is considerably increased. In the large majority of the applications, i.e., immunoblot or Western blot, the ligand used is an antibody. Separation by molecular mass followed by immunological detection provides a useful tool. Simple dot blots are often used to semi-quantify the amounts of antigens or to test the binding of uncharacterized antibodies. Colloidal-gold labeling followed by AMG significantly increases the sensitivity and the intensity of the staining signal.

As early as 1984, the use of an indirect IGSS method, more correctly the "IGS-AMG$_{Au}$" method, was suggested for immunoblotting (4,46,63). These authors showed that indirect IGSS or protein-A-gold-AMG can significantly improve the sensitivity and quality of immunoblots. The setup was assessed as being more sensitive than peroxidase anti-peroxidase (PAP) or avidin-biotin-complex (ABC) immunoperoxidase systems. The AMG amplification results in high contrast with low background staining of the gel. Because of its high detection sensitivity, even picogram amounts of antigen can be detected (46).

Nanogold and Undecagold were recently described as giving results superior to those obtained with traditional gold probes in microscopical IGSS methods (39). The new reagents have also been successfully used in Western and dot blots and microtiter plate assays. Hainfeld and collaborators have shown that even as little as 0.2 pg ($10^{-18}$ moles) of target molecules can be detected with Nanogold-IGS-AMG$_{Au}$ (39). It also appears possible to run Nanogold-labeled proteins on gels and silver amplify them directly (J. Hainfeld, Stony Brook, NY; personal communication).

A novel immunoassay, the silver enhanced gold-labeled immunoadsorbent assay (SEGLISA or IGS-AMG-IA) were recently introduced (64–66). It was based on conventional microtitration technology commonly used in enzyme-linked immunoadsorbent assays (ELISA) but employed the IGS-AMG instead of enzyme-labeled conjugates and substrate reactions. AMG produces a signal that can be read either visually or with a conventional color plate reader, measuring the apparent absorbency of the thin gold-silver layer (65). The authors claim that their technique possesses several advantages to conventional ELISA procedures, including a permanent record of the tests.

## CONCLUSION

Application of AMG as a tool in immunocytochemistry and related methods has manifold advantages over other techniques for LM and EM. Small gold particles give high labeling densities and possess good penetration properties. This is a particular advantage in preparations where only small amounts of gold-labeled molecules are present. It should be mentioned, however, that some immunogold reagents available on the market do not produce high enough labeling density and generate an unacceptable background labeling. When high-quality products are used, one still needs to check and optimize every step of the procedure. However, in the hands of a skilled technician, IGSS and other AMG procedures can be used in a variety of applications with satisfactory results and superior sensitivity.

## RECOMMENDED PROTOCOL FOR IMMUNOGOLD-SILVER STAINING

### Immunostaining

1. Mount paraffin or semithin resin sections on PLL or APES-coated glass slides and dry them for 1 h at 60°C. Before staining, deparaffinize paraplast sections through xylene and take to water through graded alcohols (isopropanol, industrial methylated spirit). Treat semithin sections with saturated sodium ethoxide for about 20 min and wash in ethanol (3 × 2 min). IGSS is not well suited for cryosections; however, to achieve optimum results, one-hour air-dried cryostat sections need to be taken to xylene using graded alcohols and then taken back to water through alcohols. Cytological material (e.g., smears, imprints, cytospins) mounted on coated glass slides and dried for at least 1 h should be postfixed in 4% neutral phosphate-buffered formaldehyde for 2–10 min.

2. Wash in distilled water (3 min).

3. Immerse in Lugol's iodine (1% iodine in 2% potassium iodide, readymade from Merck No. 9261; Darmstadt, FRG) (5 min).

4. Rinse briefly in distilled water.

5. Treat with 2.5% aqueous sodium thiosulfate until sections become colorless (up to 30 s).

6. Wash in distilled water (2 min).

7. Immerse in TBS-gelatin (Tris-buffered saline, pH 7.6, containing 0.1% cold-water fish gelatin; e.g., Aurion or Amersham) (10 min). In some cases, superior results are obtained if the buffer in this step also contains 0.1% Triton X-100 or Tween 80, and 2.5% NaCl.

8. Apply normal serum of the species providing the secondary antibody (1/10 in TBS-gelatin) (5 min) and drain off.

9. Incubate with primary antibody (90 min at room temperature or overnight at 4°C). The dilution should be optimized carefully. The suggested antibody diluent is 0.1 M TBS or PBS, pH 7.2–7.6, containing 0.1% BSA and 0.1% sodium azide.

10. Wash in TBS-gelatin (3 × 3 min).

11. Apply normal serum 1/10 as in step 8.

12. Incubate with gold-conjugated second layer antibodies (60 min at room temperature). Optimal dilution is usually between 1/25 and 1/200 and should be determined by titration.

13. Wash in TBS-gelatin (3 × 3 min).

14. Postfix in 2% glutaraldehyde in PBS pH 7.2 (2 min).

15. Rinse briefly 5 times in distilled water (about 30 s each), followed by three washes (3 min each) in the same.

16. Perform silver acetate autometallography.

## Silver Acetate Autometallography

This method can be applied for immunocytochemical IGSS, as well as for lectin histochemistry, *in situ* hybridization, *in situ* PCR and the detection of metallic gold and silver, and sulphides or selenides of mercury, silver and zinc.

1. Solutions A and B should be freshly prepared for every run. Solution A: Dissolve 80 mg silver acetate (code 85140; Fluka, Buchs, Switzerland) in 40 mL of glass double-distilled water. (Silver acetate crystals can be dissolved by continuous stirring within about 15 min.)

2. Citrate buffer: Dissolve 23.5 g of trisodium citrate dihydrate and 25.5 g citric acid monohydrate in 850 mL of deionized or distilled water. This buffer can be kept at 4°C for at least 2–3 weeks. Before use, adjust to pH 3.8 with citric acid solution.

3. Solution B: Dissolve 200 mg hydroquinone in 40 mL citrate buffer.

4. Just before use, mix solution A with solution B.

5. Silver amplification: Place the slides vertically in a glass container (preferably with about 80 mL volume and up to 19 slides) and cover them with the mixture of solutions A and B. Staining intensity can be checked in the light microscope during the amplification process, which usually takes about 5 to 20 min, depending on primary antibody concentration, incubation conditions and the amount of accessible antigen.

6. Photographic fixer (e.g., Agefix®; Agfa Gevaert, Leverkusen, Germany; diluted 1:20) can be used to stop the enhancement process immediately. (This solution can be reused for many stainings). Leave the slides in this solution not more than 10 S. Alternatively, a 2.5% aqueous solution of sodium thiosulfate can be used.

Rinse the slides carefully in tap water for at least 3 min. After silver amplification, sections can be counterstained with hematoxylin and eosin (H&E) or Nuclear Fast Red, dehydrated and mounted in DPX (BDH Chemicals, Poole, UK).

**REFERENCES**

1.**Auld, J.** 1994. Antigen unmasking in routinely processed paraffin sections by pressure cooking. UK NEQUAS Immunocytochemistry News *3*:6-9.

2. **Bastholm, L., L.Scopsi and M.H. Nielsen.** 1986. Silver-enhanced immunogold staining of semithin and ultrathin cryosections. J. Electron Microsc. Tech. *4*:175-176.

3. **Boon, M.E. and L.P. Kok.** 1988. Microwave Cookbook of Pathology. The Art of Microscopic Visualization, 2nd ed. Coulomb Press, Leyden.

4. **Brada, D. and J. Roth.** 1984. "Golden blot"-detection of polyclonal and monoclonal antibodies bound to antigens on nitrocellulose by protein A-gold complexes. Anal. Biochem. *142*:79.

5. **Brandtzaeg, P.** 1981. Prolonged incubation staining of immunoglobulins and epithelial components in ethanol- and formaldehyde-fixed paraffin-embedded tissues. J. Histochem. Cytochem. *29*:1302-1315.

6. **Breitschopf, H., G., Suchanek, G., Gould, R.M., D.R. Colman and H. Lassmann.** 1992. In situ hybridization with digoxigenin- labeled probes: sensitive and reliable detection method applied to myelinating rat brain. Acta Neuropathol. [Berl] *84*:581-587.

7. **Burry, R.W., D.D. Vandre and D.M. Hayes.** 1992. Silver enhancement of gold antibody probes in pre-embedding electron microscopic immunocytochemistry. J. Histochem. Cytochem. *40*:1849-1856.

8. **Cossu, M., A. Floris and M.S. Lantini.** 1992. An immunohistochemical study of the distribution of ABH antigens in human submandibular glands at the light and electron microscopic levels. Eur. J. Histochem. *36*:489-499

9. **Danforth, H.D., J.R. Barta and P.C. Augustine.** 1992. Localization of a low molecular weight antigen of Emeria tenella by use of hybridoma antibodies. J. Parasitol. *78*:460-465.

10. **Danscher, G.** 1981. Histochemical demonstration of heavy metals. A revised version of the sulphide silver method suitable for both light and electron microscopy. Histochemistry *71*:1-16.

11. **Danscher, G.** 1981. Localization of gold in biological tissue. A photochemical method for light and electronmicroscopy. Histochemistry *71*:81-88.

12. **Danscher, G.** 1981. Light and electron microscopic localisation of silver in biological tissue. Histochemistry *71*:177-186.

13. **Danscher, G. and J.O. Norgaard.** 1983. Light microscopic visualization of colloidal-gold on resin-embedded tissue. J. Histochem. Cytochem. *31*:1394-1398.

14. **Danscher, G.** 1984. Autometallography. A new technique for light and electron microscopical visualization of metals in biological tissue (gold, silver, metal sulphides and metal selenides). Histochemistry *81*:331-335.

15. **Danscher, G. and J.O. Norgaard.** 1985. Ultrastructural autometallography: a method for silver amplification of catalytic metals. J. Histochem. Cytochem. *33*:706-710.

16. **Danscher, G.** 1991. Applications of autometallography to heavy metal toxicology. Pharmacol. Toxicol. *68*:414-423.

17. **Danscher, G., G.W. Hacker, J.O. Norgaard and L. Grimelius.** 1993. Autometallographic silver amplification of colloidal-gold. J. Histotechnol. *16*:201-207.

18. **Danscher, G. and C. Montagnese.** 1994. Autometallographic localization of synaptic vesicular zinc and lysosomal gold, silver and mercury. J. Histotechnol. *17*:15-22.

19. **Danscher, G., G.W. Hacker, C. Hauser-Kronberger and L. Grimelius.** 1995. Trends in autometallographic silver amplification of colloidal-gold particles, p. 1-19. In M.A. Hayat (Ed.), Immunogold-Silver Staining: Methods and Applications. CRC Press, Boca Raton.

20. **Farell, C.L., J. Yang and W.M. Pardridge.** 1992. GLUT-1 glucose transporter is present within apical and basolateral membranes of brain epithelial interfaces and in microvascular endothelia with and without tight junctions. J. Histochem. Cytochem. *40*:193-199.

21. **Fujimori, O. and M. Nakamura.** 1985. Protein A gold-silver staining method for light microscopic immunohistochemistry. Arch. Histol. Jpn. *48*:449-452.

22. **Gershoni, J.M. and G.E. Palade.** 1983. Protein blotting: Principles and applications. Anal. Biochem. *131*:1-15.

23. **Grimelius, L.** 1968. The argyrophil reaction in islet cells of adult human pancreas studied with a new silver nitrate procedure. Acta Soc. Med. Ups. *73*:271-294.

24. **Grimelius, L. and E. Wilander.** 1980. Silver stains in the study of endocrine cells of the gut and pancreas. Invest. Cell Pathol. *3*:3-12.

25. **Grimelius, L., H. Su and G.W. Hacker.** 1994. The use of silver stains in the identification of neuroendocrine cell types, p. 1-8. In J. Gu and G.W. Hacker (Eds.), Modern Methods in Analytical Morphology. Plenum Press, New York.

26. **Gu, J., J. De Mey, M. Moeremans and J.M. Polak.** 1981. Sequential use of the PAP and immunogold-staining methods for the light microscopical double staining of tissue antigens. Its application to the study of regulatory peptides in the gut. Regul. Pept. *1*:465-474.

27. **Gu, J., M. D'Andrea, C.-Z. Yu, M. Forte and L.B. McGrath.** 1993. Quantitative evaluation of indirect immunogold-silver electron microscopy. J. Histotechnol. *16*:19-26

28. **Gu, J.** 1994. Microwave immunocytochemistry, p. 67-80. In J. Gu and G.W. Hacker (Eds.), Modern

51

Methods in Analytical Morphology. Plenum Press, New York.

29. **Hacker, G.W., D.R. Springall, S. Van Noorden, A.E. Bishop, L. Grimelius and J.M. Polak.** 1985. The immunogold-silver staining method. A powerful tool in histopathology. Virchows Arch. A Pathol. Anat. Histopathol. *406*:449-461.

30. **Hacker, G.W., J.M. Polak, D. Springall, J. Ballesta, A. Cadieux, J. Gu, J.Q. Trojanowski, D. Dahl and P.J. Marangos.** 1985. Antibodies to neurofilament protein and other brain proteins reveal the innervation of peripheral organs. Histochemistry *82*:581-593.

31. **Hacker, G.W., L. Grimelius, G. Danscher, G. Bernatzky, W. Muss, H. Adam and J. Thurner.** 1988. Silver acetate autometallography: an alternative enhancement technique for immunogold-silver staining (IGSS) and silver amplification of gold, silver, mercury and zinc in tissues. J. Histotechnol. *11*:213-221.

32. **Hacker, G.W.** 1989. Silver-enhanced colloidal gold for light microscopy, p. 297-321. *In* M.A. Hayat (Ed.), Colloidal-gold - Principles, Methods, and Applications, Vol. 1. Academic Press, San Diego.

33. **Hacker, G.W., A.-H. Graf, C. Hauser-Kronberger, G. Wirnsberger, A. Schiechl, G. Bernatzky, U. Wittauer, H. Su., H. Adam, J. Thurner, G. Danscher and L. Grimelius.** 1993. Application of silver acetate autometallography and gold-silver staining methods for *in situ* DNA hybridization. Chin. Med. J. *106*:83-92.

34. **Hacker, G.W., C. Hauser-Kronberger, A.-H. Graf, G. Danscher, J. Gu and L Grimelius.** 1994. Immunogold-silver staining (IGSS) for detection of antigenic sites and DNA sequences, p. 19-35. *In* J. Gu and G.W. Hacker (Eds.), Modern Methods in Analytical Morphology. Plenum Press, New York.

35. **Hacker, G.W., W.H. Muss, C. Hauser-Kronberger, G. Danscher, R. Rufner, J. Gu, H. Su, A. Andreasen, M. Stoltenberg and O. Dietze.** 1996. Electron microscopical autometallography: immunogold-silver staining (IGSS) and heavy-metal histochemistry. Methods Companion Meth. Enzymol. *10*:257-269.

36. **Hacker, G.W., I. Zehbe, J. Hainfeld, J. Sällström, C. Hauser-Kronberger, A.-H. Graf, H. Su, O. Dietze and O. Bagasra.** 1996. High-Performance Nanogold *in situ* hybridization and *in situ* PCR. Cell Vision *3*:209-215.

37. **Hacker, G.W., G. Danscher, L. Grimelius, C. Hauser-Kronberger, W.H. Muss, A. Schiechl, J. Gu. and O. Dietze.** 1995. Silver staining techniques with special reference to the use of different silver salts in light and electron microscopical immunogold-silver staining, p. 20-45. *In* M.A. Hayat (Ed.), Immunogold-Silver Staining: Methods and Applications. CRC-Press, Boca Raton.

38. **Hainfeld, J.F.and F.R. Furuya.** 1992. A 1.4-nm gold cluster covalently attached to antibodies improves immunolabeling. J. Histochem. Cytochem. *40*:177-184.

39. **Hainfeld, J.F., F.R. Furuja, K. Carbone, M. Simon, B. Lin, K. Braig, A.L. Horwich, D. Safer, B. Blechschmidt, M. Sprinzl, J. Ofengand and M. Boublik.** 1993. High resolution gold labeling. *In* G.W. Bailey and C.L. Rieder (Eds.), Proceedings of the 51st Annual Meeting of the Microscopy Society of America. San Francisco Press, San Francisco.

40. **Hauser-Kronberger, C., G.W. Hacker, E. Arrer and G. Danscher.** 1995. Non-microscopical colloidal-gold autometallography (AMGAu): use of immunogold-silver staining in blot staining and immunoassay, p. 289-297. *In* M.A. Hayat (Ed.), Immunogold-Silver Staining: Methods and Applications. CRC Press, Boca Raton.

41. **Hayat, M.A.** 1993. Stains and Cytochemical Methods. Plenum Press, New York.

42. **Hayat, M.A. (Ed.).** 1993. Immunogold-Silver Staining for Light and Electron Microscopy. J. Histotechnol. 16:197-199.

43. **Hayat, M.A. (Ed.).** 1995. Immunogold-Silver Staining: Methods and Applications. CRC Press, Boca Raton.

44. **Holgate, C.S., P. Jackson, P.N. Cowen and C.C. Bird.** 1983. Immunogold-silver staining: New method of immunostaining with enhanced sensitivity. J. Histochem. Cytochem. *31*:938-994.

45. **Huang, W.-M., S.J. Gibson, P. Facer, J. Gu and J.M. Polak.** 1983. Improved section adhesion for immunocytochemistry using high molecular weight polymers of L-lysine as a slide coating. Histochemistry 77:275-279.

46. **Jones, A. and M. Moeremans.** 1988. Colloidal-gold for the detection of proteins on blots and immunoblots. *In* J. Walker (Ed.), Methods in Molecular Biology, Vol. 3. The Humana Press.

47. **King, T.P., L. Brydon, G.W. Gooday and L.H. Chappell.** 1987. Silver enhancement of lectin-gold and enzyme-gold cytochemical labelling of eggs of the nematode Onchocerca gibsoni. Histochem. J. *19*:281-286.

48. **Komminoth, P., F.P. Merk, I. Leav, H.J. Wolfe and J. Roth.** 1992. Comparison of $^{35}$S- and digoxigenin-labeled RNA and oligonucleotide probes for *in situ* hybridization. Expression of mRNA of the seminal vesicle secretion protein II and androgen receptor genes in the rat prostate. Histochemistry *98*:217-228.

49. **Krenács, T., E. Molnár, E. Dobó and L. Dux.** 1989. Fibre typing using sarcoplasmic reticulum Ca2+-ATPase and myoglobin immunohistochemistry in rat gastrocnemius muscle. Histochem. J. *21*:145-155.

50. **Krenács, T., Z. Láslik and E. Dobó.** 1989. Application of immunogold-silver staining and immunoenzymatic methods in multiple labelling of human pancreatic Langerhans islet cells. Acta Histochem. *85*:79-85.

51. **Krenács, T., L. Krenács, B. Bozùky and B. Iványi.** 1990. Double and triple immuno-cytochemical labelling at the light microscopical level in histopathology. Histochem. J. *22*:530-536.

52. **Krenács, T., B. Iványi, B. Bozùky, Z. Lászik, L. Kranács, Z. Rázga and J. Ormos.** 1991. Postembedding immunoelectron microscopy with immunogold-silver staining (IGSS) in Epon 812, Durcupan ACM and LR-White resin embedded tissues. J. Histotechnol. *14*:75-80.

53. **Krenács, T. and L. Dux.** 1994. Silver-enhanced immunogold labeling of calcium-ATPase in sarcoplasmic reticulum of skeletal muscle. [Letter] J. Histochem. Cytochem. *42*:967-968.

54. **Kummer, W., C. Hauser-Kronberger and W.H. Muss.** 1994. Pre-embedding immunohistochemistry in transmission electron microscopy, p. 187-202. *In* J. Gu and G.W. Hacker (Eds.), Modern Methods in Analytical Morphology. Plenum Press, New York.

55. **Lackie, P.M., R.J. Hennessy, G.W. Hacker and J.M. Polak.** 1985. Investigation of immunogold-silver staining by electron microscopy. Histochemistry *83*:545-550.

56. **Lah, J.J., D.M. Hayes and R.W. Burry.** 1990. A neutral pH silver development method for the visualization of 1-nanometer gold particles in pre-embedding electron microscopic immunocytochemistry. J. Histochem. Cytochem. *38*: 503-508.

57. **Larochelle, R.and R. Magar.** 1993. The application of immunogold silver staining (IGSS) for the detection of transmissible gastroenteritis virus in fixed tissues. J. Vet. Diagn. Invest. *5*:16-20.

58. **Liesegang, R.E.** 1911. Die Kolloidchemie der histologischen Silberfärbungen, p. 1-46. *In* W. Oswald (Ed.), Kolloidchemische Beihefte (Ergänzungshefte zur Kolloid-Zeitschrift) - Monographien zur reinen und angewandten Kolloidchemie. Theodor Steinkopff Verlag, Dresden and Leipzig, Germany.

59. **Liesegang, R.** 1928. Histologische Versilberung. Z. Wiss. Mikr. *45*:273-279.

60. **Liesi, P., J.-P. Julien, P. Vilja, F. Grosveld and L. Rechardt.** 1986. Specific detection of neuronal cell bodies: *in situ* hybridisation with a biotin-labeled neurofilament cDNA probe. J. Histochem. Cytochem. *34*:923-926.

61. **Maddox, P.H. and D. Jenkins.** 1987. 3-Aminopropyltriethoxysilane (APES): a new advance in section adhesion. J. Clin. Pathol. *40*:1256-1257.

62. **Magar, R. and R. Larochelle.** 1992. Immunohistochemical detection of porcine rotavirus using immunogold-silver staining (IGSS). J. Vet. Diagn. Invest. *4*:3-7.

63. **Moeremans, M., G. Daneels, A. Van Dijck, G. Langanger and J. De Mey.** 1984. Sensitive visualisation of antigen-antibody reactions in dot and blot immune overlay assays with immunogold and immunogold-silver staining. J. Immunol. Meth. *74*:352-360.

64. **Patel, N., B.F. Rocks and M.P. Bailey.** 1991. A silver enhanced, gold labelled, immunosorbent assay for detecting antibodies to rubella virus. J. Clin. Pathol. *44*:334-338.

65. **Patel, N., B.F. Rocks and M.P. Bailey.** 1993. Sandwich silver enhanced, gold labeled immunosorbent assay for determination of human growth hormone. J. Histotechnol. *16*:259-262.

66. **Rocks, B.F., V.M.R. Bertram and M.P. Bailey.** 1990. Detection of antibodies to the human immunodeficiency virus by a silver enhanced gold-labelled immunosorbent assay. Ann. Clin. Biochem. *27*:114-120.

67. **Roth, J.** 1982. Applications of immunocolloids in light microscopy. Preparation of protein A-silver and protein A-gold complexes and their application for the localization of single and multiple antigens in paraffin sections. J. Histochem. Cytochem. *30*:691-696.

68. **Roth, J.** 1983. The colloidal-gold marker system for light and electron microscopic cytochemistry, p. 217-284. *In* G.R. Bullok and P. Petrusz (Eds.), Immunocytochemistry, Vol. 2. Academic Press, London.

69. **Roth, J., P. Saremaslani and C. Zuber.** 1992. Versatility of anti-horseradish peroxidase antibody-gold complexes for cytochemistry and *in situ* hybridization: preparation and application of soluble complexes with streptavidin-peroxidase conjugates and biotinylated antibodies. Histochemistry *98*:229-236.

70. **Roth, J., P. Saremaslani, M.J. Warhol and P.U. Heitz.** 1992. Improved accuracy in diagnostic immunohistochemistry, lectin histochemistry and *in situ* hybridization using a gold-labeled horseradish peroxidase antibody and silver intensification. Lab. Invest. *67*:263-269.

71. **Scopsi, L. and L.I. Larsson.** 1985. Increased sensitivity in immunocytochemistry. Effects of double application of antibodies and of silver intensification on immunogold and peroxidase-antiperoxidase staining techniques. Histochemistry *82*:321-329.

72. **Schmidt, J. and W. Peters.** 1987. Localization of glycoconjugates at the tegument of the tapeworms

Hymenolepis nana and H. microstoma with gold labelled lectins. Parasitol. Res. *73*:80-86.

73. **Shi, S.R., M.E. Key and K.I. Kalta.** 1991. Antigen retrieval in formalin-fixed, paraffin embedded tissues: an enhancement method for immunohistochemical staining based on microwave oven heating of tissue sections. J. Histochem Cytochem. *39*:741-748.

74. **Shimizu, H., A. Ishida-Yamamoto and R.A. Eady.** 1992. The use of silver-enhanced 1-nm gold probes for light and electron microscopic localization of intra- and extracellular antigens in skin. J. Histochem. Cytochem. *40*:883-888.

75. **Shimizu, H., T. Masunaga, A. Ishiko, T. Hashimoto, D.R. Garrod, H. Shida and T. Nishikawa.** 1994. Demonstration of desmosomal antigens by electron microscopy using cryofixed and cryosubstituted skin with silver enhanced gold probe. J. Histochem. Cytochem. *42*:678-692.

76. **Shin, R.-W., T. Iwaki, T. Kitamoto and J. Tateishi.** 1991. Hydrated autoclave pre-treatment enhances TAU immunoreactivity in formalin-fixed normal and Alzheimer's disease brain tissues. Lab. Invest. *64*:693-702.

77. **Skutelsky, E., V. Goyal and J. Alroy.** 1987. The use of avidin-gold complex for light microscopic localization of lectin receptors. Histochemistry *86*:291-295.

78. **Slater, M.** 1991. Differential silver enhanced double labeling in immunoelectron microscopy. Biotechnol. Histochem. *66*:153-154.

79. **Springall, D.R., G.W. Hacker, L. Grimelius and J.M. Polak.** 1984. The potential of the immunogold-silver staining method for paraffin sections. Histochemistry *81*:603-608.

80. **Stefanini, M., C. De Martino and L. Zamboni.** 1967. Fixation of ejaculated spermatozoa for electron microscopy. Nature *216*:173-179.

81. **Stierhof, Y.-D., B.M. Humbel, R. Hermann, M.T. Otten and H. Schwarz.** 1992. Direct visualization and silver enhancement of ultra-small antibody-bound gold particles on immunolabeled ultrathin resin sections. Scan. Electron Microsc. *6*:1009-1022.

82. **Tacha, D.E., Ph.D. Bowman and L.A. McKinney.** 1993. High resolution light microscopy and immunocytochemistry with glycol methacrylate embedded sections and immunogold-silver staining. J. Histotechnol. *16*:13-16.

83. **Towbin, H. and J. Gordon.** 1984. Immunoblotting and dot immunoblotting. Current status and outlook. J. Immunol. Methods *72*:313.

84. **Valck, V. de, W. Renmans and E. Segers.** 1991. Light microscopical detection of leukocyte cell surface antigens with a one-nanometer gold probe. Histochemistry *95*:483.

85. **Van de Kant, H.J. and D.G. De Rooij.** 1992. Periodic acid incubation can replace hydrochloric acid hydrolysis and trypsin digestion in immunogold-silver staining of bromodeoxyuridine incorporation in plastic sections and allows the PAS reaction. Histochem. J. *24*:170-175.

86. **Van de Kant, H.J., M.E. Boon and D.G. De Rooij.** 1993. Microwave application before and during immunogold-silver staining. J. Histotechnol. *16*:209-215.

87. **Van Den Pol, A.N.** 1985. Silver-intensified gold and peroxidase as dual immunolabels for pre- and postsynaptic neurotransmitters. Science *228*:332-335.

88. **Vandre, D.D. and R.W. Burry.** 1992. Immunoelectron microscopic localization of phosphoproteins associated with the mitotic spindle. J. Histochem. Cytochem. *40*:1837-1848.

89. **Varndell, I.M., J.M. Polak, K.L. Sikri, C.D. Minth, S.R. Bloom and J.E. Dixon.** 1984. Visualisation of messenger RNA directing peptide synthesis by *in situ* hybridisation using a novel single-stranded cDNA probe. Histochemistry *81*:597-601.

90. **Westermark, K., M. Lundqvist, G.W. Hacker, A. Karlsson and B. Westermark.** 1987. Growth factor receptors in thyroid follicle cells. Acta Endocrinol. (Copenh.) Suppl. *281*:252-255.

91. **Zehbe, I., G.W. Hacker, J. Sällström, E. Rylander and E. Wilander.** 1992. *In situ* polymerase chain reaction (*in situ* PCR) combined with immunoperoxidase staining and immunogold-silver staining (IGSS) techniques. Detection of single copies of HPV in SiHa cells. Anticancer Res. *12*:2165-2168.

92. **Zehbe, I., G.W. Hacker, W. Muss, J. Sällström, E. Rylander, A.-H. Graf, H. Prömer and E. Wilander.** 1993. An improved protocol of *in situ* polymerase chain reaction (PCR) for the detection of human papillomavirus (HPV). J. Cancer Res. Clin. Oncol. *119*:22S.

93. **Zehbe, I., J. Sällström, G.W. Hacker, E. Rylander, A. Strand, A.-H. Graf and E. Wilander.** 1994. Polymerase chain reaction (PCR) *in situ* hybridization: detection of human papillomavirus (HPV) DNA in SiHa cell monolayers, p. 297-306. *In* J. Gu and G.W. Hacker (Eds.), Modern Methods in Analytical Morphology. Plenum Press, New York.

94. **Zehbe, I., G.W. Hacker, J.F. Sällström, E. Rylander and E. Wilander.** 1994. Self-sustained sequence replication-bases amplification (3SR) for the *in situ* detection of mRNA in cultured cells. Cell Vision *1*:20-24.

# CHAPTER 3

# Elimination of Background Staining in Immunocytochemistry

**Jiang Gu[1] and Neera Agrawal[2]**

[1]Institute of Molecular Morphology, Mount Laurel, NJ; [2]Deborah Research Institute, Browns Mills, NJ, USA

## SUMMARY

*Background staining is a major concern for immunocytochemistry. It occurs for a number of reasons and each requires different treatment for correction. Once the cause is identified, measures can be taken to rectify the problem. One of the limitations of the widely employed indirect immunocytochemistry technique is that rabbit polyclonal antibodies cannot be used on rabbit tissue because of the strong, obscuring background that usually occurs. We experimented with a new procedure that entailed the formation of a primary and secondary antibody complex in solution prior to its application to tissue sections. In this procedure, the nonspecific binding of the secondary antibody to the endogenous rabbit immunoglobulin of the tissue was avoided. The results were very clean, completely eliminating the background staining. Various conditions for the complex formation, antibody concentrations and other steps in the procedure were tested. Results were compared with extensive controls. Antibodies on five tissue types were employed in the avidin-biotin-peroxidase complex, indirect immunofluorescent and immunogold-silver methods. The procedure was found to be universally applicable. The possible theoretical mechanisms, practical considerations, potentials and limitations of this technique are discussed. This method should find widespread application for the detection of rabbit tissue antigens using readily available rabbit polyclonal antibodies with indirect immunocytochemical techniques.*

## INTRODUCTION

For immunocytochemists, nonspecific background staining is one of the most common and bothersome problems to be dealt with. Background staining is defined as untoward labeling of tissue sections or cell preparations by the reporting markers that interfere with correct interpretation of specific immunostaining results. The ease of use and widespread applications of immunostaining have broadened the negative impact of this common problem. Without stringent controls, the unspecific background staining may be mistaken as positivity and lead to erroneous interpretation. Together with false-negative staining, they present major obstacles for successful immuno-cytochemistry.

Nonspecific background staining can manifest itself in many forms. It can be an overall staining of all or most of the tissue structures or a part of them. Some of these labelings appear very specific, confined to certain tissue or cell structures, while others are meshed over many cell and tissue types. Certain stainings are weak compared with true specific stainings, making it hard to distinguish between specific and nonspecific labelings.

Generally, background staining can be divided into three categories according to their causes: technical, conditional or inherited.

Background staining caused by technical mistakes is fairly easy to recognize and not difficult to correct. It is often related to the negligence or ignorance of the person performing the experiment. Any steps that are not carefully carried out according to established protocols may cause background staining. The most common ones include: the antibody solution or any of the solutions are dried or semi-dried on the preparation; the incubation duration is too long; the incubation temperature is too high; the composition of the buffer or any of the solutions are not correct; the washing is not long or vigorous enough, etc. Generally, this kind of background can be distinguished by the appearance of overall coloring of the sample or widespread deposits of granular chromogen aggregates without any structural preference. Certain undesirable reactions are known to occur if the protocols are not followed to their fullest extent. For example, the chromogen for peroxidase will react with intrinsic peroxidase in the tissue samples, such as in the red blood cells, if it is not completely blocked by enough hydrogen peroxide at the beginning of the immunostaining. Also, if levamisole is not used, the alkaline phosphatase labeling is likely to bind unspecifically to tissue structures and cause considerable background. These can be corrected by repeating the procedure and carefully following the protocols and conditions set forth for the particular antibodies and antigens.

Another kind of background staining is caused by suboptimal conditions of the protocols. Those problems are not easy to fix and often need systemic experimentation to identify the factor or factors that cause the problem. These may include: the concentration of the primary antibodies are too high; the chromogen developing duration is too long; the pretreatment of the samples is too harsh; over or under tissue fixation, etc. Those problems are not foreseeable or preventable even if the established protocols are followed to the letter. Every antibody-antigen reaction demands a set of optimal conditions under which the immunoreactions will take place to their fullest capacity where the maximal labelings will be retained with minimal background. Those conditions vary slightly from antigen to antigen and from antibody to antibody. They need to be adjusted individually. When such problems are suspected, a range of conditions should be tested to define the best condition for the particular antibody and antigen under study. It is recommended that one begins with testing the primary antibody, following with chromogen development and then the secondary antibody. When conditions for one parameter are tested, the other parameters should be kept constant. When the specificity of the antibody is established and preservation of the antigen is assured, finding the optimal condition is the next logical step and often the most important measure to obtain a good immunostaining.

Unspecific background staining is sometimes associated with inherited properties of particular antibodies and/or antigens in question. One in particular is the cross-reaction of antibody to untoward antigens, a common yet complicated problem that requires careful consideration and testing. Basically, there are four types of cross-reactions: first, the antibody employed specifically recognizes an antigenic epitope or set of epitopes that are different from the one of interest. These may occur when the epitopes share extensive sequence or sterile structural similarities. This problem is difficult to overcome as the immunoreaction itself is specific. One should avoid using these antibodies; or at least, one should be aware of this possibility and obtain information of known cross-reactivity of the particular antibody so that the confusion of cross-reaction might be clarified during result interpretation. One should bear in mind that not all antibody cross-reactivities are documented or known, not to mention the fact that there are many new or undiscovered antigenic molecules. There is a fine line between the discovery of new locations of the antigen of interest and the manifestation of antibody cross-reaction with unrelated antigenic epitopes. When such a situation is suspected, Western blotting or dot blotting with known antigens should clarify the situation. Second, the antibody, particularly polyclonal antibodies, may not be pure, i.e., they contain substantial amounts of antibodies to other unrelated antigens. In this case, further purification of the antibody in question would be the best strategy. One may also dilute the antibody as much as possible while retaining the specific immunostaining. In this way, the nonspecific antibodies, which often have a much lower concentration than the specific antibody, would be too diluted to cause any visible reaction. The diluted specific antibody can still achieve reasonable labeling by prolonging the incubating time or slightly raising the incubating temperature. Third, nonspecific immunostaining might be caused by excessive treatment of the tissue samples; for example, over-digestion is known to cause background staining. This might also create "specific" bonding sites for antibodies to stick to, but has nothing to do with the distribution of the particular antigen under investigation. Such problems may also occur during over-"cooking" of the paraffin sections or from excessive heating during incorrectly performed antigen retrieval procedures. Fourth, certain unspecific background stainings occur because of incompatibility of the reagents and the tissue samples used. For example, rabbit polyclonal antibodies will cause pronounced background staining when applied onto rabbit tissue samples. This is because the secondary antibodies against the primary antibody, the rabbit imunoglobulin, will also react to intrinsic immunoglobulin in the rabbit tissue. Similarly, antibodies raised in mice or rats may cause high background in tissue samples of corresponding species. Those background stainings are almost unavoidable, unless specially designed protocols are implemented. One such protocol developed by our group is presented in detail in this chapter.

The importance of performing appropriate controls for immunocytochemistry cannot be overemphasized. Without adequate controls, the specificity of immunostaining cannot be established and results could not be published no matter how beautiful or specific they may appear. The best controls include Western blotting to test the reactivity of the antibody and antigen in question, using unrelated antibodies (or nonimmune serum) and the preabsorption test. The latter involves the incubation of the primary antibody with its specific antigen in a test tube before applying this mixture on the tissue sample. A negative result will establish the specificity of the immunostaining when performed without this quenching step.

In this chapter, we present the development of an immunostaining protocol that enables the application of commonly available rabbit polyclonal antibodies on rabbit tissue samples. The principle and practical considerations for elimination of background immunostaining are thereby illustrated.

## DEVELOPMENT OF THE PRIMARY-SECONDARY ANTIBODY COMPLEX (PSC) METHOD

While immunocytochemistry has had widespread application and provides a powerful tool for morphological research and diagnosis, one limitation remains: it is difficult, if not impossible, to use rabbit polyclonal antibodies to detect antigens in rabbit tissue. Such immunostaining creates a tremendous amount of background that makes interpretation of the results nearly impossible. The background is caused by the binding of the secondary antibody to the endogenous immunoglobulin or other charged molecules in the rabbit tissue (6,10). The secondary antibody, usually goat-anti-rabbit IgG, is raised by using rabbit immunoglobulin as an immunogen. The resulting goat anti-rabbit antibodies are also polyclonal and can recognize the Fc fragment of rabbit immunoglobulin and a number of other rabbit antigens if the immunogens or the antibodies were not affinity-purified. It is only natural for these secondary polyclonal antibodies, usually labeled with biotin, fluorescein or other ligands, to also bind to the antigenic epitopes in the rabbit tissue sections in addition to the primary antibodies. This background problem has severely limited the ability to perform immunocytochemistry on rabbit tissue. Nevertheless, the rabbit remains a popular animal model for physiological and pharmacological studies because of its size, ease of handling and low cost. It would be of significant interest if a procedure can be developed that allows the detection of rabbit antigens with the readily available rabbit polyclonal antibodies without excessive background staining.

Various approaches have been used to reduce unwanted background staining on rabbit tissue. Normal serum has been used to block nonspecific binding of the secondary antibody to the tissue (10). An acidic potassium permanganate solution has been employed to destroy the endogenous immunoglobulin (11). The primary antibodies have also been directly labeled, thus omitting the need for a secondary antibody (1,7,9). However, none of these methods has given consistent and satisfactory results. Similar problems exist when murine antibodies are used to detect murine antigens or human antibodies to detect human antigens. An approach employed by Tuson and colleagues using human monoclonal antibody on human tissue (12) and by Fung et al. using mouse monoclonal antibody on transgenic mouse tissue (3) has had some success in eliminating background activity. In this study, we employed a similar approach that enabled us to use polyclonal rabbit antibodies on rabbit tissues. Various conditions using a range of antibodies on different tissue types in three immunostaining methods were used. An optimal procedure was established. The background staining in all these tissue types was greatly reduced and the specific staining became clearly visible.

## Specimen Preparation

Eight New Zealand White rabbits (Charles River, Wilmington, MA, USA) weighing approximately 2 kg were used. The rabbits were anesthetized with intravenous sodium pentobarbital (40 mg/kg) injection. The heart, pancreas, duodenum and kidney were removed, cut into small pieces and fixed for 3 to 4 h in 4% paraformaldehyde. The tissue was processed for paraffin embedding. Seven-micrometer-thick paraffin sections were cut. Additional tissue sections were also snap-frozen and 10-μm thick cryostat sections were postfixed in 4% paraformaldehyde for 2 min or 0.1% glutaraldehyde. The same organs from rats were used as controls.

## Immunostaining Procedures Tested

The types of antibodies used in this study and their characteristics are presented in Table 1. A number of experiments were performed to test the new procedure for detecting rabbit antigens using rabbit antibodies. They included the following.

1. Conventional avidin-biotin-peroxidase (ABC) method (8), indirect immunofluorescence method (4) and immunogold-silver method (5) were performed on rat tissue sections using the polyclonal antibodies against various antigens as listed in Table 1. These were used as positive controls and were carried out as follows.

**The ABC method (8).** Reagents used were from a kit (Zymed, San Francisco, CA, USA). Deparaffinized and rehydrated tissue sections were treated with 3% hydrogen peroxide solution for 30 min to exhaust intrinsic peroxidase. Tissue sections were then incubated with 10% normal goat serum for

**Table 1. Antibody Type, Concentration, Source, Tissue Type and Methods Used**

| Antibodies | Concen- tration | Tissue Type | Detection Method | Source |
|---|---|---|---|---|
| ANP | 1:1000 | atria | ABC, IF, IGSS | Peninsula Laboratories Belmont, CA |
| Insulin | 1:100 | pancreas | ABC, IF, IGSS | Biogenex Laboratories San Ramon, CA |
| NPY | 1:1000 | atrium, duodenum | ABC | Peninsula Laboratories Belmont, CA |
| S-100 | 1:4000 | duodenum, kidney | ABC | Accurate Chemical Westbury, NY |
| PGP 9.5 | 1:1000 | duodenum | ABC | Accurate Chemical Westbury, NY |
| PKC | 1:100 | ventricle | ABC | Research and Diagnostic Antibodies, Berkeley, CA |

Note: ABC = Avidin-Biotin-Peroxidase complex method; ANP = atrial natriuretic peptide; IF = Indirect immunofluorescent method; IGSS = Immunogold-silver staining method; NPY = Neuropeptide Y; PGP 9.5 = Protein gene product 9.5; PKC = Protein kinase C.

The specificities of the antibodies have been well established with previous experiments and publications. The specificities of the antibodies to PKC were further confirmed with Western blotting assay using rabbit tissue antigens.

30 min. Sections were incubated with primary antibody in a humid chamber for 18–20 h at 4°C. The dilution of each antibody was optimized in preliminary experiments. The sections were washed with phosphate-buffered saline (PBS; 0.1 M, pH 7.2) and incubated with biotinylated secondary antibody for 1 h; rinsed with PBS and incubated with streptavidin-biotin-peroxidase conjugate for 30 min; rinsed with PBS and incubated with substrate chromogen mixture. The slides were monitored for color development. Color development time varied from 5 min to 15 min for various antibodies. The sections were then counterstained with hematoxylin and coverslipped.

**Indirectimmunofluorescence method (4).** This method is essentially the same as the ABC method except for the following. Blocking of the intrinsic peroxidase step was omitted. The secondary goat anti-rabbit antibody was labeled with fluorescein-isothiocyanate (FITC) or rhodamine instead of biotin. Following incubation with the labeled secondary antibodies, the slides were washed in PBS, mounted and viewed with a Nikon transmission light microscope with a UV light source and appropriate UV filters.

**Immunogold-silver staining (IGSS) method (5).** A previously established procedure of immunogold-silver staining method was used (5). Paraffin sections were de-waxed, rehydrated, washed in tap water and then immersed in Lugol's iodine solution for 5 min (1% iodine in 2% potassium iodate; Merck #9261, Darmstadt, Germany). After rinsing briefly in tap water, the slides were further rinsed with 2.5% sodium thiosulfate until the sections became colorless (usually less than 30 s). After washing in tap water, the slides were immersed in IGSS buffer consisting of 0.1 M PBS with 0.1% fish gel, 0.1% Triton® X-100 and 2.5% NaCl. Ten percent normal goat serum was then applied to the tissue sections for 30 min. After draining excess normal goat serum, the primary antibody was applied to the tissue section and incubated overnight at 4°C. The antibody was diluted in 0.1 M PBS or Tris-buffered saline (TBS 0.1 M Tris and 0.15 M NaCl) containing 0.1% bovine serum albumin (BSA) and 0.1% sodium azide. The slides were then washed in PBS with 0.1% fish gel and incubated in 5-nm colloidal gold-labeled secondary antibody (goat anti-rabbit) for 1 h at room temperature. After washing as before, the slides were immersed in 0.1 M PBS for 2 min and post-fixed in 2% glutaraldehyde in PBS for 2 min. They were then washed well in distilled water and developed in a silver enhancement solution consisting of silver acetate (100 mg silver acetate in 50 mL distilled water), and hydroquinone (250 mg hydroquinone in 50 mL citric buffer, pH 3.8). Citric buffer (pH 3.8) was prepared by dissolving 23.5 g trisodium citrate dihydrate and 25.5 g citric acid monohydrate in 1 liter of distilled water. The silver acetate solution and hydroquinone solution were mixed immediately prior to the enhancement. The slides were monitored periodically under a light microscope until the desired darkness of reaction was achieved. They were then immersed in photographic fixer for 1 min, washed with tap water and counterstained with hematoxylin and eosin.

2. The above methods and antibodies were applied to rabbit tissues. They were used as negative controls to demonstrate the amount of background stainings produced when using rabbit polyclonal antibodies on rabbit tissue with the conventional protocols.

3. The newly designed PSC procedures were performed with all the above antibodies on rabbit tissue. The principles of the new procedure are illustrated in Diagram 1. The details of this procedure are described below.

In this immunohistochemical detection system, primary and secondary antibodies were incubated in a test tube at 37°C for 1 h. The pre-diluted biotinylated secondary antibody was used as the diluent. Undiluted nonimmune rabbit serum was then added to the primary-secondary antibody mixture at a ratio of 1:100. The mixture was incubated at 37°C for 1 h, then chilled on ice for an additional hour.

While the antibody mixture was incubating, the sections were deparaffinized, rehydrated, treated with 3%–5% hydrogen peroxide solution for 30 min, rinsed with TBS three times for 2 min each and incubated with 10% nonimmune goat serum or 5% nonfat dry milk and 1% fetal bovine serum in TBS for 1 h at room temperature. Sections were washed in TBS as before and immersed in prechilled 2% fetal bovine serum in TBS for 15 min at room temperature. Sections were incubated with the antibody complex at 4°C for 12–18 h in a humid chamber, washed three times with TBS, 5 min each, and then treated with TBS with 1% fetal bovine serum for 10 min (this step was later found to be unnecessary). Sections were then incubated with strept-avidin-biotin-peroxidase complex for 1 h at room temperature. After washing twice in TBS, 3 min each, sections were incubated with substrate-chromogen mixture. The color development was monitored under a microscope. The

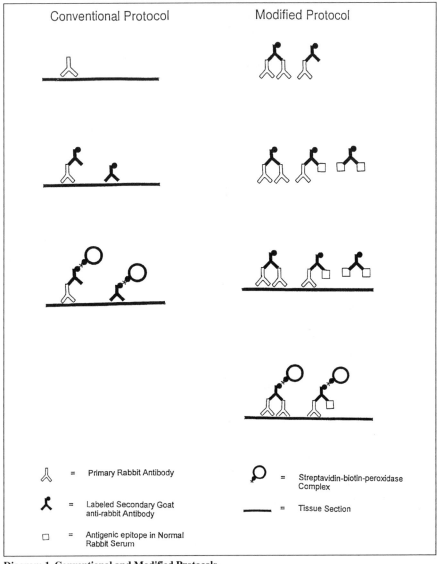

**Diagram 1. Conventional and Modified Protocols.**

slides were then washed with distilled $H_2O$, counterstained with hematoxylin and mounted in mounting media.

4. The modified procedure described above was also applied to rat tissue to further evaluate whether it causes any side effects, i.e., weakening the specific staining, causing other nonspecific reactions, etc.

5. Various conditions for the new procedure were further evaluated in an attempt to optimize the new protocol. These included changing the concentrations of the primary antibody, secondary antibody and normal rabbit serum, varying the incubation period and temperatures of the complex formation reaction, omitting each background blocking step and using different buffer systems.

6. Other possible methods to block the background were also tested. These included using unlabeled goat anti-rabbit IgG antibodies or their Fab fragments to block the intrinsic bonding sites on the rabbit tissues prior to applying the primary antibodies, using normal rabbit serum, goat serum or nonfat dry milk at various concentrations, varying lengths of time for incubations and combinations of the above.

7. The new PSC procedure was also incorporated into the indirect immunofluorescence method and the IGSS method. Standard procedures were used, except that the primary and respective secondary antibodies were incubated in test tubes with normal rabbit serum for the formation of PSC complex.

8. The results were evaluated with a computerized image analyzer (Universal Imaging, Image I™, West Chester, PA, USA) as well as manually by at least two investigators independently, and treated with Student's $t$ test.

## OBSERVATIONS

All the immunostaining methods using the employed antibodies on rat tissue gave strong positive immunostaining in their known locations with little or no background staining in the nonspecific regions. Insulin was detected exclusively in the β cells of the pancreatic islets, ANP in the atrial myocardiocytes, S-100 in the Schwann cells of all the organs (particularly the gut), PGP 9.5 in the nerves and PKC isoforms in the myocardiocytes. When those antibodies were used on the rabbit tissue with conventional immunostaining procedures, heavy background staining obscured specific antibody-antigen reactions, making result interpretation problematic (Figures 1a, 2a, 3a and 4a; See color plates in Addendum, p. A-4 and A-5). When the primary antibodies were omitted, the background staining still occurred in the extracellular spaces. When the new procedure in which the primary and secondary antibody complexes were first formed in solution was employed the background staining was completely removed, and the specific immunostaining was clearly visible in the appropriate locations (Figures 1b, 2b, 3b and 4b). The immunostaining results with rabbit tissue were generally weaker than those with the rat tissue (Figures 2 and 5). However, when the PSC and the conventional procedures were performed on the rat tissues, no apparent differences in staining intensities between the two methods were observed (Figure 5, see color plate in Addendum, p. A-5).

Image analysis revealed that the difference between the background stainings of the PSC method and the conventional method was statistically extremely significant ($P$ <0.0001). The staining intensities between the specific and nonspecific stainings with the conventional method were not significant ($P = 0.14$). With the PSC method, they became extremely significant ($P$ <0.0001).

The same beneficial effects of the PSC method were obtained with all three immunostaining methods, i.e., ABC, indirect immunofluorescence and immunogold-silver staining methods (Figure 4).

Using two to three times more concentrated primary antibodies in the PSC method did not appear to increase the specific staining or increase the background staining.

The optimal temperature to form the primary-secondary antibody complexes was 37°C for 1 h for the primary-secondary antibody incubation and 1 h for the normal rabbit serum incubation. The immunostaining was generally stronger when the complex was formed at 37°C than at room temperature. The 4°C cooling did not make a significant difference in the staining results. When the complex formation reaction was shortened to 30 min for each step instead of 1 h, the staining intensity decreased slightly.

On comparing different background blocking methods, the unlabeled goat anti-rabbit antibody or the unlabeled Fab fragment of goat anti-rabbit IgG was only partially effective in removing the background staining on the rabbit tissue. However, there was still a substantial amount of background staining remaining on the tissue section, thus making the quality of the immunostaining far inferior to that obtained by the PSC procedure. No significant difference was observed when the dry milk blocking procedure was replaced by normal goat serum blocking. The second blocking step following the primary antibody incubation did not appear to further improve the quality of the immunostaining. One background blocking step with normal goat serum prior to the application of the primary antibody or the primary-secondary antibody complex to the tissue section appeared to be sufficient.

Both PBS and TBS were effective as the washing solutions and the diluent, although the staining quality appeared to be slightly better when TBS was used as the buffer.

## DISCUSSION

In this chapter, we demonstrate that the PSC method can effectively remove the background staining while permitting specific detection of rabbit antigens in rabbit tissue. The specificity of the staining achieved with this procedure was established with various controls.

The key to removing the excessive background of conventional immuno-staining when using rabbit antibodies on rabbit tissue is to prevent the secondary antibody raised against rabbit immunoglobulin from binding to the rabbit immunoglobulins endogenous to the tissue. These immunoglobulins are normally present in plasma cells, lymphocytes and in the plasma. The latter is diffused throughout most tissue types and particularly in the connective tissues that are often filled with body fluids containing immunoglobulin. As a result, the secondary goat anti-rabbit IgG would normally bind to those antigens and create heavy background staining. For this reason, it has been very difficult to investigate the morphological distribution of tissue antigens in rabbit tissue using the readily available rabbit antibodies and the indirect immunostaining method.

In theory, there should be at least four approaches that could be undertaken to eliminate or circumvent the nonspecific reactions in the rabbit tissue. The first is to form a PSC complex as we did in the present study. In this way, all the secondary antibody binding sites would be occupied by either the primary antibody or the normal rabbit serum, making them unavailable to bind to the intrinsic tissue antigens. Second, it should be possible to block all the intrinsic immunoglobulin reaction sites by using unlabeled secondary antibodies or their Fab fragments that recognize rabbit immunoglobulin (6,10). This reaction should be performed prior to the primary antibody application so that the labeled goat-anti-rabbit antibodies will have no free intrinsic binding sites to react with. This direct blocking approach was tested in our study but was found to be ineffective. This could be related to the competition between the labeled secondary antibodies and the Fab fragments or the unlabeled goat anti-rabbit antibodies used in the blocking reaction. In addition, unless the blocking Fab fragment or unlabeled antibodies were from exactly the same batch as the labeled secondary antibodies, the blocking would not be complete and background staining would occur. Even if it worked, this approach would be more expensive because it requires one additional reagent. Third, the primary antibody might be directly labeled with biotin or other markers that would make the secondary goat-anti-rabbit antibody unnecessary (1,7,9). This approach is more expensive and laborious as each individual primary antibody would have to be labeled separately and is less sensitive than the indirect methods. Fourth, primary antibodies raised in other species, such as murine monoclonal antibodies, may be used to detect rabbit tissue antigens. Those antibodies, however, are more difficult and expensive to develop and generally give immunostainings that are weaker than their rabbit polyclonal counterparts. Therefore, the first approach, i.e., development of a primary and secondary antibody complex as used in this study, possibly in combination with the blocking method, would be the method of choice.

The concentrations of the primary and secondary antibodies and normal serum used in this new procedure are important factors in the success of the method. Should the secondary antibody have a concentration less than that of the primary antibody, the detecting sensitivity would be reduced, as there would be primary antibodies reacting to tissue antigens but not being picked up by the labeled secondary antibodies. If there are too many secondary antibodies, the normal rabbit serum may not be able to occupy all the available binding arms, resulting in binding of secondary antibodies to tissue antigens and nonspecific staining. The concentrations of the three reagents should be approximately matched with slightly more secondary antibody molecules than the primary and a slightly higher amount of rabbit immunoglobulins contained in the normal rabbit serum to form a complex that allows maximum detection of specific antigenic sites in the tissue section without too much left over for the normal rabbit serum. This requires experimentation of different concentrations of the reagents prior to performing the immunostaining.

The detecting sensitivity of the new procedure at the primary antibody dilutions used in this study was unchanged or slightly decreased from the routine procedure. Because appropriate controls were not available, assessment of the detecting sensitivity of the PSC method was not feasible. The large sizes of the primary-secondary antibody complex did not appear to interfere with the mobility of the molecular diffusion and tissue antigen recognition. However, in the PSC method, we recommend using a slightly higher concentration of the primary antibody than would be used in the conventional procedure. The dry milk used in the blocking solution by Fung et al. (3) can be replaced by normal goat serum, normally included in immunostaining kits, without causing any problem. Blocking for a second time with the same solution before subsequent immunostaining steps did not cause any noticeable benefit in our experiments, nor is it theoretically necessary. The complex formation step can be performed at 37°C for a total of 2 h (1 h for priming and secondary antibody reaction and 1 h for normal rabbit serum incubation). The additional 1 h incubation at 4°C, as previously reported (3) was not necessary. Incubation at 37°C appeared to be better than at room temperature or 4°C. This is in agreement with previous experience that higher temperature accelerates and strengthens immunoreaction (6,10). The modified procedure performed on rat tissue produced identical, if not better, results when compared to the conventional procedure, indicating that the new procedure neither caused untoward side effects nor weakened the immunostaining.

Our experiments demonstrate that this new procedure is universally applicable. Antibodies on five different tissue types were tested. Those antigens ranged from peptides to proteins, and their locations included both cytoplasm and nucleus. In each case, the use of the PSC method gave much improved immunostaining without any nonspecific background. The tissue type did not seem to have an effect on the procedure either. All the immunostaining and labeling procedures including the avidin-biotin-peroxidase complex, immunogold-silver and indirect immunofluorescent methods gave much improved results. The modified procedure should also be applicable to other species such as mouse, rat or human using primary antibodies raised in the respective species, as demonstrated by previous studies (3,12). It should also be applicable for immunogold electron microscopy. The different conditions and precautions established in this study should also be applicable. In theory, the new procedure should be applicable to all immunostaining procedures as an alternative approach to prevent background formation caused by the secondary antibodies. Nevertheless, we recommend that, depending on the abundance of the tissue antigens and the affinity and avidity of the antibodies, the PSC method should be tested and optimized for each individual system. In a way, the PSC method is similar to the direct labeling immunostaining method with the added advantage of an increased detecting sensitivity that is comparable to the indirect method, and the convenience of using the widely available indirect immunostaining reagents and kits. In conclusion, this new procedure permits the use of polyclonal rabbit antibodies on rabbit tissue to achieve clean and specific immunostainings.

## RECOMMENDED PROTOCOL OF PSC METHOD

For the biotin-labeled secondary antibody and strepavidin-peroxidase complex methods:

1. Mix primary and secondary antibodies in a test tube to their appropriate dilutions using TBS as the diluent, and incubate at 37°C for 1 h.

2. Add nonimmune rabbit serum 1:100 to the above mixture and incubate for 1 h at 37°C.

3. Chill the mixture before applying it to tissue sections. It can be stored at 4°C for 10 min to overnight.

4. Deparaffinize and rehydrate slides.

5. Immerse slides in 5% hydrogen peroxide in TBS for 30 min at room temperature.

6. Wash 3 times in $dH_2O$, 2 min each.

7. Wash in TBS for 5 min.

8. Incubate in 10% blocking serum (nonimmune goat serum) for 30 min at room temperature.

9. Incubate the sections with primary-secondary antibody complex in a humid chamber at 4°C or room temperature for 10–20 h overnight.

10. Wash 3 times in TBS, 3 min each.

11. Incubate with streptavidin-horseradish peroxidase complex for 1 h at room temperature.

12. Wash 3 times in TBS, 2 min each.

13. Incubate with chromogen and monitor for color development. Color develops in 5–20 min (Zymed kit).

14. Counterstain with hematoxylin.

15. Mount and coverslip.

**Note:** Other indirect immunostaining methods, e.g., immunofluorescence method, peroxidase anti-peroxidase (PAP) method, IGSS method, can be performed in the same manner by forming the PSC followed by their own established protocols.

## ACKNOWLEDGMENTS

The authors wish to thank Dr. P.P. Wang, Dr. M.V. Cohen and Dr. J. Dow for their assistance in performing the experiments, providing the PKC antibody, and comments on the chapter; Nancyleigh Carson and Michele Forte for their technical assistance; Dr. Gene Wood Gupta and Dr. Tak-Shun Choi for their critical review and comments; and Gayle Englund for preparing the manuscript. This work was supported in part by Deborah Hospital Foundation (OC-94-70), Browns Mills, NJ, and in part by a grant to J. Gu from the National Institute of Health, Heart, Lung and Blood Institute, HL-50688.

**REFERENCES**

1. **Baetge, E.E., R.R. Behringer, A. Messing, R.L. Brinster and R.D. Palmiter.** 1988. Transgenic mice express the human phenylethanolamine N-methyltransferase gene in adrenal medulla and retina. Proc. Natl. Acad. Sci. USA *85*:3648-3652.
2. **Foulds, S. and O. Eremin.** 1990. The use of Fab fragments in a screening method for the detection of human antitumor monoclonal antibodies. Hybridoma *9*:91-96.
3. **Fung, K-M., A. Messing and V.M.-Y. Lee.** 1992. A novel modification of the avidin-biotin complex method for immunohistochemical studies of transgenic mice with murine monoclonal antibodies. J. Histochem. Cytochem. *40*:1319-1328.
4. **Gu, J., K.N. Islam and J.M. Polak.** 1983. Repeated application of first layer antiserum improves immunostaining, a modification of indirect immunofluorescence procedure. J. Histochem. *15*:475-483.
5. **Hacker, G.W., C. Hauser-Kronberger, A-H. Graf, G. Danscher, J. Gu and L. Grimelius.** 1994. Immunogold-silver staining (IGSS) for detection of antigenic sites and DNA sequences, p. 19-36. *In* J. Gu and G.W. Hacker (Eds.), Modern Analytical Methods in Morphology, Plenum Publishing, New York.
6. **Harlow, I. and D. Lane.** 1988. Antibodies. A Laboratory Manual. Cold Spring Harbor Press, Cold Spring Harbor, NY.
7. **Haspel, M.V., R.P. McCab, N. Pomato, N.J. Janesch, J.V. Knowlton, L.C. Peters, H.C. Hoover Jr. and J.G. Hanna Jr.** 1985. Generation of tumor cell-reactive human monoclonal antibodies using peripheral blood lymphocytes from actively immunized colo-rectal carcinoma patients. Cancer Res. *45*:3951.
8. **Hsu, S.M., L. Raine and H. Fanger.** 1981. A comparative study of peroxidase-antiperoxidase method and an avidin-biotin complex method for studying polypeptide hormones with radioimmunoassay antibodies. Am. J. Pathol. *75*:734-738.
9. **Imam, A., M.M. Drushella and C.R. Taylor.** 1986. A novel immunoperoxidase-procedure of using human monoclonal antibodies for the detection of cellular antigens in tissue sections. J. Immunol. Methods *86*:17-20.
10. **Jennette, J.C.** 1989. Immunohistology in Diagnostic Pathology. CRC Press, Boca Raton, FL.
11. **Johansson, O., T. Hökfelt, B. Pernow, S.L. Jeffcoate, N. White, H.W.M. Steinbusch and A.A.J. Vethofstad.** 1981. PCE: Immunohistochemical support for three putative transmitters in one neuron coexistence of 5-hydroxy-tryptamine, substance P- and thyrotropin releasing hormone-like immunoreactivity in medullary neurons projecting to the spinal cord. Neuroscience *6*:1857-1881.
12. **Tuson, J.R., E.W. Pascoe and D. Jacob.** 1990. A novel immunohistochemical technique for demonstration of specific binding of human monoclonal antibodies to human cryostat tissue sections. J. Histochem. Cytochem. *38*:923-926.

# CHAPTER 4

# Principles, Applications and Protocols of Microwave Technology for Morphological Analysis

Anthony S-Y. Leong and F. Joel Leong

Division of Tissue Pathology, Institute of Medical & Veterinary Science, Adelaide, South Australia

## ABSTRACT

*Since the introduction of microwave irradiation for morphologic diagnosis and research just over 20 years ago, the applications of microwaves have expanded considerably beyond their initial use as a safe, clean and rapid method of tissue fixation. Microwaves can now be employed for the fixation of large specimens, including brain, to accelerate the action of cross-linking fixatives, as well as to greatly reduce the various stages of tissue processing so that a paraffin-embedded block can be produced in about 30 min. Almost every histochemical stain can be greatly accelerated by microwaves, and immunohistochemical staining procedures can be completed in 20 min by employing microwaves. Cellular antigens are distinctly better preserved in tissues fixed by microwaves than by conventional cross-linking fixatives, and the cytomorphology of cryostat sections irradiated in Wolman's solutions is vastly improved. Microwaves can similarly be employed in transmission and electron microscopy both for fixation and staining of ultrathin sections, and they have the potential to greatly accelerate the polymerization of resins. They also have applications in newer molecular techniques; they not only increase the speed of the procedures but also improve sensitivity. This wide range of applications makes microwave technology invaluable in the pathology laboratory.*

## HISTORICAL PERSPECTIVE

The earliest report to highlight the potential applications of microwave (MW) energy in histology is attributable to Mayers (59) at the University of Edinburgh Medical School, Edinburgh, Scotland, who employed a 630-W generator used by physiotherapists as a heat device to produce rapid and uniform heating of biological specimens as a means of preserving tissue architecture for light microscopy. Direct exposure to MWs for 90 s produced satisfactory fixation of 1-cm cubes of human and mouse tissues. Almost concurrently, Stavinoha et al. (73), at the University of Texas in San Antonio, employed MW energy to stop enzymatic degradation of acetylcholine by acetylcholinesterase in rodent brain to permit chemical analysis. Similarly, Maruyama (57), employing a 5.0-kW magnetron at 2.45 GHz, confirmed that significantly higher levels of acetylcholine, beta-endorphin and dopamine were demonstrable in small animals killed by microwaves. Focused microwave irradiation is an accepted method of killing laboratory animals, producing brain temperatures in mice in excess of 80°C after 200 ms and death within 5 s, so that many of the enzymes which catalyze biochemical reactions are completely and irreversibly deactivated, allowing more accurate measurements of many neurosubstances (71).

The first detailed study of MWs for tissue fixation was done by Bernard (1), who examined the effects of different temperatures from 60° to 85°C, on various organs in four adult hairless mice and concluded that each organ had an optimal temperature for histological fixation. He also suggested that microwaves were potentially useful for electron microscopical studies, although he himself was unable to achieve optimal organelle preservation.

Other important landmarks in the development of MW fixation techniques include the contributions of Login (48), who introduced a method for standardization of exposure time and solution volume in domestic MW ovens; Patterson and Bulard (66), who developed a protocol for immunofluorescence staining of cell monolayers stabilized by MW irradiation; and those of Leong et al. (35) and Leong and Duncis (37), who adapted MW fixation for routine use in a diagnostic laboratory and for the fixation of large biopsy specimens and whole organs.

In the last 20 years, through the pioneering efforts of several groups of investigators in different parts of the world (4,32,34,35,49–51), there has been a slow but increasing acceptance of microwave technology in many aspects of morphologic diagnosis and research. The number of publications describing innovative applications of MWs in the pathology laboratory continue to increase. There are currently two books on the subject (4,53), and a microwave newsletter has been published as a supplement to the *European Journal of Morphology*, clearly attesting to the increasing interest in the subject.

This review will discuss the applications of MW technology in the pathology laboratory both in the area of light microscopy as well as ultrastructural examination.

## CURRENT APPLICATIONS

### Fixation

**Fixation of biopsies and large specimens.** The applications of MW technology have come a long way since the first experimentation with 630-W generators used for physiotherapy (59). With the ready availability of domestic microwave ovens, there was renewed interest in the laboratory applications of such devices. Domestic MW ovens operating at 2.45 GHz and at 600-W output can be used to produce satisfactory fixation of most tissues by irradiating in normal saline to a temperature of 58°–70°C. A variety of immersion fluids, including 10% formalin, water and various buffers, have been employed, but in our experience, normal saline produces the best and most consistent results (13,35). When only a small volume of tissue is irradiated, fixation can be accomplished in a 600-W oven in about 120 s. MWs can thus be used in the diagnostic setting as the primary method of routine fixation, not only providing increased speed but also eliminating the use of formaldehyde, which is a noxious reagent with a potential for toxicity and carcinogenicity. Although any domestic oven may be employed, the use of one with a rotating plate or carousel provides more even distribution of the electromagnetic waves, and a digital timer allows greater accuracy. It is easy to calibrate domestic ovens because the temperature attained is proportional to the duration of electromagnetic flux (35) and the availability of temperature probes in some models of domestic ovens makes operation even easier. Much more expensive laboratory ovens offering slightly greater convenience are also available. These are precalibrated and have the added feature of fume extraction, which is particularly useful when irradiating various chemicals and dyes employed in tissue staining.

MWs do not have a deleterious effect on special stains, and a wide range of tissue antigens are preserved in materials fixed by MWs, as demonstrated with immunohistochemical staining (35). Other workers have confirmed these findings and showed that histochemical techniques for mucous substances, various enzymes and lipids remain effective in tissues fixed by MW irradiation (22).

Whole mice and rabbit fetuses were successfully fixed with MWs (1,68). Leong and Duncis (37) demonstrated the feasibility of fixing large biopsy specimens and viscera with MWs. MWs have limited penetration in solid tissues and temperature gradients of up to 15°C were observed between the surface and core of large pieces of solid tissues such as the spleen, liver, kidney and breast. As it was not possible to raise the core temperature to the optimal level to accomplish fixation without overheating the surface, irradiation had to be done in two phases. Initial exposure to MWs was performed with the intent of rendering the tissue sufficiently firm to allow easy dissection. This was applied to large specimens such as stomach, solid organs and segments of bowel, which are conventionally opened and emptied before pinning out in the case of viscera, or sliced up, in the case of solid specimens, and fixed in formaldehyde for 18–24 h before dissection and sampling. By completely immersing the specimen in a sufficient volume of normal saline and irradiating to a saline temperature of 68°–74°C, the large specimens were rendered sufficiently hard and firm to allow easy dissection. In addition, they had the added advantage of retaining much of their natural color and pliability, allowing better observation of pathologic changes. Lymph node dissection was also easier because they became opaque after irradiation and contrasted sharply against the background bright yellow adipose tissue. Because of the limited penetration of MWs and the temperature gradient between the surface and deeper tissues, fixation was uneven, and further exposure to MWs was required. This was performed by immersing the 2-mm thick tissue samples in saline and irradiating for a further period to attain a temperature of 58°–70°C.

In diagnostic surgical pathology, speed of turnaround is essential and the use of MWs to accomplish fixation of both large and small biopsy specimens, as described above, is a major asset. Specimens can continue to be transported to the laboratory in 10% buffered formalin. Indeed, it may be necessary to retain this procedure to avoid autolysis, which results from delays and other mishaps that may occur during transportation of fresh specimens, particularly if material is received from distant sites. The procedure adopted in our laboratory, which has an accession of about 30 000 biopsies a year, is used for specimens that will be immediately examined after laboratory accessioning and sampled as 2-mm thick blocks before placing in plastic cassettes. These cassettes are completely immersed in normal saline solution and irradiated to a temperature of 62°C. It is convenient to contain 20 cassettes in each of three glass beakers of saline, which are located equidistantly at the periphery of the rotating plate in the floor of the microwave oven. Although morphologic preservation is slightly better at higher temperatures not exceeding 72°C, a range of 60°–68°C produces satisfactory cytomorphologic preservation and antigen preservation. On attaining the optimal temperature, it is maintained for an additional 5 min before the tissue blocks are transferred to a vacuum-assisted automated processor and processed through cycles of absolute alcohol, chloroform and wax with the complete exclusion of formaldehyde from the processing cycles. Cycles of 1.5 h and 3.5 h are used, the former for smaller endoscopic and needle biopsies and the latter for virtually all other tissue blocks. For convenience, an overnight cycle of 16.5 h may be used, whereby MW-fixed blocks are processed through several changes of absolute alcohol, chloroform and wax (Tables 1–4) (33). During the interim between MW irradiation and commencement of tissue processing, the tissue samples can be held in 70% alcohol, Carnoy's solution (60% ethanol, 30% chloroform and 10% glacial acetic acid), 10% buffered formalin or normal saline. The first two solutions are particularly useful for tissues that contain large quantities of fat; however, we have found it convenient to routinely employ 10% buffered formalin because this is generally universally applicable for all tissues. This interim period will depend very much on the schedule of the individual laboratory. In our setup, the interim period is generally about 30–60 min, although with late afternoon specimens, this time may extend to 90 min. This protocol allows rapid preparation of high-quality diagnostic sections (Figures 1 and 2, see color plates in Addendum, p. A-6), possibly with the exception of large pieces of skin and dense connective tissues such as my-

Table 1. Processing Cycle for Microwave-Fixed Tissues: Ultrashort Cycle - 1 h 41 min

| Step | Reagent | Temperature | Vac | Immersion |
|------|---------|-------------|-----|-----------|
| 1 | 100% ethanol | 45 | Yes | 5 min |
| 2 | 100% ethanol | 45 | Yes | 5 min |
| 3 | 100% ethanol | 45 | Yes | 5 min |
| 4 | 100% ethanol | 45 | Yes | 10 min |
| 5 | chloroform | 45 | No | 10 min |
| 6 | - | - | - | - |
| 7 | chloroform | 45 | Yes | 10 min |
| 8 | wax | 62 | Yes | 10 min |
| 9 | wax | 62 | Yes | 10 min |

ometrium, which require emersion in formalin for about 30 min before microwave irradiation. This extra step can be achieved without much disruption to the procedure outlined above by simply cutting up these specimens last so that they have a longer exposure to formalin before being sampled (33).

Formaldehyde is the most common chemical fixative used for light microscopy but its noxious odor and concerns over its potential toxicity, as well as suggestions of increased risk of cancer in humans and other animals exposed directly to airborne formaldehyde, have made the search for alternative methods of tissue fixation necessary. While a primary requirement of fixation is the stabilization of tissue proteins to withstand processing for paraffin-embedding, an ideal fixative also has to maintain the tissue in as near a lifelike state as possible. In addition, diagnostic laboratories are also concerned with costs and speed of turnaround. MW irradiation satisfies all these considerations and is a suitable alternative to formaldehyde as the primary method of fixation. It also allows the exclusion of formaldehyde from the laboratory and from the autoprocessor, eliminating the use of this noxious and potentially toxic agent. MW irradiation fulfills all the requirements of a good fixative, is cheap and clean, and has the added advantage of speed.

While MW irradiation as a method of fixing neural tissue for neurochemical analysis is well established (57,71,74), it has more recently also been employed for morphologic studies of neural tissue. Brain tissue, previously perfused with physiologic saline or formaldehyde, when irradiated, produced excellent tissue sections without the morphologic changes that result from autolysis and those that occur during dehydration and impregnation for routine paraffin processing (56). Compared with cryostat sections of untreated brain tissue, MW-irradiated sections showed fewer freeze-thaw artifacts and produced superior microscopic sections. The results were better in tissues

**Table 2. Processing Cycle for Microwave-Fixed Tissues: Short Cycle - 3 h 31 min**

| Step | Reagent | Temperature | Vac | Immersion |
|------|---------|-------------|-----|-----------|
| 1 | 100% ethanol | 45 | Yes | 20 min |
| 2 | 100% ethanol | 45 | Yes | 10 min |
| 3 | 100% ethanol | 45 | Yes | 10 min |
| 4 | 100% ethanol | 45 | Yes | 15 min |
| 5 | 100% ethanol | 45 | Yes | 20 min |
| 6 | chloroform | 45 | No | 20 min |
| 7 | chloroform | 45 | No | 10 min |
| 8 | chloroform | 45 | No | 20 min |
| 9 | wax | 63 | Yes | 30 min |
| 10 | wax | 63 | Yes | 20 min |

that were exposed to MWs following perfusion with saline than with formaldehyde. In addition, the Bodian stain and immunohistochemical staining of neurofilaments in axonal structures were more exquisite and distinct in saline-perfused, irradiated tissues (56).

The entire fresh brain can be hardened sufficiently for convenient handling by irradiating the immersed brain in a bucket of saline. This procedure is tedious and labor-intensive because the saline has to be replaced as soon as the temperature of 68°C is attained. Nonetheless, it is possible to cut the brain satisfactorily after this procedure, which takes up to half an hour. Besides employing MW irradiation as the primary method of fixation, the non-ionizing radiation can also be employed to accelerate fixation by cross-linking aldehydes and by alcohols. Boon et al. (5) developed a three-step method employing MW irradiation to produce microscopic slides of fresh brain tissue within one working day. The procedure comprised exposure of the whole brain, sprinkled with physiologic saline, to MWs for 30 min, followed by irradiation of the brain slices for 15 min before immersion in 10% formalin for 3.5 h; the final step was a period of further MW irradiation in formalin for 6 min. The quality of the microscopic sections obtained was assessed to be excellent, with good preservation of brain tissue equivalent to that of conventionally fixed and processed sections.

**Fixation of cryostat sections.** MW irradiation can be used to produce vastly improved cytomorphology in cryostat sections. The exposure of freshly cut cryostat sections immersed in Wolman's (95% ethanol and 5% glacial acetic acid) or Kryofix (EM Science, Fort Washington, PA, USA) (27) to MWs produced noticeable improvement in the quality of the microscopic image without any significant delay in the frozen section procedure. This can

**Table 3. Processing Cycle for Microwave-Fixed Tissues: Overnight Cycle - 11 h 36 min**

| Step | Reagent | Temperature | Vac | Immersion |
|------|---------|-------------|-----|-----------|
| 1 | 70% ethanol | 45 | Yes | 30 min |
| 2 | 100% ethanol | 37 | Yes | 1 h |
| 3 | 100% ethanol | 37 | Yes | 20 min |
| 4 | 100% ethanol | 37 | Yes | 20 min |
| 5 | 100% ethanol | 37 | Yes | 20 min |
| 6 | 100% ethanol | 37 | Yes | 30 min |
| 7 | 100% ethanol | 37 | Yes | 1 h |
| 8 | chloroform | 37 | No | 1 h |
| 9 | chloroform | 37 | No | 1 h |
| 10 | chloroform | 37 | No | 1 h |
| 11 | wax | 60 | Yes | 2 h 30 min |
| 12 | wax | 60 | Yes | 1 h 30 min |

be achieved by microwaving the cryostat sections immersed in a Coplin jar of Wolman's solution or Kryofix for 15 s with the oven at "high" setting. The cytomorphology is clearly improved compared to sections conventionally fixed in 95% ethanol, 10% buffered formalin or formalin vapor.

**MW fixation for electron microscopy.** Irradiation of specimens in normal saline does not produce optimal preservation of fine structural features. However, MW-irradiation can be applied to accelerate fixation in glutaraldehyde and other cross-linking agents such as Karnovsky's fixative (0.05% glutaraldehyde and 2% formaldehyde) (49). Irradiation can be employed in the following way: tissue samples are contained in a vial with 2 mL of the fixative and located centrally in the microwave oven, positioned on a polystyrene block, 1.5 cm above the floor of the rotating plate. The sample is irradiated to 50°C, usually taking 5–10 s. On completion of irradiation, the tissue is removed immediately from the warm fixative and stored in 0.1 M sodium cacodylate buffer with 0.02% sodium azide for up to two weeks or immediately processed (13,42,43,51).

MW irradiation can also be employed to accelerate aldehyde fixation for scanning electron microscopy. After irradiating in 2.5% glutaraldehyde, the preservation of the fine structural features was considered to be at least equal to that of routinely processed control tissues (69). MWs have been successfully applied in the fixation of renal biopsies for both light microscopy and electron microscopy with irradiation in glutaraldehyde producing good

Table 4. Processing Cycle for Microwave-Fixed Tissues: Ultralong Cycle - 16 h 36 min

| Step | Reagent | Temperature | Vac | Immersion |
|------|---------|-------------|-----|-----------|
| 1 | 70% ethanol | 37 | Yes | 1 h |
| 2 | 100% ethanol | 37 | Yes | 1 h |
| 3 | 100% ethanol | 37 | Yes | 1 h |
| 4 | 100% ethanol | 37 | Yes | 30 min |
| 5 | 100% ethanol | 37 | Yes | 30 min |
| 6 | 100% ethanol | 37 | Yes | 1 h |
| 7 | 100% ethanol | 37 | Yes | 1 h |
| 8 | chloroform | 37 | No | 2 h |
| 9 | chloroform | 37 | No | 1 h 30 min |
| 10 | chloroform | 37 | No | 1 h 30 min |
| 11 | wax | 62 | Yes | 2 h 30 min |
| 12 | wax | 62 | Yes | 2 h 30 min |

preservation of glomeruli for both scanning electron microscopy and transmission electron microscopy. In addition, MW-fixed renal biopsies were suitable for immunofluorescence studies; the results were considered equivalent to those of tissues that had been directly snap-frozen (30).

Login et al. (50) developed a MW irradiation device that maximizes the power output and coupling of the electromagnetic energy to the tissue sample, making ultrafast MW fixation possible. Excellent preservation of ultrastructural morphology was obtained by irradiating in cross-linking aldehyde solutions for as brief a period as 26 ms. Similarly, they have extended the use of MWs to include rapid primary osmium fixation for electron microscopy (52).

Unlike the situation with light microscopy, speed is not as important a consideration in the preparation of specimens for routine electron microscopy. However, the ultrastructural examination of fine needle aspiration samples requires rapid processing to maintain the expediency of this diagnostic procedure. Rapid fixation and processing for such specimens can be accomplished by exposing the aspirated tissue contained in an 8–10 mL of 1% glutaraldehyde-4% formaldehyde mixture in cacodylate buffer, for 25 s in a 650-W domestic oven, followed by staining and processing through graded ethanol and embedding in epoxy resin. This allows a sample to be available for ultrastructural examination within two hours, allowing excellent preservation of ultrastructural morphology and allowing for rapid diagnosis.

As in the case of paraffin-embedded tissues, MW-stimulated fixation in electron microscopy also appears to allow the preservation of cellular enzymes and proteins that are otherwise difficult to conserve with conventional methods. Login et al. (51) demonstrated the preservation of antigenic sites of rat muscle chymase, using an ultrastructural post-embedding immunogold technique in MW-fixed cell suspensions, and Ohtani (65) showed that MW-stimulated fixation of frozen sections for pre-embedding immuno-electron microscopy improves the preservation of subcellular structures without any deleterious effect on antigenicity.

**Proposed Classification for MW Fixation**

It is evident that MWs can be employed as the primary mode of fixation with the tissues immersed in saline or buffer, or it can be used to accelerate the effects of various chemical fixatives. Login and Dvorak (53) proposed a system to classify MW fixation. They suggested that MWs can be employed in five ways for fixation:

1. MW stabilization: refers to the preservation of structure by MW irradiation alone without the superimposed effects of chemical fixatives.
2. Fast and ultrafast, primary MW-chemical fixation: specimens are irradiated by MW energy in a chemical environment for milliseconds (ultrafast) or for seconds to minutes (fast). In this approach, the tissues are not exposed to chemical fixatives before or after MW irradiation.
3. MW irradiation followed by chemical fixation: specimens are irradiated initially followed by immersion in a chemical environment for minutes to hours after MW irradiation.
4. Primary chemical fixation followed by MW irradiation: specimens are immersed at room temperature in chemicals for minutes before MW irradiation.
5. Combinations of freezing and MW fixation: methods such as that described above for cryostat sections.

While such a classification system allows a clear distinction between the various applications of MWs for fixation of tissues, the basic problem remains of not knowing exactly how MWs act on tissues to stabilize proteins. MWs are a form of non-ionizing radiation, commonly generated by domestic ovens at a frequency of 2.45 GHz. When dipolar molecules such as water or the polar side chains of proteins are exposed to the rapidly alternating electromagnetic fields, they oscillate through $180°$ at the rate of 2.45 billion cycles per second. The molecular movement or kinetics thus induced results in instantaneous heat that is proportional to the energy flux and continues until radiation ceases. Therefore, heating by MWs offers a method of overcoming the limitations imposed by the normally poor heat-conducting properties of biological tissues. The 2.45-GHz MWs penetrate several centimeters into biological material and the heat produced within the tissue can be controlled by adjustments of the energy levels and the duration of exposure to the non-ionizing rays. Only recently was it recognized that other molecules with an uneven distribution of electrical charge, such as inorganic material and copper oxides, can also be rotated in the electromagnetic field. Therefore, microwaves are used to melt inorganic materials and to produce mixed copper oxides for the making of superconductors.

Heat is currently considered the primary factor responsible for many of the applications of MWs in tissue fixation, as well as for the processing and staining of tissue to be described below. However, the rapid movement of molecules in the electromagnetic fields may itself have a direct role to play. While heat or thermal energy will increase molecular kinetics and hasten chemical reactions, the rapid oscillation of molecules directly induced by the alternating electromagnetic fields will give rise to increased collision of molecules. This in turn will directly accelerate chemical reactions, and the heat generated may be a phenomenon secondary only to the kinetics. Hjerpe et al. (21) examined the MW stimulation of CEA/anti-CEA reaction in an enzyme-linked immunosorbent assay system. Despite continuous cooling by ice, MW stimulation increased reaction rates by a factor of 1000, allowing the investigators to conclude that such rate increases were far too large to be explained solely by the modest increase in temperature. In contrast, other workers have suggested that heat is the most important element of MW fixation of tissues. Hopwood et al. (23) examined the effects of MW irradiation on lysosome and hemoglobin in formaldehyde and observed that MWs did not produce cleavage or polymerization of the proteins. Irradiation resulted in an electrophoretic pattern that was similar to that obtained when lysosome or hemoglobin was heated in formaldehyde to 60°C for 30 min.

It is clear that MW irradiation of tissues in saline without any fixative can produce sufficient stabilization of protein to allow subsequent tissue processing with good cytomorphologic preservation. It is possible, with other methods in which there is brief immersion of tissues in a fixative prior to MW exposure, that MWs act in a different manner. MWs appear to enhance rapid penetration of the fixative, e.g., it has been shown that radiolabeled formaldehyde is homogeneously distributed throughout tissue blocks when they are exposed to MWs for 10 s (62). The action of MWs in the third method of fixation when irradiation is followed by exposure to a chemical fixative is less certain as the period of exposure to the chemical is only very brief and, on its own, the duration of chemical immersion is quite insufficient to produce fixation.

MWs stimulate the fixation of tissues with cross-linking aldehyde solutions, not only by causing rapid penetration as shown by Mizuhira and Hasegawa (62), but probably also by accelerating the degradation of these fixatives. Common chemical fixatives generally exist as oligomers, and they are degraded to monomers or dimers; the diffusion of these monomers or dimers is enhanced by exposure to MWs. A physical law states that viscosity, and hence diffusion rates, in liquids can vary considerably with temperature. It is a well-accepted fact that temperature has an influence on tissue penetration by fixing reagents. Indeed, it has been shown that the fixation of tissues with glutaraldehyde can be hastened with increasing temperatures up to 45°C (67), and a method of rapid fixation for light microscopy has been described in which the tissues were immersed in a mixture of ethanol, formaldehyde and acetic acid, and heated to 80°C (64). Fixation in formaldehyde at 70°C has been reported to give good ultrastructural preservation of some zoologic specimens (82).

A number of other hypothetical physical mechanisms may also play a role in the actions of MW irradiation. Field-induced alterations in macro-molecular hydrogen bonding, proton tunneling and disruption of bound water, may induce alterations in biologic systems. Although the proton energy generated by MWs is too small to alter covalent bonding, low-intensity MW fields may readily affect the integrity of the non-covalent secondary bonding, including hydrophobic interactions, hydrogen bonds and van der Waal's interactions that make up the precise steric interactions at the cell membrane (10).

## Microwave-Stimulated Tissue Processing

Tissue processing can be greatly accelerated by irradiating the specimens sequentially in the dehydrating and clearing reagents before a final irradiating in molten wax. Fixed tissues of not more than $5 \times 5 \times 2$ mm, when irradiated sequentially for 5 min each in 70% alcohol, Histoclear (Agar Aids Ltd., Essex, England) and melted Paramat (EM Science), allow the preparation of embedded tissue blocks after a total 15 min (2). With minor variations in the duration of irradiation in each solution, depending on the output of the microwave oven, the quality of microscopic sections produced as a result of this rapid processing is excellent. Morphometric comparisons have shown no differences in nuclear size of several types of cells in the sections obtained by this method, compared with those prepared by much longer conventional procedures (Figure 3, see color plate in Addendum, p. A-7). By employing this technique, we have been able to process as many as nine tissue blocks at a time in a compartmentalized perspex container of $6- \times 6- \times 6$-cm size. The number of specimens that can be processed in this manner is limited because current models of domestic MW ovens have very uneven flux within the oven cavities.

## Microwave-Stimulated Staining of Tissue Sections

Brinn (7) was probably the first to employ MWs to produce acceleration of histochemical stains. He irradiated the periodic acid oxidation step to accelerate the methenamine silver staining procedure that normally takes 90–180 min to less than 20 min. He also accelerated Pascual's modification of Grimelius', Masson-Fontana and Perls' procedures by exposure to MWs. In all instances, staining times were markedly reduced and background precipitation was also decreased. Since this pioneering work, there has been extensive use of MWs to produce acceleration of many other histochemical stains in smear preparations as well as paraffin and plastic sections. Giammara et al. (15) and Hanker and Giammara (19) have successfully employed MWs to produce accelerated and precise staining with methenamine silver in both light and electron microscopic preparations. Other reported MW-stimulated techniques include the accelerated staining for the Alcian Blue-Periodic acid-Schiff stain (58), the Steiner procedure for spirochetes and non-filamentous bacteria (75), Grocott's methenamine silver for *Pneumocystis carinii* (54), stains for acid and alcohol fast organisms (18), and the Warthin-Starry method for spirochetes and bacteria (9). In addition, we have been able to accelerate a wide variety of other histochemical procedures by exposure to MWs and these include the methenamine silver stain for fungi, ammoniacal silver stain for reticulin fibers and the PAS stain (unpublished). We have also developed a MW-stimulated colloid silver nitrate stain for melanin, which is completed in 45 s (40). A compilation of such MW-accelerated procedures has been prepared by Boon and Kok (4).

Besides producing significant reductions in staining times, MW-stimulation often results in less nonspecific precipitation in the preparations. In the case of acid and alcohol-fast organisms, MW-irradiation eliminated the tedious stages of dewaxing and avoided the hazards of an open or naked flame in the laboratory. It has been suggested that the mycobacteria are killed by irradiation for 30–60 s at 640 W and rapid staining is facilitated by increased permeability of the organisms, resulting in easy passage of bacterial enzymes and staining reagents (18). MW irradiation can also be employed to stimulate staining in plastic sections (41) and can be used to accelerate immunogold staining with monoclonal antibodies in plastic-embedded tissues (80).

Estrada et al. (14) employed MWs to accelerate the staining of ultrathin sections for transmission electron microscopy. Irradiation of grids in uranyl acetate for 15 s, followed by a 15-s burst of irradiation in lead citrate, resulted in an overall staining quality that was better than that obtained with the conventional method that takes up to 40 min. The staining is more consistent and shows selective localization with less background precipitation.

**Technical aspects of MW-stimulated staining.** MW-accelerated staining can be performed in two ways. The tissue section can be immersed in a jar of stain and a temperature probe employed for precise temperature control (70). The ability to control temperature in MW-accelerated staining is an important factor because there appears to be a well-defined optimal staining temperature for each procedure. For example, in the Romanowsky-Giemsa stain this is 55°C (3,24), which is also the optimal temperature for the Alcian Blue method. The optimal temperature for the Southgate mucicarmine method is 60°C and for the Grimelius method it is 75°C. The optimal temperature for the Grocott, Jones and Masson-Fontana methods is 80°C (4). In the case of other staining procedures, such as the Oil Red O stain, the temperature is not so critical and the dye solution can even be brought to boil. In the Rio-Hortega method for staining dendrites and axons, the tissue blocks are first soaked in the silver solution and the chemical reaction is achieved in the MW oven; this approach avoids crust formation and produces outstanding results (55).

Besides accurate temperature control, physical agitation of the solution or agitation by air bubbling produces a more even temperature distribution, and this appears to be of particular importance when high power levels are used in the impregnation and reducing steps of histochemical stains (9).

An alternative procedure is to irradiate the tissue section that is covered with the staining solution. In this situation precise temperature control is impossible and this method can be adopted only when temperatures are not critical and excessively high temperatures are not detrimental. The method is very well suited for staining single slides. In some situations, the staining procedure is enhanced by allowing the tissue section to be bathed or immersed in the hot dye solution for a period of time after irradiation has ceased.

## MWs for Accelerated Immunostaining

Leong and Milios (36) demonstrated that MWs can be used to accelerate the antibody-antigen reactions in the staining of labile lymphocyte membrane antigens in cryostat sections. MW irradiation was applied to cryostat sections both for the purpose of fixation and drying of the sections as well as to accelerate the incubation of the primary antibody, which was achieved by irradiating for a few seconds. The immunostaining obtained in this manner was intense, with good morphologic preservation. More recently, we have extended the use of MWs in immunostaining for the acceleration of the entire immunohistochemical staining procedure. Irradiation can be applied to the primary, secondary and tertiary stages of the standard streptavidin-biotin peroxidase staining procedure, resulting in completion of the procedure within 20 min, with excellent results that equal those of conventional techniques (44). Similarly, Chiu and Chan (8) were able to accelerate immunofluorescence staining of renal biopsy specimens with MWs.

Immunogold-staining in plastic sections can also be accelerated by exposure to MWs (80). MW irradiation has also been employed to immobilize various biological substances such as interleukin-1 and tumor necrosis factor (20) and suppressor gene p53 protein (26), allowing them to be stained in paraffin sections with immuno-enzyme techniques. MW stimulation seems particularly promising in the immunogold-silver technique using Immunogold-labeled antibodies and protein A (6,25,78), antibody dilutions increased by a factor of 10 over conventional staining procedures and there was low or negligible nonspecific background staining (78). The gold-labeling step used for DNA *in situ* hybridization can also be performed in the MW oven (4,77).

## MW Irradiation and Antigen Preservation

With the increasing use of immunohistochemistry as an adjunct to microscopic diagnosis, the optimal preservation of tissue antigens in paraffin-embedded sections is an important requirement in diagnostic pathology. Despite the preservation of a large number of antigens following formaldehyde fixation and routine processing, many lymphocyte membrane antigens and other labile antigens in particular are lost with increasing durations of exposure to formaldehyde (39). Formaldehyde exposure also has a deleterious effect on the immunostaining of the intermediate filaments, particularly neurofilaments, vimentin, desmin and cytokeratins, all of which are important in the analysis and characterization of human tumors.

Leong and Gilham (39) demonstrated that tissues fixed in formaldehyde for varying periods from 6 h to 30 days show a distinct and progressive loss in the staining of many of the 25 antigens commonly studied in histopathology. There was a distinct decrease in staining intensity of some antigens after 3 days of exposure to formaldehyde, and many antigens were lost after 7 days. In particular, the intermediate filament proteins, vimentin, desmin and neurofilaments, failed to be labeled by monoclonal antibodies after 1 day of formaldehyde fixation. With the exception of leukocyte common antigen (CD45), lymphocyte antigens detected by LN1 (CDw75), LN2 (CD74), LN3 (not clustered) and UCHL1 (CD45RO) were universally lost after 3 days. However, some markers remained weakly positive after 14 days of formaldehyde fixation, and these included S100 protein, prostate-specific antigen, thyroglobulin and carcino-embryonic antigen. In contrast, a wide range of tissue antigens were preserved in MW-irradiated tissues. When compared with formaldehyde-fixed tissues, immunostaining in MW-irradiated tissues was found to be clearly superior, both more intense and more extensive (32,39). Another finding was that proteolytic digestion was not uniformly necessary for MW-irradiated tissues, except in the case of cytokeratins and desmin; and it was also possible to stain for CD22 and CD5, lymphocyte antigens that are lost during formaldehyde fixation (38). Earlier work had demonstrated that MW fixation of cells grown in tissue culture resulted in total retention of cellular matrix protein, in contrast to the 40%–50% protein loss that occurred in conventional formaldehyde-fixed cells (66).

More recently, there has been tremendous interest in an antigen retrieval technique employing MW irradiation of tissue sections. Shi et al. (72) described the MW heating of tissue sections in the presence of heavy metal solutions, such as lead thiocyanate, up to temperatures of 100°C to "unmask" a wide variety of tissue antigens for immunostaining. We (45) and others (12,17) have shown that MW irradiation of tissue sections in 10 mmol citrate buffer solution produced, with few exceptions, increased intensity and extent of immunostaining of a wide variety of tissue antigens (Figures 4 and 5, see color plates in Addendum, p. A-8 and A-9), including estrogen and progesterone receptors (46). Moreover, proteolytic enzyme digestion was not necessary in many instances and some primary antibodies could be used at higher working dilutions (45). The use of citrate buffer eliminates the need to employ heavy metal solutions, such as lead thiocyanate, sold as an Antigen Retrieval Solution (Biogenex Labs, San Ramon, CA, USA), which is toxic so that irradiation procedures must be performed in a chemical hood. Momose et al. (63), furthermore, did not find MW irradiation in lead thiocyanate to be superior over protease digestion prior to immunostaining. Other commercial unmasking reagents used in conjunction with MW irradiation are also available, but excellent results can generally be obtained with the citrate buffer solution. Irradiation in 4 M urea is an excellent method for the immunostaining of cellular and membranous immunoglobulin (Figure 6, see color plate in Addendum, p. A-10) (61), and 0.05 M glycine HCl solution has been advocated for nuclear antigens (76). Lan et al. (31) describe the use of MWs to block antibody cross-reactivity and retrieve antigens during multiple immunoenzyme staining, allowing the discrete simultaneous staining of cell surface, cytoplasma and nuclear antigens in the same cell.

**MWs for *in situ* hybridization.** Coates et al. (11) demonstrated the use of MWs to achieve the high temperatures necessary for denaturation of probe and tissue DNA in the *in situ* hybridization detection of cytomegalovirus nuclei acid sequences in tissue sections, emphasizing the ease of control and rapidity of MW irradiation and the increased sensitivity of the method. MWs have also been employed to reduce optimal antibody incubation times in a three-step indirect labeling method using peroxidase and diaminobenzidine for *in situ* hybridization (77,79).

**Resin Polymerization**

Resin polymerization is one of the longest steps in the preparation of tissues for ultrastructural examination, and any reduction in this stage of processing would reduce the entire preparation time significantly. McLay et al. (60) demonstrated that exposure to MWs produced polymerization of epoxy resin within a matter of seconds, in comparison with control polymerization, which was achieved by conventional oven heating for 16 h at 70°C. Giammara (16) has performed extensive studies on microwave embedding methods, employing a wide variety of polymers and formulations with success.

**FUTURE DEVELOPMENTS**

Despite the current extensive use of microwave technology for morphologic diagnosis and research, basic aspects of the action of MWs remain poorly understood. L.P. Kok, a theoretical physicist (28), has discussed the physical properties of microwaves and has studied some of the effects of microwaves in different media. He highlights the uneven nature of the heating effect of microwaves because of the differences in depth of penetration in different media and provides a thoughtful discussion of the variation in fields within the cavity of the microwave oven. Such studies are particularly important if we aim to produce a microwave histo-processor. Such a machine has the potential of producing tissue blocks within 30–40 min and, if as many as 20 blocks can be produced in one processor, the entire routine and work patterns of diagnostic histopathology laboratories would be radically altered. A MW-stimulated tissue processor would immensely increase turnaround times, as current processing schedules, however rapid (such as those employed in our laboratories), still require a minimum of 1.5 to 3.5 h, depending on the size of the tissue block. It is currently possible to produce tissue blocks within 45 min by applying MWs to all stages of tissue processing, but the process is labor-intensive and only a limited number of blocks may be handled in this manner (2,32). A microwave-assisted histo-processor would be a significant contribution to the practice of histopathology, not only because of its rapidity but also because it will represent a major step towards formalin-free histopathology laboratories, eliminating the use of a noxious, potentially toxic and carcinogenic agent.

Recently, Kok and Boon (29) showed that it is possible to achieve ultrarapid tissue processing through a vacuum-microwave method of histoprocessing. They passed fixed tissue blocks of 2- to 4-mm thickness through 100% ethanol under vacuum to a pressure of 250 hPa and stimulated by microwaves to a temperature of 40°C. This step took 11 min and was followed by a further 11 min of clearing in isopropanol at 700 hPa heated to 60°C by microwaves and the final embedding stage of 18 min at 700 dropping to 100 hPa with microwave irradiation at variable power outputs. The entire process took approximately 40 min. They also suggested that the approach could be used to produce large sections of giant blocks of $4 \times 6 \times 1$ cm, which can be cut easily on a routine microtome due to the optimal paraffin impregnation.

This report by Kok and Boon (29) is particularly exciting in view of our anticipation of its possibilities (47). By employing a prototype device designed by Milestone (Sorisole, Italy), we found it possible to process up to 60 blocks simultaneously in this manner. We have had initial success with this prototype device kindly made available by Franco Visinoni, President of Milestone.

As with many laboratory stains and procedures, much of our knowledge concerning the actions of MWs is empirical. For example, the staining of tissue sections and cell preparations appears to be based on at least two important factors, namely the diffusion of the dye or antibody into cells and its binding to the substrate or antigen. Diffusion is a physical process that can be accelerated significantly by microwave irradiation, but the influence of MWs on binding mechanisms is much more complicated and less understood.

While this report has reviewed the many applications of MWs in the areas of histochemistry and immunohistochemistry, many further potential applications require exploring. For example, MWs can be combined with other physical modalities, such as ultrasound, to enhance its action in tissue sections (81), and there needs to be further development and refinement of laboratory microwave ovens with provisions for fume extraction, accurate programmable power level, cycle time and temperatures. In addition, microwave fields in the oven cavity should be more even with the elimination of "hot" spots and there should be provision for focusing the irradiation over a smaller area. Such developments will require the combined input of microwave engineers and morphologists.

## PROTOCOLS

### Microwave Fixation and Processing

1. Ensure that the fresh or partially fixed tissue blocks do not exceed 2 mm in thickness.
2. Place tissue blocks into plastic cassettes and retain in 10% buffered formalin until sufficient blocks are accumulated for microwave irradiation.
3. Remove cassettes from formalin and completely immerse the cassettes in 1-L glass beakers containing $N$ saline. It is convenient to place 20 cassettes into each of three beakers, which are located equidistantly at the periphery of the rotating plate of a microwave oven (650 W, 2.5 GHz).
4. When using an oven with a temperature probe, set the temperature to 68°C and turn the oven on "high" output. If a probe is not available, the oven must be previously calibrated so that the duration of irradiation required to attain 68°C is known. Once this temperature is attained, maintain at this temperature for an additional 5 min.
5. Remove all cassettes and transfer to a vacuum-assisted automated tissue processor through cycles of absolute alcohol, xylene or chloroform and wax. Note the exclusion of formaldehyde from the tissue processor.
6. Employ the processing cycles tabulated in the text above. Ultrashort cycles of 1 h 41 min can be used to process endoscopic and needle biopsies. Routine tissue blocks are processed through the short cycle of 3 h 31 min with the exception of thick or large tissue blocks or samples that contain a large component of fatty tissue. These latter blocks are best processed through the overnight cycle of 11 h 36 min, which can also be conveniently employed at the end of the working day, if laboratory staff are not rostered on duty during the night. Similarly, the ultralong cycle of 16 h 36 min can be used for processing over the weekend.

## ULTRARAPID PROCESSING OF SMALL TISSUE BLOCKS

1. Place tissue sample, not exceeding $5 \times 5 \times 2$ mm, in a 100-mL beaker of $N$ saline. Locate the beaker in the center of the rotating dish in a microwave oven and irradiate at 650 W until 68°C is attained. Maintain at 68°C for 5 min.

2. Transfer the tissue into a 100 mL of 70% ethanol and similarly irradiate until the ethanol boils at about 70°C. Maintain at simmering for 5 min.
3. Transfer tissue into a 100-mL beaker of Histoclear and repeat irradiation to boiling. Similarly, maintain at simmering for 5 min.
4. Transfer to molten Paramat and irradiate at 650 W for 5 min or until tissue is completely penetrated by wax.
5. Remove tissue and embed in Paramat.

## COMMENTS

It may be necessary to employ a container with a lid so that the boiling alcohol and Histoclear do not spill over. Solid wax is microwave transparent; however, when wax is molten it is opaque to microwaves and continues to heat up and melt.

## MICROWAVE EPITOPE RETRIEVAL

1. Five-micrometer paraffin-embedded sections are mounted on amino-alkylsilane-treated slides.
2. Deparaffinized with xylene and pass through graded alcohols. Rinse in deionized water followed by phosphate-buffered saline (PBS).
3. Endogenous peroxidase is blocked with 0.5% hydrogen peroxide-methanol for 30 min.
4. Immerse the deparaffinized and rehydrated sections in 10-mmol citrate buffer at pH 6.0 (2.1 g citric acid monohydrate per 1 L water, pH adjusted with NaOH).
5. The sections are washed in PBS and immersed in a closed plastic container (Kartell, Milan, Italy) filled with the appropriate amount of citrate buffer. The container permits stacking of 20 slides in 250 mL of buffer.
6. The immersed sections are placed in the center of the rotating dish of a microwave oven and irradiated at maximum power setting until the solution boils. As soon as boiling point is attained, normally within 5 min, the power is adjusted so that the solution simmers.
7. Continue simmering for 10 min.
8. The irradiation is turned off and the sections are allowed to remain in the hot solution for an additional 25 min.
9. The sections are blocked in 3% nonimmune horse serum for 20 min.
10. Proteolytic enzyme digestion may be employed as appropriate to the primary antibody used (see reference 45).
11. Stain with a standard streptavidin-biotin peroxidase technique.

## ACKNOWLEDGMENT

I am most grateful to Margaret Elemer for the transcription of this manuscript.

# REFERENCES

1.**Bernard, J.R.** 1974. Microwave irradiation as a generator of heat for histological fixation. Stain Technol. *49*:215-224.
2.**Boon, M.E., L.P. Kok and E. Ouwerkerk-Noordan.** 1986. Microwave-stimulated diffusion for fast processing of tissue: reduced dehydrating, clearing and impregnating times. Histopathology *10*:303-309.
3.**Boon, M.E., L.P. Kok, H.E. Moorlag, P.O. Gerrits and A.J.H. Suurmeijer.** 1987. Microwave-stimulated staining of plastic-embedded bone marrow sections with the Romanowsky-Giemsa stain: improved staining patterns. Stain Technol. *62*:257-264.
4.**Boon, M.E. and L.P. Kok.** 1988. Microwave Cookbook of Pathology. The Art of Microscopic Visualisation, 2nd ed., Coulomb Press, Leyden.
5.**Boon, M.E., E. Marani, P.J.M. Adriolo, J.W. Steffelaar, G.T. Bots and L.P. Kok.** 1988. Mirowave irradiation of human brain tissue: production of microscopic slides within one day. J. Clin. Pathol. *41*:590-593.
6.**Boon, M.E., L.P. Kok, H.E. Moorlag and A.J.H. Suurmeijer.** 1989. Accelerated immunugold-silver staining of paraffin sections using microwave irradiation: factors influencing results. Am. J. Clin. Pathol. *91*:137-143.
7.**Brinn, N.T.** 1983. Rapid metallic histological staining using the microwave oven. J. Histotechnol. *6*:125-129.
8.**Chiu, K.Y. and K.W. Chin.** 1987. Rapid immunofluorescence staining of human renal biopsy specimens using microwave irradiation. J. Clin. Pathol. *40*:689-692.
9.**Churukian, C.J. and E.A. Schenk.** 1988. A Warthin-Starry method for spirochetes and bacteria using a microwave oven. J. Histotechnol. *11*:149-151.
10.**Clearly, S.F.** 1978. Survey of microwave and radio frequency: biological effects and mechanisms. D.H.E.W. Publication (FDA) *78-8055*:1-33.
11.**Coates, P.J., P.A. Hall, M.G. Butler and A.J. D'Ardenne.** 1987. Rapid technique of DNA-DNA *in situ* hybridisation on formalin fixed tissue sections using microwave irradiation. J. Clin. Pathol. *40*:865-869.
12.**Cuevas, E.C., A.C. Bateman, B.S. Wilkins, P.A. Johnson, J.H. Williams, A.H.F. Lee, D.B. Jones and D.H. Wright.** 1994. Microwave antigen retrieval in immunocytochemistry: a study of 80 antibodies. J. Clin. Pathol. I:448-452.
13.**Daymon, M.E. and A.S-Y. Leong.** 1984. Microwave fixation for surgical and autopsy tissues. Pathology 16:418.
14.**Estrada, J.C., N.T. Brinn and E.H. Bossen.** 1985. A rapid method of staining ultra-thin sections for surgical pathology TEM with the use of the microwave oven. Am. J. Clin. Pathol. *83*:639-641.
15.**Giammara, B.L., J. Pierce, A. Rustioni, M. Borowitz, D. Chandler, P. Yates and J. Hanker.** 1985. Microwave acceleration of the metallization and intensification of cytochemical and immunocytochemical reaction products with silver methenamine, p. 704-705. Proceedings of the 43rd Annual Meeting of the Electron Microscopy Society of America.
16.**Giammara, B.L.** 1992. Microwave embedding methods. Scanning (Suppl. I) *14*:60-61.
17.**Gown, A.M., N. de Wever and H. Battifora.** 1993. Microwave-based antigenic unmasking. A revolutionary new technique for routine immunohistochemistry. Appl. Immunohistochem. *1*:256-266.
18.**Hafiz, S., R.C. Spencer, M. Lee, H. Gooch and B.I. Duerden.** 1985. Use of microwaves for acid and alcohol fast staining. J. Clin. Pathol. *38*:1073-1076.
19.**Hanker, J.S. and B.L. Giammara.** 1992. Microwave-accelerated cytochemical stains for the electron microscopic examination and the image analysis of light microscopy diagnostic slides. Scanning (Suppl.II) *14*:61-64.
20.**Haruna, N., T. Monden and H. Morimoto.** 1990. Use of a rapid microwave fixation technique for immunocytochemical demonstration of tumor necrosis factor, interleukin-1a, and interleukin-1b in activated human peripheral mononuclear cells. Acta Histochem. Cytochem. *23*:563-572.
21.**Hjerpe, A., M.E. Boon and L.P. Kok.** 1988. Microwave stimulation of an immunological reaction (CEA/anti-CEA) and its use in immunohistochemistry. Histochem. J. *20*:388-396.
22.**Hopwood, D., G. Coghill, J. Ramsay, G. Milne and M. Kerr.** 1984. Microwave fixation. Its potential for routine techniques: Histochemistry, immunocytochemistry and electron microscopy. Histochem. J. *16*:1171-1191.
23.**Hopwood, D., G. Yeaman and G. Milne.** 1988. Differentiating the effects of microwave and heat on tissue proteins and their cross linking by formaldehyde. Histochem. J. *20*:341-346.
24.**Horobin, R.W. and M.E. Boon.** 1988. Understanding microwave-stimulated Romanowsky-Giemsa staining of plastic embedded bone marrow. Histochem. J. *20*:329-334.

25. **Jackson, P., E.N. Lalani and J. Boutsen.** 1988. Microwave stimulated immunogold-silver staining. Histochem. J. *20*:353-358.

26. **Kawasaki, Y., T. Monden, H. Morimoto, M. Murotani, Y. Miyoshi, T. Kobayashi, T. Shimano and T. Mori.** 1992. Immunohistochemical study of p53 expression in microwave fixed, paraffin-embedded sections of colorectal carcinoma and adenoma. Am. J. Clin. Pathol. *97*:244-249.

27. **Kok, L.P., M.E. Boon and A.J.H. Suurmeijer.** 1987. Major improvement in microscope-image quality of cryostat sections combining freezing and microwave-stimulated fixation. Am. J. Clin. Pathol. *88*:620-623.

28. **Kok, L.P. and M.E. Boon.** 1990. Microwaves for microscopy. J. Microsc. *158*:191-322.

29. **Kok, L.P. and M.E. Boon.** 1995. Ultrarapid vacuum-microwave histoprocessing. Histochem. J. *27*:411-419.

30. **Lai, F.M., K.N. Lai, E.C. Chew and J.C.K. Lee.** 1987. Microwave fixation in diagnostic renal pathology. Pathology *19*:17-21.

31. **Lan, H.Y., W. Mu, D.J. Nikolic-Paterson and R. Atkins.** 1995. A novel, simple, reliable and sensitive method for multiple immunoenzyme staining; use of microwave oven heating to block antibody cross-reactivity and retrieve antigens. J. Histotech Cytochem *43*:97-102.

32. **Leong, A.S-Y.** 1988. Applications of microwave irradiation in histopathology. Pathol. Annu. 2:213-234.

33. **Leong, A.S-Y.** 1991. Microwave fixation and rapid processing in a large throughput histopathology laboratory. Pathology *23*:271-273.

34. **Leong, A.S-Y.** 1991. Microwave irradiation-applications in tissue fixation: processing and staining for light microscopy and electron microscopy, p. 47-60. *In* World Health Organisation Bi Regional Training Course on Electron Microscopy in Biomedical Research and Diagnosis of Human Diseases. University of Adelaide Press, Adelaide.

35. **Leong, A.S-Y., M.E. Daymon and J. Milios.** 1985. Microwave irradiation as a form of fixation for light and electron microscopy. J. Pathol. *146*:313-321.

36. **Leong, A.S-Y. and J. Milios.** 1986. Rapid immunoperoxidase staining of lymphocyte antigens using microwave irradiation. J. Pathol. *148*:183-187.

37. **Leong, A.S-Y. and C.G. Duncis.** 1986. A method of rapid fixation of large biopsy specimens using microwave irradiation. Pathology *18*:222-225.

38. **Leong, A.S-Y., J. Milios and C.G. Duncis.** 1988. Antigen preservation in microwave-irradiated tissues: a comparison with formaldehyde fixation. J. Pathol. *156*:275-282.

39. **Leong, A.S-Y. and P.N. Gilham.** 1989. The effects of progressive formaldehyde fixation on the preservation of tissue antigens. Pathology *21*:266-271.

40. **Leong, A.S-Y. and P. Gilham.** 1989. A new, rapid microwave-stimulated method of staining melanocytic lesions. Stain Technol. *64*:81-87.

41. **Leong, A.S-Y. and S. Pulbrook.** 1989. Microwave-stimulated staining of reticulin fibres in plastic sections by an ammoniacal silver nitrate method. J. Histotechnol. *12*:289-292.

42. **Leong, A.S-Y. and D.W. Gove.** 1990. Microwave techniques for tissue fixation, processing and staining. EMSA Bull. *20*:61-65.

43. **Leong, A.S-Y. and D.W. Gove.** 1990. Applications of microwave irradiation in electron microscopy. cf Ref. 62. Proc. Int. Congr. Electron. Microsc. *12*:140-141.

44. **Leong, A.S-Y. and J. Milios.** 1990. Accelerated immunohistochemical staining by microwaves. J. Pathol. *161*:327-334.

45. **Leong, A.S-Y. and J. Milios.** 1993. An assessment of the efficacy of the microwave antigen-retrieval procedure on a range of tissue antigens. Appl. Immunohistochem. *1*:267-274.

46. **Leong, A.J.-Y. and J. Milios.** 1993. A comparison of two antibodies to entrogen and progesterone from commercial sources and the effects of microwave antigen retrieval. Appl. Immunohistochem. *1*:282-288.

47. **Leong, S.S.-Y.** 1994. Microwave technology for morphological analysis. Cell Vision *1*:278-288.

48. **Login, G.R.** 1978. Microwave fixation versus formalin fixation of surgical and autopsy tissue. Am. J. Med. Technol. *44*:435-437.

49. **Login, G.R. and A.M. Dvorak.** 1985. Microwave energy fixation for electron microscopy. Am. J. Pathol. *120*:230-243.

50. **Login, G.R., W.B. Stavinoha and A.M. Dvorak.** 1986. Ultrafast microwave energy fixation for electron microscopy. J. Histochem. Cytochem. *34*:381-387.

51. **Login, G.R., S.J. Galli, E. Morgan, N. Arizono, L.B. Schwartz and A.M. Dvorak.** 1987. Microwave fixation of rat mast cells. 1. Localization of granule chymase with an ultrastructural post-embedding immunogold technique. Lab. Invest. *57*:592-599.

52. **Login, G.R., B.K. Dwyer and A.M. Dvorak.** 1990. Rapid primary microwave-osmium fixation. I.

Preservation of structure for electron microscopy in seconds. J. Histochem. Cytochem. *38*:755-762.

53. **Login, G.R. and A.M. Dvorak.** 1994. Methods of Microwave Fixation for Microscopy. A Review of Research and Clinical Applications: 1970-1992. Gustav Fischer Verlag, Stuttgart.

54. **Loughman, N.T.** 1989. Pneumocystis carinii: rapid diagnosis with a microwave oven. Acta Cytol. *33*:416-417.

55. **Marani, E., J.M. Guldemond, P.J.M. Adriolo, M.E. Boon and L.P. Kok.** 1987. The microwave Rio-Hortega technique: A 24 hour method. Histochem. J. *19*:658-664.

56. **Marani, E., M.E. Boon, P.J.M. Adriolo, W.J. Rietveld and L.P. Kok.** 1987. Microwave-cryostat technique for neuro-anatomical studies. J. Neurosci. Methods 22:97-101.

57. **Maruyama, Y.** 1981. Rapid tissue fixation by microwave irradiation. Trends Pharmacol. Sci. 2:239-241.

58. **Matthews, K. and J.K. Kelly.** 1989. A microwave oven method for the combined alcian blue-periodic acid-Schiff stain. J. Histotechnol. *12*:295-303.

59. **Mayers, C.P.** 1970. Histological fixation by microwave heating. J. Clin. Pathol. *23*:273-275.

60. **McLay, A.L.C., J.D. Anderson and W. McMeekin.** 1987. Microwave polymerisation of epoxy resin. Rapid processing technique in untrastructural pathology. J. Clin. Pathol. *40*:350-352.

61. **Merz, H., O. Richers, S. Schrimel, K. Orscheschek and A.C. Feller.** 1993. Constant detection of surface and cytoplasmic immunoglobulin heavy and light chain expression in formalin-fixed and paraffin-embedded material. J. Pathol. *170*:257-264.

62. **Mizuhira, V. and H. Hasegawa.** 1990. Opening images of synaptic vesicles and calcium localisation by means of microwave fixation method. Proc. Int. Congr. Electron Microsc. *12*:182-183.

63. **Momose, H., P. Mehta and H. Battifora.** 1993. Antigen retrieval by microwave irradiation in lead thiocyanate: A comparison with protease digestion retrieval. Appl. Immunohistochem. *1*:77-82.

64. **Ni, C., T.C. Chang, S.S. Searl, E. Coughlin-Wilkinson and D.M. Albert.** 1981. Rapid paraffin fixation for use in histological examination. Ophthalmology (Rochester) 88:1372-1376.

65. **Ohtani, H.** 1991. Microwave-stimulated fixation for pre-embedding immunoelectron microscopy. Eur. J. Morphol. *29*:64-67.

66. **Patterson, M.K., Jr. and R. Bulard.** 1980. Microwave fixation of cells in tissue culture. Stain Technol. *55*:71-75.

67. **Peracchia, C. and B.S. Mittler.** 1972. New glutaraldehyde fixation procedures. J. Ultrastruct. Res. *39*:57-64.

68. **Petrere. J.A. and J.L. Schardein.** 1977. Microwave fixation of foetal specimens. Stain Technol. *52*:113-114.

69. **Riches, D.J. and E.C. Chew.** 1984. The use of microwaves for fixation in electron microscopy, p. 257-260. *In* M.F. Chung (Ed.), Proceedings of the 3rd Asia-Pacific Conference on Electron Microscopy, Hentexco Trading Co., Hong Kong.

70. **Schaffner, R.** 1986. The Perls' iron staining procedure for use in the microwave oven using a temperature probe. J. Histotechnol. *9*:107-108.

71. **Schneider, D.R., B.T. Felt and H. Goldman.** 1982. On the use of microwave irradiation energy for brain tissue fixation. J. Neurochem. *38*:749-752.

72. **Shi, S., M.E. Key and K.L. Kaira.** 1991. Antigen retrieval in formalin-fixed, paraffin-embedded tissues: an enhancement method for immunohistochemical staining based on microwave oven heating of tissue sections. J. Histochem. Cytochem. *39*:741-748.

73. **Stavinoha, W.B., B. Pepelko and P.W. Smith.** 1970. Microwave radiation to inactivate cholinesterase in the rat brain prior to analysis for acetylcholine. Pharmacologist 12:257.

74. **Stavinoha, W.B., J. Frazer and A.T. Modak.** 1977. Microwave fixation for the study of acetyl choline metabolism, p. 167-179. *In* D.J. Jenden (Ed.), Cholinergic Mechanisms and Cycle Pharmacology. Plenum Press, New York.

75. **Swischer, B.L.** 1987. Modified Steiner procedure for microwave staining of spirochetes and non-filamentous bacteria. J. Histotechnol. *10*:241-243.

76. **Taylor, C.R., S-R. Shi, B. Chaiwun, L. Young, S.A. Imam and R.J. Cote.** 1994. Strategies for improving the immunohistochemical staining of various intranuclear prognostic markers in formalin fixed paraffin sections. Hum. Pathol. 25:263-270.

77. **van den Brink, W.J., H.J.M.A. Zijlmans, L.P. Kok, P. Bolhuis, H.H. Volkers, M.E. Boon and H.J. Houthoff.** 1990. Microwave irradiation in labile-detection for diagnostic DNA-*in situ* hybridisation. Histochem. J. 22:327-334.

78. **van de Kant, H.J.G., M.E. Boon and D.G. de Rooij.** 1988. Microwave-added technique to detect BrdU in S-phase cells using immunogold-silver staining and plastic-embedded sections. Histochem. J. *20*:335-340.

79. **Volkers, H.H., W.J. van den Brink, R. Rook and F.M. van den Berg.** 1992. Microwave label detec-

tion technique for DNA *in situ* hybridisation. Eur. J. Morphol. *29*:59-62.

80. **Wannakrairot, P. and A.S-Y. Leong.** 1989. Microwave stimulated immunogold-silver staining. Presented at the National Scientific Meeting of the Australian Institute of Laboratory Scientists, Adelaide.

81. **Yasuda, K., S. Yamashita, M. Shiozawa, S. Aiso and Y. Yasui.** 1992. Application of ultrasound for tissue fixation: Combined use with microwave to enhance the effect of chemical fixation. Cytochemistry (Kyoto) *28*:237-244.

82. **Zeikus, J.A. and H.C. Aldridge.** 1975. Use of hot formaldehyde fixatives in processing parasitic nematodes for electron microscopy. Stain Technol. *50*:219-223.

# CHAPTER 5

# Microwave Immunohistochemistry: Advances in Temperature Control

**Tak-Shun Choi[1], Michael M. Whittlesey[2], Steven E. Slap[2], Virginia M. Anderson[1] and Jiang Gu[3]**

[1]Department of Pathology, SUNY Health Science Center at Brooklyn, Brooklyn, NY; [2]Energy Beam Sciences, Agawam, MA; [3]Institute of Molecular Morphology, Mount Laurel, NJ, USA

## ABSTRACT

*The advantages and limitations to employing a fiber-optic temperature probe for feedback control of a microwave oven designed for immunohistochemistry were tested. The probe was designed to be submerged in a drop of immunoreagent on a glass slide, and the measured temperature of the droplet was used to regulate the output power and the time needed for the magnetron to reach and maintain the desired incubation temperature. A three-step streptavidin-biotin-peroxidase microwave method, employing a range of antibodies, particularly anti-insulin on rat pancreas, was used as the model system to illustrate the effect of temperature and duration of incubation in microwave immunostaining. Insight into the factors affecting a droplet temperature under microwave irradiation and the relationship among temperature, duration, concentrations of antibodies and immunoreaction for immunostaining were obtained. The optimal temperature and duration for microwave immunostaining in our model system were 37 °C for 3 min at 75% power level followed by a 2-min incubation inside the oven for all three stages of immunostaining. Use of droplet temperature-regulated microwave standardizes conditions used in microwave immunohistochemistry. Other potential applications where uniform heating of small amounts of reagent and control of reagent temperature are important may benefit from this technique.*

## INTRODUCTION

Microwave (MW) technology has been employed in accelerating immunohistochemistry for about a decade (1,3,5,6,9,11,15,16). MW irradiation shortens the time of immunostaining procedures by at least 20-fold while still retaining staining quality (9). There are currently two ways of performing MW immunostaining. One approach is to irradiate small droplets of immunoreagents on individual microscopic slides for a given empirical period of incubation and power level of a particular MW oven. However, before the development of the H2900 Immunohistochemistry Microwave oven (Energy Beam Sciences, Agawam, MA, USA), precise temperature control of the reagent droplets during MW irradiation was not possible. The droplet typically has a volume of approximately 40–120 µL, and the previously available thermocouple thermometer was too large to be inserted into the small droplet volume. Moreover, being metal, i.e., a conductor, the thermocouple probe sometimes is heated at a different rate than the reagent being microwaved and causes inaccurate temperature measurement (14), although some laboratory microwave ovens have built-in mechanisms to compensate for this effect. An excessive increase in temperature causes evaporation of the droplet and interferes with the antibodies' activities. These situations can lead to erratic results and poor reproducibility (1,5,6).

In order to control the incubation temperature, Boon et al. (4) used a thermocouple temperature probe in their MW immunostaining. They immersed the slides in antibody solution contained in a jar, with the probe placed in the reagent to regulate the amount of MW irradiation and maintain the desired incubation temperature for a period of time. They applied this approach in detecting human papillomavirus (HPV) in skin biopsies and leukocyte common antigen (LCA) in lymph node tissues using anti-HPV and anti-LCA, respectively. The dilutions of these antibodies were increased by 20-fold to make the cuvette incubation economically feasible. However, the large amount of antibodies (at least about 100 mL of diluted antibodies per jar) used still hinders its routine diagnostic use.

A composite approach to the existing strategies for the thermocouple probe is to irradiate the slides, which are covered with droplets of reagents, and to use a fiber-optic probe to measure the droplet temperature. This reading is then conveyed to a built-in controller to provide a feedback control of the amount of MW energy needed to reach and maintain the desired incubation temperature for a designated period of time. The fiber-optic probe we developed overcomes the above limitations of the thermocouple probe. Its diameter is small enough to insert into a droplet of liquid, and it is manufactured from a microwave-transparent material, i.e., an insulator, so that the fiber-optic probe itself will not be heated by the microwaves and affect the temperature reading. We have adopted this new approach in the H2900 Immunohistochemistry Microwave oven to provide a precise control of the droplet's incubation temperature in MW immunostaining (7). The fiber-optic probe is coupled with a control panel that is capable of programming the duration, the power output, and the on/off cycles of the oven. This has enabled us to demonstrate the major factors affecting the droplet temperature and gain insight into the relationships among incubation temperature, duration, antibody concentration and heating interval of a few antibodies on a number of tissue types, particularly insulin in the rat pancreas.

As with other laboratory procedures and staining, much of the knowledge concerning the actions of MWs on immunostaining is empirical. The precise, continuous measurement of reagent droplet temperature and a feedback control of the MW irradiation strength and duration had not been possible until the advent of the newly designed H2900 Immunohistochemistry Microwave oven. In order to understand the effects of MW on immunostaining, we will first discuss the basic properties of the MWs and the factors affecting a temperature droplet with MW irradiation. Then, we will illustrate the relationships among incubation temperature, duration, antibody concentration and heating interval by using a three-step streptavidin-biotin-peroxidase microwave method on paraformaldehyde-fixed, paraffin-embedded sections of rat pancreas, duodenum and atrium as our model system.

## Microwave Energy

MWs are electromagnetic waves with a frequency range of 300 MHz to 300 GHz. In vacuum, MWs travel at the speed of light and the two frequencies correspond to wavelengths of 1 m and 1 mm, respectively. In any MW oven, MWs are generated by a magnetron through interaction between strong magnetic and electric fields. The magnetron emits MWs into a wave guide that opens into the oven cavity. Typically, MWs have a standard frequency of 2.45 GHz, a wavelength of 12.2 cm and a photon energy of $10^{-5}$ electron volts. They are capable of forcing dipolar molecules, as well as the polar side chain of proteins, to rotate through 180 degrees at the rate of 2.45 billion cycles per second. The increased intermolecular and intramolecular motion results in instantaneous heat. Molecules with an uneven distribution of electrical charge, such as inorganic materials and copper oxides, can also be oscillated by MW irradiation. MW energy is not strong enough to alter covalent bonds or to break hydrogen bonds. MWs can, however, redistribute the hydrogen bonds, which may be essential for many biochemical reactions (1,9,17).

## Temperature Measurements

A newly designed H2900 Immunohistochemistry Microwave oven was used (Figures 1A and 1B),with a maximum-power limit of 800 W. The power output level can be set from 0% to 100%. Cycle times can be selected to be 1, 2, 4, 8, 16, 32 or 64 s. An optimum cycle time of 2 s was used in our study. In addition, the rate of droplet temperature increase, from a lower temperature to a higher temperature (the ramp-up rate), can be adjusted by setting the proportional band. A narrow proportional band will increase the ramp-up rate, but the droplet temperature may overshoot the set-point temperature before the controller of the oven has time to react and compensate by adjusting the amount of MWs emitted. A proportional band of 0.5 in our experiments allowed both even and rapid ramp-up rates, and the maintenance of temperatures within 1°C of the set point.

The temperature profiles of droplets of solution were recorded during MW irradiation. The measurements were performed with the conditions shown in

Table 1. Various Combinations of Conditions to Study the Temperature Profile of a Droplet of Solution with MW Irradiation

| Test Number. | Power Output Level (%) | Water Load (mL) | Droplet Solution | Droplet Size (μL) | Droplet Position | Type of Water Load |
|---|---|---|---|---|---|---|
| 1 | 100,87,80,75,63,50,25 | 200 | PBS | 100 | Center | Incubator |
| 2 | 75 | 200/100/0 | PBS | 100 | Center | Incubator |
| 3 | 75 | 200 | dH$_2$0,BLS,BSA,EC,PBS | 100 | Center | Incubator |
| 4 | 75 | 200 | PBS | 75/100/200/300 | Center | Incubator |
| 5 | 75 | 200 | PBS | 100 | Center,RL,RR,FL,FR | Incubator |
| 6 | 75 | 200 | PBS | 100 | Center | Incubator, Beaker(RL,RR) |

Note: PBS = 0.1 M phosphate-buffered saline, pH 7.3. BLS = blocking serum. BSA = biotinylated secondary antibody. EC = enzyme conjugate. Position of droplet or beaker inside the oven: RL = rear left, RR = rear right, FL = front left, FR = front right.

Table 1 over a "total time" of 2–3 min with the temperature set point at 100°C. These included variation in: 1) power output level, 2) amount of water load, 3) type of droplet solution, 4) volume of the droplet, 5) location of the droplet in the oven and 6) the types and locations of water load.

## Results of Temperature Measurement

The temperature curve of a droplet rose initially and reached a plateau level. The rapid heat loss of the droplet, due to the large surface area-to-volume ratio, balances the energy supplied by the MWs, and results in a plateau of the temperature. The other cooling mechanisms involved are conduction, convection and infrared radiation of heat to the cooler environment of the oven cavity. The ramp-up rate and level of the temperature plateau of the droplet depended on several factors that are discussed as follows.

## Cycle Time and Power-Level Control

Cycle time of a MW oven is the total time of irradiation per cycle when the oven is set at 100% power. For a given cycle time, when a power level is selected at less than 100%, e.g., 75%, the time of MW irradiation will be 0.75 of the cycle time. For example, a cycle time of 2 s at 75% power-level setting means the magnetron will emit MW energy for 1.5 s (so-called duty cycle) and will be off for the remaining 0.5 s of the 2-s cycle. Thus, when we adjust the power-level setting of the H2900, we are changing the duration of a duty cycle for a given cycle time.

The output power of the MW oven affected the ramp-up rate and level of the temperature plateau of the droplet (Figure 2). A higher power level gave a steeper rise and reached a higher temperature level. The temperature of a 100-µL phosphate-buffered saline (PBS) droplet was 45°C and 36°C after 1 min, and 57°C and 47°C after 3 min, with 100% and 75% power level, respectively. Furthermore, with the H2900, we chose a cycle time of 2 s to allow for both even ramp-up rates and maintenance of temperatures within 1°C of the set point. When a kitchen microwave oven is employed at less than full power, it emits MWs for a selected percentage of a fixed cycle, usually 12–20 s. This sequence of heating and cooling steps results in the temperature rising in a characteristically jagged, "sawtooth" pattern (2), and the temperature can easily overshoot the set point.

### Hot Spots

The level of the temperature plateau depended on the position of the droplet in the oven (Figure 3). A droplet placed in the rear of the oven reached higher temperature level in 3 min than in the front. In addition, droplets placed at the left of the oven had a higher temperature level in 3 min than that at the right. This uneven distribution of MW energy within a MW oven is known as "hot spots".

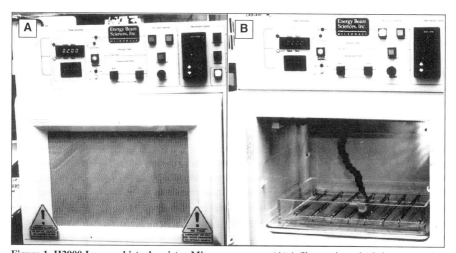

**Figure 1. H2900 Immunohistochemistry Microwave ovens.** (A) A fiber-optic probe is incorporated to measure droplet temperature and to provide a feedback control of the amount of MW energy emitted to maintain a desired incubation temperature for a designated period of time. (B) Inside view showing the fiber-optic probe submerged in a droplet of solution. Note that the slide is placed on the bars of an incubating chamber and a water load is in the bottom of the chamber.

To obtain MW uniformity throughout the MW oven cavity, three strategies have been used. First, the dimensions and material composition of the cavity itself are engineered for maximum energy efficiency. Second, a "microwave stirrer" mounted on the ceiling of the cavity turns with the movement of air within the cavity and moves the MWs around the cavity. Third, a rotator can be used to move the material to be heated relative to the MWs within the cavity, passing the specimen through potentially hotter or cooler areas to even out the exposure to MWs. However, whatever strategies are employed, the distribution will never be perfect, and introduction of anything at all into the oven, even the specimen itself, will change the distribution of MWs. Therefore, the only certain way to achieve reproducibility is to use a probe to measure the temperature of the droplet directly. With a fiber-optic probe interacting with a temperature controller, a droplet on a slide can be placed anywhere within the oven cavity without worrying about hot and cold spots. The H2900 will produce just the right amount of MW energy to reach and maintain the desired incubation temperature (Figure 4).

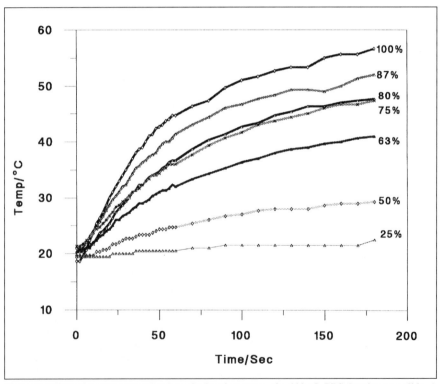

**Figure 2. Effect of power-output level on the heating curve of a 100-μL PBS droplet on a slide.** A high-power setting resulted in a steeper rise and a higher plateau of the droplet temperature. Cycle time 2 s. Diameter on the slide drawn by a PAP Pen® (Kiyota International, Eik Grove Village, IL, USA) = 15 mm. 200 mL of distilled water was placed in the bottom of the incubating chamber to absorb excessive MW energy and prevent drying of the droplet during MW irradiation.

## Water Load

MW energy will be absorbed by any object in the cavity that is not MW transparent. When a volume of microwavable material is very small, e.g., 100 µL of immunoreagent, the power density can be very high. This can cause very rapid increases in temperature that can be hard to control, resulting in excessive evaporation of the droplet. To solve this problem, a "water load" or "dummy load" has been used. A container of microwavable liquid, generally water, of known temperature and volume is placed into the cavity. This load absorbs some of the MW energy and reduces the power density to the droplet (Figure 5). Without a water load, the droplet temperature rose from 21°C to 67°C in 1 min and reached 82°C in 3 min. Moreover, there was an 80% evaporative water loss of the droplet after the 3-min incubation. With a 100-mL water load, the droplet temperature reached 37°C (from 21°C) in 1 min and 49°C in 3 min, and the evaporative water loss of the droplet was reduced to

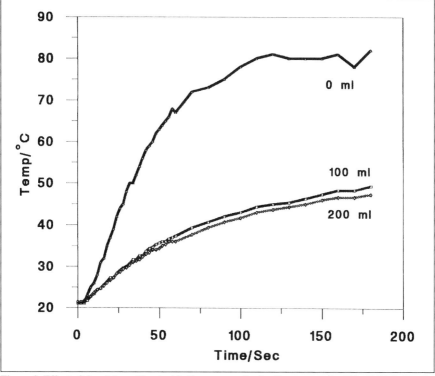

**Figure 3. Effect of droplet position inside the oven on the heating curve of a 100-µL PBS.** Droplet placed in the rear of the oven (RL and RR) had similar rate of rise in temperature but reached higher temperature levels than that placed in the front (FL and FR) or the center. In addition, droplet placed on the left of the oven (RL and FL) reached higher temperature levels in 3 min than that on the right (RR and FR), respectively. Cycle time 2 s, power level 75%. Diameter on the slide 15 mm. Water load 200 mL in an incubating chamber.

about 20% of its initial volume. In addition, the water load in a chamber can serve as a humid chamber. This prevents excessive evaporation of the droplet, due to the air flow generated by the microwave stirrer and the raised droplet temperature.

The type and location of the water load also influenced the temperature profile of a droplet with MW irradiation (Figure 6). A water load in a beaker resulted in a steeper rise and a higher droplet temperature in 3 min than the same amount in an incubating chamber. Moreover, the beaker water load placed in the rear right (BRR) resulted in a steeper rise and a higher droplet temperature than that placed in the rear left (BRL). The final temperature of the water load was 45°C, 53°C and 57°C for BRR, BRL and incubating chamber, respectively. Evaporative water loss of the droplet was 45%, 30% and 15% for BRR, BRL and incubating chamber, respectively. Therefore, the amount and the type of water load used for immunostaining should be stan-

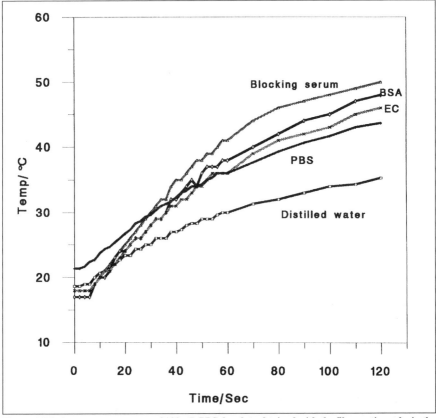

**Figure 4. Ideal temperature curves of 100-μL PBS droplets obtained with the fiber-optic probe in the H2900 oven, providing a feedback mechanism for controlling the amount of MW irradiation.** Various positions of the droplets inside the oven did not significantly affect the droplet temperature. Cycle time 2 s; power level 75%. Water load 200 mL in an incubating chamber. Temperature set point 37°C.

dardized to reproduce a similar temperature profile of droplet reagents. When the same amount of water load is placed in a hot spot in an oven, the load absorbs more MW energy and results in a reciprocal decrease in MW energy available to the droplet reagent, and, hence, the droplet reaches a lower temperature. In addition, the same amount of water in the bottom of an incubating chamber serves as a better MW energy absorbent than the water contained in a beaker (Figure 6). With the droplet-temperature feedback mechanism in H2900, the amount of water load does not affect the temperature profile of a droplet significantly (Figure 7). This has the advantage of not having to refresh the water load in between irradiations to maintain a constant amount of energy uptake (1). However, without the water load, the desired temperature is reached rapidly and the temperature tends to overshoot the set point temperature. Therefore, it is necessary to have a water load in an incubating chamber to avoid overshoot of the droplet temperature and excessive droplet evaporation.

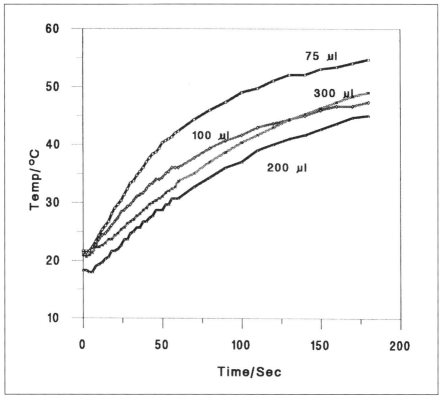

**Figure 5. Effect of water load on the heating curve of a 100-µL PBS droplet on a slide.** The presence of water load prevented excessive increase in droplet temperature and evaporation of the droplet. A larger water load resulted in a slower rise and a lower plateau of the droplet temperature at a given power level and time duration. Cycle time 2 s; power level 75%. Diameter on the slide = 15 mm.

## Type, Size and Shape of a Droplet and Its Heating Curve

The temperature curve depended on the type of the droplets (Figure 8). The reagents used in the ABC method, i.e., PBS, biotinylated secondary antibody (BSA), and enzyme conjugte (EC), all had steeper rises and higher plateaus than that of the distilled water. The droplet temperature of 100 µL distilled water rose from 19°C to 30°C in 1 min and reached 35°C in 3 min, whereas that of BSA rose from 17°C to 38°C in 1 min and reached 48°C in 3 min. This represented an approximately 200% difference in change in droplet temperature for the same amount of MW energy.

The temperature curve also depended on the size and shape of the droplet (Figure 9). For a fixed diameter of 15 mm on a slide, the temperature of a relatively flat droplet of 75 µL PBS rose from 22°C to 42°C and reached 55°C in 1 and 3 min, respectively. However, for the 200-µL and 300-µL PBS

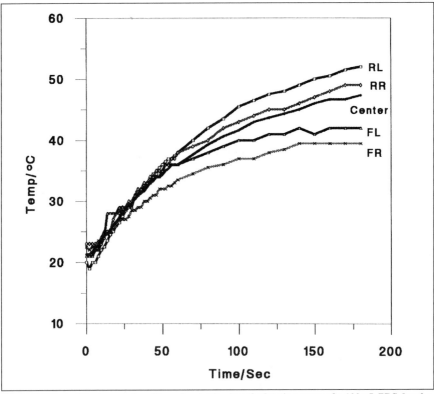

**Figure 6. Effect of the type and position of water load on the heating curve of a 100-µL PBS droplet.** Water load placed in a beaker resulted in a steeper rise and reached a higher droplet temperature in 3 min than that placed in an incubating chamber. Beaker water load placed on the right (BRR) resulted in a steeper rise and a higher temperature plateau of the droplet than that placed on the left (BRL). Final temperature of the water load was 45°C, 53°C and 57°C for BRR, BRL and incubator, respectively. Evaporative water loss of the droplet was 45%, 30% and 15% for BRR, BRL and incubator, respectively.

droplets, the temperature profiles were different. The temperature of a 300-µL droplet rose from 21°C to 34°C and reached 49°C in 1 and 3 min, respectively, whereas that of a 200-µL droplet rose from 18°C to 30°C and reached 45°C in 1 and 3 min, respectively. This can be explained in terms of the "antenna effect" of electomagnetic radiations.

Absorption of MW energy by an object depends on the shape of the object (14). The absorption is maximal when the linear size of the object is on the order of half a wavelength, i.e., 6.1 cm (so-called "antenna effect"). Therefore, smaller droplets take up MW energy less effectively than larger ones. Smaller droplets also have a larger surface area-to-volume ratio, and thus cooling due to evaporation takes place much faster. The composite effects of both poor antenna effect and efficient cooling make the temperature of smaller sizes of droplets attain a lower plateau for a given MW irradiation setting. However, if the droplet size is fixed by a given diameter on a slide, the flatter shape of a droplet gives a steeper rise and a higher plateau. This is because the effective

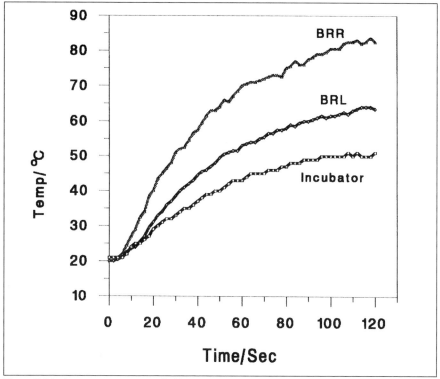

**Figure 7. Ideal temperature curves of 100-µL PBS droplets obtained with the fiber-optic probe in the H2900 oven.** Various amounts of water load in an incubating chamber did not significantly affect the droplet temperature. The temperature ramp-up rates were different slightly between 100 and 200 mL water load. The final set-point temperature was reached without being overshot. Without a water load, the set-point temperature was reached rapidly, often overshooting. Cycle time 2 s, power level 75%. Temperature set point 37°C.

diameter of a flat droplet is larger than a dome-shaped droplet and hence has a better antenna effect and less effective cooling (Figure 9).

## MW Immunostaining

Based on our preliminary data on the temperature profile of a droplet with H2900 MW irradiation, we chose a power output of 75% to allow a rapid ramp-up time to reach a desired temperature without excessive evaporation or drying of the sections. For each of the three incubation steps, we started with an irradiation time of 60 s (1) at 37°C and then incubated at room temperature outside the oven for an additional 5 min. We compared this result to the intensity of staining obtained with the conventional method. An extensive range of variations in the power output of the oven, incubation durations, irradiation mode (on/off intervals vs. continuous), antibody concentrations, and the number of immunostaining steps being irradiated were tested individually and in combinations. The results were evaluated for specific label-

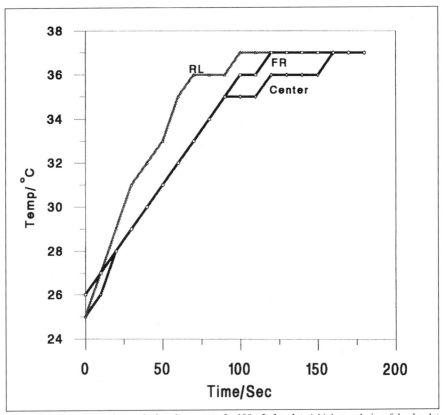

**Figure 8. Effect of molarity on the heating curve of a 100-μL droplet.** A higher molarity of the droplet resulted in a steeper rise and a higher plateau of the droplet temperature. Cycle time 2 s; power level 75%. Diameter on the slide 15 mm. Water load 200 mL in an incubating chamber.

ing and background staining. Specific staining was graded on a scale of 0 to 3, where 0 = no staining, 1 = weak staining, 2 = strong staining and 3 = very strong staining. Nonspecific background was graded in terms of none (N), slight (S), moderate (M) or intense (I). After evaluating the relationships among temperature, duration and droplet evaporation, we optimized the incubation time and then the incubating temperature sequentially for each of the ABC steps.

## Temperature Control

The ability to control the incubation temperature in immunostaining is of the utmost importance. Too high a temperature will deactivate the antibody and results in no antibody-antigen complex formation. Too low a temperature will retard the rate of diffusion and chemical reaction, and result in increased time needed for acceptable staining. The temperature need not be very high,

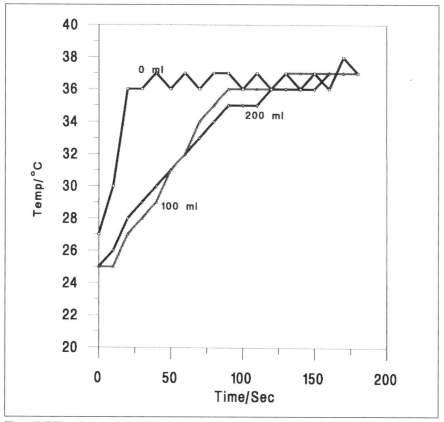

**Figure 9. Effect of size and shape of a droplet on its heating curve**. A relatively flat droplet resulted in a steeper rise and a higher temperature plateau. Cycle time 2 s; power level 75%. Diameter on the slide 15 mm. Water load 200 mL in an incubating chamber.

because the diffusion rate and chemical reaction rate increase exponentially with temperature (14). Moreover, there may be an optimal incubation temperature and duration for each antibody-antigen interaction, the type of tissue and fixative as well as duration of fixation. Thus, it is necessary to have control of the incubation temperature to find the optimal conditions for MW immunostaining.

## Immunohistochemistry Results

For rat pancreas using monoclonal anti-insulin antibodies, we found that for primary antibody (PA), BSA and EC steps, a 3-min MW incubation with power level at 75% and temperature set point at 37°C followed by a 3-min room temperature incubation (23°C) outside the oven gave very strong staining and no background (test 1 of Table 2). However, a "total time" of 60 s MW irradiation followed by a 5-min incubation at room temperature outside the oven for each of the three stages, resulted in weak staining when compared with the conventional overnight ABC method (Figures 10 and 11, See color plate in Addendum, p. A-11). Moreover, repeating the 5 cycles of 1-min MW irradiation followed by 1-min post-MW incubation inside the oven, and then followed by a 3-min incubation at room temperature, gave no additional specific staining or increased background. In addition, we found that using a washing bottle containing PBS to jet rinse the sections between each step was adequate in removing excess reagents and giving a clean background. With a 200-mL water load in an incubating chamber and the proportional band set at 0.5, the temperature of the reagent droplet rose from 25°C to 37°C within 1 min. The evaporative water loss of the droplet was about 10%–15%. Therefore, these conditions were then used as the starting conditions to further optimize the duration of incubation and then the temperature sequentially.

The post-MW incubation was essential to obtain very strong staining. When we only irradiated the PA, BSA, and EC steps for 3 min each without post-MW incubation, the staining was weak (Figure 14, see color plate in Addendum, p. A-13). It was important to post-incubate all three steps. When we omitted the post-MW incubation of either the PA or BSA and EC steps, the staining was strong. However, it was weaker than when post-MW incubation was done on all three layers (Figure 15, test 3 and 4, see color plate in Addendum, p. A-13).

We found that the length of post-MW incubation could be shortened to 2 min and still resulted in very strong staining (test 5). Moreover, the post-incubation could be performed inside the MW oven with power off for the same duration, resulting in very strong staining (Figure 12, p. test 6, see color plate in Addeneum, p. A-12). The temperature of the reagent droplet cooled exponentially, from 37°C to 29°C, with an initial drop from 37°C to 34°C in 10 s.

The optimal length of MW incubation in our model system was 3 min. Further reduction of incubation time to 1 or 2 min resulted in weaker staining (Figure 16; test 7 and 8, see color plate in Addendum, p. A-14). Therefore, in our model system, MWs have an additive effect on the primary, secondary and tertiary steps of the ABC staining procedure, and post-incubation after MW irradiation is important to facilitate affirmative antibody-antigen bindings.

## Power Level and Droplet Evaporation

In an attempt to shorten the incubation time by increasing the incubation temperature, we found that the limiting factors were the ramp-up time of the droplet temperature, the problem of temperature overshoot and the problem of excessive evaporative water loss of the droplet if a higher power level and/or a smaller proportional band was used. Excessive evaporation of the droplet occurs if the power level is too high and the incubation time is too long (more than 5 min). Drying of reagents during immunostaining aborts the procedure. Evaporation can also lead to chemical condensation, which decreases immunoreaction efficiency (13). Thus a balance of power level and the ramp-up time is necessary to maintain the composition of the reagents by avoiding droplet evaporation. In addition, a power level should be chosen such that the ramp-up rate is fast and the MW controller has time to react and compensate for the temperature increase in the droplet. The higher the power level, the more the duty cycle time and thus more energy is delivered per 2-s cycle. The increased power causes an increased tendency for the droplet temperature to overshoot and oscillate, making an ideal temperature curve difficult to be achieved.

## Maximal Allowable Temperature

Since the 3-min MW irradiation, followed by 2-min post-incubation inside the oven for each step of ABC method was an acceptable duration of time (1,10,19), we decided to keep the 3-min MW incubation and 2-min post-incubation as our starting points to determine the maximal allowable temperature for antibody-antigen interactions.

In order to eliminate the possibility of enzyme-complex denaturation at temperatures above the physiologic level, the EC incubation step was performed at 37°C. We found that an incubation temperature of 55°C for the PA and BSA steps resulted in weak specific immunostaining and increased background (Figure 17, see color plate in Addendum, p. A-14).

## Optimal MW Incubation Conditions

In this study the optimal conditions for immunostaining are 37°C for 3 min at 75% power level followed by a 2-min incubation inside the oven for all three stages of the ABC method. When compared with the conventional overnight ABC method, the staining of the MW method at optimal conditions was as strong (Figures 10 and 12). When the same duration, i.e., 5 min, of incubation for each step was performed at room temperature without MW irradiation, the procedure resulted in much weaker staining (Figure 13, test 11, see color plate in Addendum, p. A-12). Overall, we found that in our model

Table 2. Effect of MW Irradiation, Incubating Temperature and Duration on Each of the Three Stages of ABC immunostaining on Rat Pancreas Using Monoclonal Anti-Insulin Antibodies

| Test | Tsp/°C | PA | | BSA | | EC | | Staining Results | |
|---|---|---|---|---|---|---|---|---|---|
| | | MWI/S | PMWI/S | MWI/S | PMWI/S | MWI/S | PMWI/S | Specific | Background |
| 1 | 37 | 180 | 180 | 180 | 180 | 180 | 180 | 3 | N |
| 2 | 37 | 180 | 0 | 180 | 0 | 180 | 0 | 1 | N |
| 3 | 37 | 180 | 0 | 180 | 180 | 180 | 180 | 2 | N |
| 4 | 37 | 180 | 180 | 180 | 0 | 180 | 0 | 2 | N |
| 5 | 37 | 180 | 120 | 180 | 120 | 180 | 120 | 3 | N |
| 6 | 37 | 180 | 120* | 180 | 120* | 180 | 120* | 3 | N |
| 7 | 37 | 120 | 120* | 120 | 120* | 120 | 120* | 2 | N |
| 8 | 37 | 60 | 120* | 60 | 120* | 60 | 120* | 1 | N |
| 9 | 45 | 180 | 120* | 180 | 120* | 180*1 | 120* | 2 | S |
| 10 | 55 | 180 | 120* | 180 | 120* | 180*1 | 120* | 1 | I |
| 11 | 23 | 0 | 300 | 0 | 300 | 0 | 300 | 1 | N |

Chromogen development was 5 min at room temperature. Jet rinsing the sections was done using a washing bottle containing PBS between each step. Optimal incubation condition was 3-min MW incubation, followed by a 2-min post-MW incubation inside the oven. Tsp = Temperature set point. MWI = MW incubation. PMWI = Post-MW incubation in a humid chamber at room temperature. * = PMWI done inside the oven with power off. *1 = Tsp at 37°C. Specific staining: 1 = weak, 2 = strong and 3 = very strong. Background: N = none, S = slight and I = intense.

Table 3. List of Antibodies, Dilution and Sources

| Antibodies to | Origin/Clonity | Conventional | MW | Source |
|---|---|---|---|---|
| Insulin | Mouse IgG/M | 1:40 | 1:40 | Biogenex Laboratories, Sam Ramon, CA, USA |
| S100 | Rabbit IgG/P | 1:3000 | 1:3000 | Accurate Chemical, Westbury, NY,USA |
| α-ANF | Rabbit IgG/P | 1:1000 | 1:1000 | Peninsula Laboratories, Belmont, CA, USA |
| PGP 9.5 | Rabbit IgG/P | 1:2500 | 1:2500 | Accurate Chemical, Westbury, NY, USA |
| PKC | Rabbit IgG/P | 1:100 | 1:100 | Research and Diagnostic Antibodies, Berkeley, CA,USA |

S100 = S100 protein; α-ANF = α-atrial naturetic factor; PGP 9.5 = protein gene product 9.5 and PKC = protein kinase C.

system with anti-insulin antibody, the whole procedure of ABC immunostaining could be completed within 20 min with excellent results. These results were summarized in Table 3.

The optimal incubation conditions of our model system have to be modified for other antibodies. Identical incubation conditions used on the pancreas resulted in weak or no staining of S100 and PGP 9.5 on rat duodenum, α-ANF on rat atrium and PKC on rat ventricle. When the incubation conditions were prolonged for a few more minutes at 37°C for the primary antibodies, the optimal staining was obtained.

## DISCUSSION

In this study MW irradiation was applied to the primary, secondary, and tertiary stages of the standard streptavidin-biotin-peroxidase staining procedure. The entire ABC procedure took about 20 min and results were comparable to the conventional overnight technique.

There are two ways to control the temperature of droplets heated by MW irradiation. One is to determine the empirical relation between the incubation temperature and power level, irradiation time and pattern, object size and composition for a given MW oven (3,16). In addition, several approaches have been attempted to maintain the droplet temperature near physiological levels to prevent destruction of the antibodies. These include *(i)* using a water load in the oven to absorb some of the MW energy to reduce the power density to the droplet (1,7,10,15,16); *(ii)* refreshing the water load in between irradiations to maintain a constant amount of energy uptake (1); *(iii)* stopping MW irradiation at intervals (11); and *(iv)* using cooling during irradiation (10). However, the temperature in the droplets of reagents might occasionally overshoot, even when the necessary precautions are applied. Moreover, the type and amount of water load, molarity of the reagent, droplet size, position of the slides in the MW oven and location of hot spots in the oven all play a role in the temperature buildup of the droplet during MW irradiation (Figures 3 to 7). Furthermore, the aging of the magnetron will decrease the strength of the electromagnetic field produced. This may affect the staining result if the old set of parameters is used. Therefore, the empirical power levels and time setting obtained by different investigators for immunostaining can result in completely different temperatures in the reagent droplets when either another type or an aged microwave oven is used. Hence, it is not possible to pay the required attention to such small, but important, details to achieve a constant incubating temperature during MW immunostaining. This makes the procedures neither operator- nor machine-independent, and therefore unsuitable for routine diagnostic use or research purposes. However, this has been the method of choice since a better method was not possible before the development of the H2900 oven.

In the H2900, we use a fiber-optic probe that is directly linked to a controller so that the fiber-optic probe can both measure and control the temperature of very small volumes of immunoreagents. Because of the negative-feedback feature design of the H2900, the problems of temperature overshoot, hot spots and the disturbance caused by the aging magnetron can be overcome such that reproducible immunostaining results can easily be obtained.

In our experience with the H2900 in the model system tested, the temperature profile of a droplet is highly reproducible (Figures 4 and 7), and the staining procedure can be standardized in terms of the duration of incubation at an optimal temperature. However, it should be stressed that, although the MW irradiation can significantly shorten the procedure and improve the staining, the optimal procedure for each antibody-antigen reaction varies. Our results clearly indicate that the best incubation temperature and duration depend largely on the abundance and availability of the tissue antigens and the affinity and avidity of the antibodies. They need to be established individually. Once they are established, the MW procedure will be highly reproducible and controllable with the H2900 oven.

## "Microwave Effect" on Immunostaining

Heat is currently considered the primary factor responsible for the accelerated effect of MWs on immunostaining (1). However, the rapid oscillation of molecules in the electromagnetic fields may itself play a direct role.

Diffusion is a physical process that can be accelerated tremendously by MW irradiation. While thermal energy will increase molecular kinetics and hasten chemical reactions, the rapid oscillation of molecules directly induced by the electromagnetic fields will result in increased collision of molecules. This increased chance of forming antibody-antigen complexes will directly accelerate the reaction rate. The heat generated may be an epiphenomenon, secondary only to the kinetics. Hjerpe et al. (10) concluded that a 1000-fold increased reaction rate of carcinoembryonic antigen (CEA)/anti-CEA reaction in an enzyme-linked immunosorbent assay system by MW irradiation was far too large to be explained solely by the modest increase in temperature.

The effect of MWs on binding mechanisms is much more complicated and less understood. Although the photon energy generated by MWs is too small to alter covalent bonding, the low intensity of MW fields may readily affect the integrity of the noncovalent secondary bonding, such as hydrophobic interactions, hydrogen bonds and Van der Waal's interactions (8). The resulting change in the three-dimensional conformation of the paratopes of antibodies and epitopes of antigens may increase the steric interaction of the antibody-antigen complex. Thus, this would result in a decrease in the yield of reaction product (85%–90% decrease) as demonstrated by Hjerpe et al. (10). In addition, this may also explain the fact that once the antibody-antigen complex is formed, the MW photon energy may not be strong enough to alter the binding of the complex. Overall, this might explain the increase in reaction rate, due to the increased chances of possible antibody-antigen interactions, and the decrease in product yield, due to alteration of reactive sites by MW effects. Furthermore, the fact that this inhibition of yield can be compensated for by increasing the concentration of antibody, also supports this explanation of the result observed by Hjerpe et al. More importantly, this may be played to our advantage in reducing the background unspecific binding, which is usually a weaker binding than the specific antibody-antigen reaction.

In order to elucidate the existence of a "microwave effect" in immunostaining, we applied the MW optimal conditions to immunostaining using heat via a thermal cycler (PTC-100™; MJ Research, Watertown, MA, USA) and monitored the droplet temperature via a temperature probe (Mon-a-therm Model 6510; Mallinckrodt Medical, St. Louis, MO, USA). The rate of droplet temperature increase in the thermal cycler was similar to that achieved by MW irradiation. We found that a 3-min incubation at 37°C followed by 2-min incubation without heating gave a much weaker staining than the MW method. The result suggests that thermal effect alone cannot account for the accelerated effect of MW immunostaining. In addition, the previously reported success of using a 7-s MW irradiation following by a 5-min room temperature incubation for each step of the ABC methods by Takes et al. also supports our observation (19). In our experience, a 7-s MW irradiation at 100% power level of an 800-W MW oven will raise a droplet temperature by no more than 5°C. All these results strongly suggest that, in addition to the thermal effect on accelerating immunostaining, the rapid oscillation of molecules (2.45 billion cycles per second) in the electromagnetic fields definitely plays a role in the accelerated effect of MWs on immunostaining. The increased oscillation of molecules in the electromagnetic field might enhance the penetrating ability and diffusion rate of antibodies and thus increase chances of forming antibody-antigen complexes, resulting in increased reaction rate. In summary, we conclude that there is a microwave effect, in addition to the thermal effect, of MW on immunostaining.

## OTHER POTENTIAL APPLICATIONS

If a cooling system could be incorporated into the H2900, we might have a way to utilize MWs for the *in situ* polymerase chain reaction (ISPCR). This MW-ISPCR apparatus could perform ISPCR and *in situ* hybridization in the same oven. It may even be possible to automate the entire ISPCR procedure from the amplication step through the detection step by using such an oven. Nevertheless, there are many obstacles to be overcome before the MW-ISPCR apparatus can become a reality.

## CONCLUSION

A rapid, cost-effective, reproducible and simple immunohistochemical procedure is beneficial to the increasing importance of immunohistochemistry in diagnostic anatomic pathology and biomedical research. With this new dimension of temperature control in MW immunostaining, the future protocols in MW immunostaining will have an accelerated turnaround time and reproducible parameters for the optimal performance of each antibody.

## PROTOCOL

### Equipment

The H2900 Immunohistochemistry Microwave oven is used, with maximum-power limit of 800 W. The power output level can be set from 0% to 100%. Cycle times can be selected to be 1, 2, 4, 8, 16, 32 or 64 s. An optimum cycle time of 2 s should be used. In addition, a proportional band of 0.5 should be used to allow both even and rapid ramp-up rates, and maintenance of temperatures within 1°C of the set point.

The fiber-optic temperature-measurement system in the H2900 microwave oven consists of three elements: 1) a fiber-optic probe, 2) a probe support arm and 3) an electronic module for measurement and control. The probe itself consists of a glass fiber that terminates in a temperature-sensing element. It is jacketed on the inside with a supporting sleeve and coated on the outside with Teflon®. The probe support arm is used to position and hold the probe in the desired location and to protect all but the exposed temperature-sensing end. The electronic module is used to excite the temperature sensors, measure its response, compute and report the temperature in analog and digital formats.

The temperature-sensing element at the end of the glass fiber contains a phosphor that fluoresces in the near-infrared region when excited by a red light-emitting diode (LED). The system works by sending a pulse of light to the sensor and measuring the decay time of its fluorescent response to the pulse. This decay time is a direct function of temperature and the module uses curve-fitting techniques to find the temperature in a calibration table. This information is then output as a serial data report and is also used to drive an analog output whose voltage is linearly related to temperature.

The H2900 temperature control subsystem accepts the analog voltage as an input and converts it to a display of the measured temperature. The controller also uses the information to adjust the power output of the magnetron so as to achieve and maintain the desired incubation temperature.

The H2900 has three modes of operation: 1) "control override", in which the microwave simply emits at full power for a set period of time and the temperature feedback control mechanism is bypassed; 2) "total time", in which the countdown timer starts as soon as the run button is pushed, hence the time contains both the ramp-up and temperature maintenance periods; and 3) "time at temperature", in which mode the countdown timer starts when the set point temperature is reached, excluding the ramp-up period.

### Control-Panel Settings of H2900

1. Select a power output level of 75%.
2. Use an optimum cycle time of 2 s.
3. Set the proportional band to 0.5.
4. Set the "total time" to 3 min.
5. Set the set point temperature to 37°C.

### Tissue Fixation and Treatment of Sections

1. Fix freshly collected normal rat pancreas, duodenum and atrium in 4% paraformaldehyde for 2 h before processing through alcohol and xylene to wax in an automated histoprocessor (Shandon Hypercenter XP, Cheshire, England).
2. Stain consecutive sections (7-μm thick) from all blocks according to a conventional streptavidin-biotin-peroxidase complex (ABC) method and the

MW-stimulated method described below.

## Conventional Streptavidin-Biotin-Peroxidase Staining Method

1. Deparaffinize paraffin sections and rehydrate to PBS at pH 7.3.

2. Quench endogenous peroxidase with 3% hydrogen peroxide in PBS for 30 min at room temperature.

3. Wash with PBS three times for 2 min each.

4. Apply non-immune goat serum to each section for 30 min to block non-specific binding of the bridging antibodies to charged molecules in the section (12).

5. Drain off excess reagent from the section.

6. Incubate the PA overnight at 4°C in an humid chamber.

7. Incubate the BSA for 60 min at room temperature.

8. Incubate the streptavidin-horseradish peroxidase conjugate (enzyme conjugate) for 30 min at room temperature.

9. Following steps 6, 7 and 8, rinse the slides with PBS three times for 2 min each and drain off excess fluid from the sections.

10. Incubate with 3-amino-9-ethylcarbazole (AEC), the substrate-chromogen, for 5 min.

11. Rinse the sections in distilled water, counter-stain with Mayer's hematoxylin and mount with GVA mounting medium.

## Microwave-Stimulated Streptavidin-Biotin-Peroxidase Staining Method

1. Deparaffinize paraffin sections and rehydrate to PBS.

2. Perform quenching of endogenous peroxidase and blocking with non-immune serum.

3. Drain off excess serum from the slides, add the appropriately diluted antibody and incubate in the H2900 microwave oven. Incubations are performed in an incubating chamber with inner dimensions of $34.5 \times 21.5 \times 5$ cm, and with 200 mL of distilled water in the bottom of the chamber. Perform incubations in the chamber without its lid on. This allows the temperature probe to gain access to the droplet on the slide.

4. Incubate each of the PA, BSA and EC steps for 3 min at 37°C at 75% power level, followed by a 2-min post-MW incubation inside the oven.

5. Jet rinse the sections using a washing bottle containing PBS between each step.

6. Incubate with the chromogen and counterstain the sections as above.

### Immunoreagents

1. The primary antibodies, their dilutions for the two methods of staining, and their sources are listed in Table 2.

2. Biotinylated goat anti-mouse and anti-rabbit immunoglobulins and

streptavidin peroxidase conjugate (Zymed Laboratories, South San Francisco, CA, USA) are used for both methods of staining.

3. Reagent control is obtained by replacing the primary antibody with non-immune serum.

## COMMENTS

### Probe Placement

The temperature of a droplet on a slide can be precisely controlled with the H2900; however, it is impractical to irradiate one slide at a time if a large number of slides are to be processed. Boon et al. (4) had employed a cuvette incubation method with temperature control provided by a thermocouple probe in the H2500 Microwave Processor (Energy Beam Sciences); however, this approach is limited because of the cost of reagents. The problem can be solved if the fiber-optic probe is mounted on a specially designed turntable. The turntable may have 20–50 slots in which individual slides can be inserted. One of the slots will have the probe fixed in such a way that the droplet on that slide can be measured. In this way, the probe will measure the time-average of a tested droplet for the rest of droplets on the other slides placed on the turntable. Moreover, the drag effect provided by the turntable will minimize the effect of hot spots and make the measured temperature a true representative of different droplets on the various slides (2). In addition, if the probe is not submerged in the droplet but just touching it, the Teflon coating may interfere with the heat conduction to the temperature-sensing element, resulting in inaccurate measurement of the droplet temperature. This can cause an excessive MW energy to be delivered to the reagent droplet. Therefore, by having the probe mounted on the turntable, we will eliminate the need to position and to submerge the probe into a small droplet.

### Evaporation Control

Removal of the lid of the incubating chamber holding the water load is necessary for the probe to gain access to the droplet, but this can result in as much as 15%–20% loss of droplet volume after 3 min of MW irradiation at 75% power level. Evaporation and drying of the tissue affects the efficiency of staining and intensifies the background. The water loss from evaporation changes the ionic strength and increases hydrogen ion concentration of the incubating medium, resulting in a decrease in antibody-antigen interaction (13). This problem may be overcome if an evaporation control device is incorporated either into the incubating chamber or on the individual slide, e.g., a chamber device like the Gene Cone™ (Gene Tec, Durham, NC, USA) (18). In our preliminary testing, the device showed promise in reducing evaporation in MW immunostaining. However, a modified and reusable device would be more practical.

## ACKNOWLEDGMENT

This work was performed at Deborah Research Institute, Browns Mills, NJ, USA.

### REFERENCES

1. **Boon, M.E. and L.K. Kok (Eds.).** 1992. Microwave irradiation in immunostaining, p.256-285. *In* Microwave Cookbook of Pathology: The Art of Microscopic Visualization, 3rd ed. Columbia Press, Leiden.
2. **Boon, M.E. and L.K. Kok.** 1992. Microwave oven and temperature control. p.74-115. In Microwave Cookbook of Pathology: The Art of Microscopic Visualization. 3rd ed. Columbia Press, Leiden.
3. **Boon, M.E., L.P. Kok, H.E. Moorlag and A.J.H. Suurmeijer.** 1989. Accelerated immunogold silver and immunoperoxidase staining of paraffin sections with the use of microwave irradiation. Factors influencing results. Am. J. Clin. Pathol. *91*:137-143.
4. **Boon, M.E., F.C.J. Hendrikse, P.G. Kok, P. Bolhuis and L.P. Kok.** 1990. A practical approach to routine immunostaining of paraffin sections in the microwave oven. Histochem. J. *22*:347-352.
5. **Chiu, K.Y.** 1987. Use of microwaves for rapid immunoperoxidase staining of paraffin sections. Med. Lab. Sci. *44*:3-5.
6. **Chiu, K.Y. and K.W. Chan.** 1987. Rapid immunofluorescence staining of human renal biospy specimens using microwave irradiation. J. Clin. Pathol. *40*:689-692.
7. **Choi, T-S., M.M. Whittlesey, S.E. Slap, V.M. Anderson and J. Gu.** 1995. Advances in temperature control of microwave immunohistochemistry. Cell Vision 2:151-164.
8. **Clearly, S.F.** 1978. Survey of microwave and radio frequency biological effects and mechanisms. D.H.E.W. Publication (FDA) 78-8055:1-33.
9. **Gu, J.** 1994. Microwave immunocytochemistry, p. 67-80. *In* J. Gu and G.W. Hacker (Eds.), Modern Methods in Analytical Morphology. Plenum Press, New York.
10. **Hjerpe, A., M.E. Boon and L.P. Kok.** 1988. Microwave stimulation of an immunological reaction (CEA/anti-CEA) and its use in immunohistochemistry. Histochem. J. *20*:388-396.
11. **Jackson, P., E.N. Lalani and J. Bowtsen.** 1988. Microwave stimulated immunogold silver staining. Histochem. J. *20*:353-358.
12. **Jennette, J.C. and M.R. Wick.** 1990. Immunohistochemical techniques, p.1-28. *In* J.C. Jennette (Ed.), Immunohistology in Diagnostic Pathology. CRC Press, Boca Raton, FL.
13. **Klein, J. (Ed.).** 1982. Lymphocyte-activating substances, p.347-414. *In* Immunology: The Science of Self-Nonself Discrimination. John Wiley & Sons, New York.
14. **Kok, L.P. and M.E. Boon.** 1990. Physics of microwave technology in histochemistry. Histochem. J. *22*:381-388.
15. **Leong, A.S.-Y. and J. Milios.** 1986. Rapid immunoperoxidase staining of lymphocyte antigens using microwave irradiation. J. Pathol. *148*:183-187.
16. **Leong, A.S.-Y. and J. Milios.** 1990. Accelerated immunohistochemical staining by microwaves. J. Pathol. *161*:327-334.
17. **Leong A.S.-Y.** 1994. Microwave technology for morphological analysis. Cell Vision *1*:278-288.
18. **Stapleton, M.J., M.C. Levin and S. Jacobson.** 1994. DNA amplification within cells on slides: advances in evaporation and temperature control. Cell Vision *1*:177-181.
19. **Takes, P.A., J. Kohrs, R. Krug and S. Kewley.** 1989. Microwave technology in immunohistochemistry: application to avidin-biotin staining of diverse antigen. J. Histotechnol. *12*:95-98.

# CHAPTER 6

# Applications of *In Situ* Hybridization and Immunocytochemistry for Localization and Quantification of Peptide Gene Expression — A Lesson From Islet Amyloid Polypeptide

**Hindrik Mulder[1], Bo Ahrén[2] and Frank Sundler[1]**

[1]Department of Physiology and Neuroscience, Section for Neuroendocrine Cell Biology; [2]Department of Medicine at Malmö University Hospital, University of Lund, Sweden

## SUMMARY

*The identification of the messenger phenotype of neuroendocrine cells is crucial for the understanding of neuroendocrine mechanisms. Neuroendocrine cells generally express several neurohormonal peptide messengers—sometimes together with non-peptide messengers—and this expression may be subject to alterations during ontogeny and under pathological conditions. In addition, the level of messenger expression is regulated in response to physiological stimuli. At present, several techniques can be used to assess messenger expression in neuroendocrine cells. Immunocytochemistry, using antibodies directed at these messengers and applied as single, double or triple immunostaining, reveals the localization and co-localization of neurohormonal peptides. In situ hybridization, using probes designed to hybridize with a target mRNA in tissue sections, demonstrates the cellular localization of peptide mRNA; in situ hybridization can be combined with immunocytochemistry and, thus, simultaneously demonstrate the site of expression of both mRNA and peptide. Furthermore, in situ hybridization can, by the aid of computerized image analysis, be used to quantitate levels of mRNA in tissue sections and thereby monitor alterations in gene expression under different conditions. Collectively, these techniques have greatly expanded the present-day knowledge of the neuroendocrine system. In this review, we focus on the recently discovered islet hormone candidate—islet amyloid polypeptide (IAPP or amylin)—to illustrate the use and relevance of these techniques. Protocols for in situ hybridization are given and discussed from a practical standpoint. IAPP is structurally related to calcitonin gene-related peptide (CGRP) and adrenomedullin and is predominantly co-expressed with insulin in pancreatic islets; in some species, such as rat, IAPP is also expressed in islet somatostatin cells. Moreover, IAPP is expressed in gastrointestinal neuroendocrine cells, most of which co-express somatostatin, and in a population of CGRP-containing sensory neurons.*

115

*Under diabetes-like experimental conditions induced by dexamethasone or strepto-zotocin, the islet expression of IAPP and insulin dissociate, in that IAPP is over-expressed relative to insulin. The implications of the dissociated expression of IAPP and insulin remain unclear but may be relevant to the pathogenesis and course of diabetes, due to the metabolic and paracrine effects of IAPP-restraining insulin action and release. The role of IAPP in the gastrointestinal tract and in sensory neurons remains to be clarified.*

## INTRODUCTION

For a thorough understanding of neuroendocrine cell function, it is important to establish the morphological basis of the production, storage and release of neurohormonal messengers. The present-day knowledge in this field has been expanded greatly by the development of various cytochemical techniques, and there has been remarkable progress in the optimization and utilization of these techniques during the last decade. In particular, immunocytochemistry, with the use of single-, double- and triple-staining procedures, has allowed the precise cellular localization and co-localization of neuroendocrine messengers. The immunocytochemical techniques have also been successfully transferred to electron microscopy, providing information on the subcellular localization of neuroendocrine messengers.

Neuroendocrine cells, which are known to utilize the regulated pathway of messenger production, with the inherent formation of secretory granules and the storage of such granules in large numbers, are particularly suitable for immunocytochemistry, due to the high concentration of the messenger within a limited space. One limitation of immunocytochemistry within the field of neuroendocrine research concerns the problem of cross-reactivity of the antibodies used. Neurohormonal peptides, constituting the bulk of these messengers, are often members of families of related peptides, providing potential candidates for cross-reactivity in tissue sections. Even though such potential cross-reactivity can be controlled and checked (see below), it is a distinct possibility that there are still other related peptides in the tissue that have not yet been identified.

In recent years, molecular biology has provided science with a number of new techniques that have further deepened and broadened our understanding of the neuroendocrine system and of biology in general. A prerequisite for these techniques is the cloning of genes, and at present, the list of genes that have been cloned is extended daily. Methods for monitoring gene expression are currently available and include Northern hybridization and nuclease protection assay (solution hybridization); these techniques allow quantification of mRNA levels in extracts from tissues or cell cultures. In 1969, Gall and Pardue demonstrated that labeled nucleic acid sequences, or probes, could hybridize to specific chromosomal DNA sequences in tissue sections (27). Thus, a new field of research in cytochemistry, designated *in situ* hybridization, was opened. In neuroendocrine research, *in situ* hybridization has been used for the localization of gene expression, i.e., for the demonstration of mRNA, while detection of chromosomal DNA *in situ* primarily has been of use in molecular genetic research. Since autoradiography is the most common detection method for *in situ* hybridization and there is a long tradition of quantification of autoradiography in other fields, quantification of *in situ* hybridization was soon made feasible. This approach was initially adopted by neurobiologists (45), because in a complex structure such as the brain a tissue extract from a discrete nucleus may not be possible to obtain, whereas such a structure can readily be identified in tissue sections. Hence, quantification of gene expression in a tissue section using *in situ* hybridization may be advantageous. At present, quantitative *in situ* hybridization is also emerging as a useful tool when studying the regulation of gene expression in peripheral neuroendocrine systems, such as the pancreatic islets (12,13,49,50).

Our approach to study neuroendocrine gene expression, using cytochemical techniques, includes localization of neuroendocrine messengers with immunocytochemistry, *in situ* hybridization and a combination thereof. In addition, we monitor the level of gene expression by quantitative *in situ* hybridization. In this review, we have chosen to illustrate the use of these techniques by referring to our studies on islet amyloid polypeptide (IAPP or amylin) and its expression in neuroendocrine cell systems under normal and experimental conditions. Finally, protocols for *in situ* hybridization are described in detail and the relevance and use of the different protocols are discussed from a practical standpoint.

## ISLET AMYLOID POLYPEPTIDE—A BACKGROUND

It has been known for almost a century that islets of Langerhans from patients with non-insulin-dependent diabetes mellitus contain amyloid deposits (59), the amount of which correlates with the duration and severity of the disease (74). In 1986, Westermark and associates were able to extract a 37-amino acid peptide from the amyloid of an insulinoma (78); the following year, it was shown by the same group (76) as well as by Cooper and associates (16) that this peptide was also the main constituent of islet amyloid in patients suffering from non-insulin-dependent diabetes. Interestingly, the sequence of the extracted peptide showed an approximately 50% homology to calcitonin gene-related peptide (CGRP; 16,78), a ubiquitously expressed neuropeptide derived from an mRNA that arises from alternative processing of the calcitonin gene transcript (34). Recently, the novel peptide adrenomedullin was isolated from a pheochromocytoma. The peptide, normally expressed predominantly in the adrenal medulla, is a 52-amino acid polypeptide, the C-terminus of which bears a structural resemblance to IAPP and CGRP (38). The gene for IAPP and IAPP cDNA have been cloned in humans and several animal species, demonstrating that the peptide has been

117

well conserved through evolution (8,14,22,36,42,48,55,56,68). In humans, the IAPP gene encodes an 89-amino acid prepropeptide (47,56), which is subsequently cleaved to the mature 37-amino acid peptide, containing an N-terminal disulfide bridge and an amidation of the C-terminus (16). The gene comprises three exons, of which the third encodes IAPP 1–37 (47,56).

IAPP is a product of the normal B cell (35,77), where it is stored in the same secretory granules as insulin (44). The expression of IAPP, however, is not restricted to the islets, since IAPP-like immunoreactivity has been demonstrated in the gastrointestinal tract (46,52,66) and IAPP mRNA has been detected in the stomach, dorsal root ganglia and lung (23,51,52,54). The islet expression of IAPP is subject to some interspecies variation, in that D (somatostatin)-cells in rodents also express IAPP (3,19,53).

The localization of IAPP in insulin secretory granules implies that IAPP is released together with insulin; in fact, under most conditions, IAPP is released in parallel with insulin in response to glucose and other secretagogues (37,40,65). However, data are accumulating which suggest that the secretion and expression of IAPP and insulin can be dissociated. For instance, in hyperglycemia, the relative increase in secretion of IAPP is greater than that of insulin (58); the same phenomenon has been observed in dexamethasone- and streptozotocin-treated rats (30,33,58,60). Also, the level of IAPP mRNA has been demonstrated to be dissociated from that of insulin mRNA in several experimental models (11,49,50,60) and in human islets (57). The concept of a differential regulation of IAPP and insulin gene expression would be supported by the existence of regulatory elements in the upstream 5'-region of the IAPP and insulin genes, which are not common to the genes; this issue is, however, at present unresolved. (14,29,68).

Initially, great hope was attributed to IAPP as a possible pathogenetic factor in non-insulin-dependent diabetes mellitus, since IAPP was shown to restrain the peripheral actions of insulin in skeletal muscle (25,43,83), thereby contributing to, or possibly causing, insulin resistance. These actions have been proposed to be mediated by an activation of glycogen phosphorylase and inhibition of glycogen synthase (41). A major problem, though, in the understanding of these actions of IAPP, is that they occur at concentrations supposedly higher than those of endogenous circulating IAPP and, thus, the physiological relevance of these findings has been questioned (for reviews, see References 15, 75 and 80). It is possible, however, that the actual concentration of exogenous IAPP under experimental conditions has been overestimated, due to the tendency of the peptide to form insoluble fibrillar aggregates (73). Also, a paracrine/autocrine mode of IAPP action in islets has been discussed, since IAPP inhibits insulin secretion (4,26,71). Again, in most reports, these effects are seen at fairly high concentrations of IAPP, but the local concentration of IAPP upon release within islets may reach the levels required for the effects observed. Furthermore, the use of IAPP-antagonists supports a physiological role of endogenous IAPP, since they modulate both the peripheral actions of IAPP on fuel metabolism (82) as well as its effects on insulin secretion (7,71,81).

A problem in the understanding of the physiology of IAPP is that the IAPP-receptor has yet to be cloned. Such receptors appear to exist, since an IAPP-binding site with a 10- to 150-fold higher affinity for IAPP than CGRP is present in several peripheral tissues and some areas of the brain (10). Furthermore, a high-affinity IAPP-binding site in the brain has recently been characterized (6,63,69) and is possibly identical to the type C3 CGRP receptor previously described (62). A common feature of these binding sites is that other peptides than IAPP, namely CGRP and adrenomedullin, interact with high potency at these sites.

A potential diabetogenic mechanism of IAPP is its amyloidogenic properties in islets, which is incompletely understood at present; however, as deduced by amino acid substitutions and synthetic peptide experiments, it is known that the amino acid sequence 25–29 is crucial for amyloid formation (73). The sequence divergence in this region probably explains to a great extent why some species, such as humans and cats, form islet amyloid made up of IAPP, whereas other species, such as rodents, lack these amyloid deposits, even in diabetes (9). In this context it is noteworthy and somewhat intriguing that mice, carrying a transgene containing a human IAPP gene construct and overexpressing human IAPP, still do not develop islet amyloid (24,32,70). Nevertheless, in islets from such mice cultivated in a high glucose medium (18) or in human islets transplanted under the kidney capsule of nude mice (72), IAPP fibrils are formed. Evidently a factor other than the IAPP sequence and the local concentration of IAPP—a factor that has not yet been identified—is required for amyloid formation. To speculate, this factor may be related to a decreased clearance of IAPP from islets, which prevails in both these experimental settings.

Our studies on IAPP have focused on the cellular expression of IAPP in the pancreas, gastrointestinal tract and sensory nervous system. In particular, we have examined the expression of IAPP following dexamethasone- and streptozotocin-treatment of rats with special emphasis on its correlation to insulin gene expression.

## PATHOPHYSIOLOGICAL MODELS FOR IMPAIRED ISLET FUNCTION IN THE RAT

### Induced Peripheral Insulin Resistance

Non-insulin-dependent diabetes mellitus is often preceded by a state of reduced insulin sensitivity (20). A commonly held view is that this reduced sensitivity to insulin is compensated by an increased secretion of insulin, which will counteract the development of hyperglycemia. However, in subjects with poor islet function, there is an inability to fully compensate this with an exaggerated insulin secretion, which will result in hyperglycemia and diabetes.

Although the reduced insulin sensitivity often precedes diabetes, the development of disease is dependent on impaired B cell function (61). Therefore, in studies focusing on the pathogenesis of non-insulin-dependent diabetes mellitus, states with a reduced sensitivity to insulin are of great interest, since they may give insight into the islet adaptation to the increased demand, the process which fails in diabetes. One model for reduced insulin sensitivity is dexamethasone-treatment of the rat, in which dexamethasone induces insulin resistance (64). Dexamethasone adminstered daily at a fairly high dose for one to two weeks results in reduced insulin sensitivity, compensatory islet hypertrophy and hyperplasia, as well as hyperinsulinemia with preservation of normal or near normal plasma glucose levels. The compensatory alterations induced in the islets are thus clearly reminiscent of the early changes of non-insulin-dependent diabetes mellitus and can serve as an experimental model for this disease.

### Impaired B Cell Function

In severe diabetes, on the other hand, the islet B cell mass is reduced, which inevitably impairs insulin secretion and causes massive hyperglycemia. Also this phase of diabetes is of interest to study since it will give us information on the islet function in fully developed diabetes. Experimentally, diabetes with reduced B cell mass is usually studied in rodents given the B cell toxins streptozotocin or alloxan (1,2,5). The mechanism of action is thought to involve a breakage of DNA strands and depletion of B cell NAD-levels, which cause B cell destruction and consequently a long-standing diabetes (79). However, the duration of the evolving diabetes is dose-dependent, since compensatory repair processes are initiated, counteracting the diabetogenic effects of the drugs at low dose levels. Thus, when given at a high dose, an irreversible form of diabetes is induced, whereas at a low dose, the diabetes is transient (5). Consequently, the models of diabetes induced by a higher dose of B cell toxins will yield information on B cell function when irreversible damage has occurred, whereas those induced by a lower dose will provide data on B cell repair mechanisms iniated early in the onset of diabetes.

## CELLULAR EXPRESSION OF IAPP—IMMUNOCYTOCHEMISTRY AND *IN SITU* HYBRIDIZATION

### Immunofluorescence

For the cellular localization of IAPP in the pancreas, gastrointestinal tract and sensory nervous system, we have used indirect immunofluorescence with polyclonal antibodies (52,53) and, recently, a monoclonal antibody directed at human IAPP (51). Since our studies have included rats, an affinity-purified secondary antibody specific for mouse immunoglobulin G and coupled to fluorescein isothiocyanate (FITC) has been used for the experiments with the monoclonal antibody; the affinity purification is necessary in order to avoid high background staining in rat. In simultaneous double and triple immuno-fluorescence, two and three primary antibodies directed against different islet peptides and raised in different species were applied. As secondary antibodies served anti-mouse, anti-rabbit and anti-guinea pig immunoglobulin G coupled to either FITC, tetramethyl rhodamine isothiocyanate (TRITC) or 7-amino-4-methyl coumarin-3-acetic acid (AMCA). These fluorophores emit fluorescence at different wavelengths and by shifting the microscope filters, the localization of up to three primary antibodies in a section can be determined.

Using triple immunofluorescence, the cellular localization of IAPP, pro-insulin and somatostatin in islets from dexamethasone- and streptozotocin-treated rats can be visualized simultaneously (Figure 1, see color plate in Addendum, p. A-15). In dexamethasone-treated rats, the centrally located bulk of cells—B cells—harbor both proinsulin and IAPP (Figure 1, A and B). In addition, a small number of peripherally located islet cells—D cells—harbor somatostatin and IAPP (Figure 1, A and C); co-localization of proinsulin and somatostatin is not seen. In streptozotocin-treated rats, the islet architecture is disrupted; the number of B cells is reduced and the cells appear scattered (Figure 1, D and E). IAPP still occurs predominantly in B cells but also to some extent in D cells (Figure 1, E and F).

In the stomach of rats, IAPP predominantly occurs in somatostatin cells (Figure 1, G and H), but also to a lesser degree in G (gastrin) cells. In rat sensory ganglia, such as the dorsal root ganglia and trigeminal ganglia, IAPP is expressed in a moderate number (approximately 25% of the total number of neurons) of small to medium-sized nerve cell bodies (Figure 2, A and B), constituting a subpopulation of neurons expressing CGRP (51). Furthermore, nerve fibers containing IAPP are seen in the superficial layers of the dorsal horn in the spinal cord (Figure 2C) and in small subsets of nerve fibers in tissues known to receive sensory innervation, such as the tongue (Figure 2D).

Thus, using immunofluorescence it is possible to verify the co-localization of several different neurohormonal peptides. The ability of neuroendocrine cells to express multiple messengers greatly increases their potential to affect the target tissue or cells. Furthermore, the co-expression of neurohormonal peptide messengers (sometimes referred to as the chemical coding) is also subject to changes during ontogeny and under altered functional and/or pathological conditions. The specificity of immunocytochemical staining can be checked, e.g., by preabsorption tests. Here, the antigen or related compounds, with a potential for cross-reactivity with the antibody, is added in excess to the antibody at working dilution. Quenching of the immunocytochemical signal, using preabsorbed antibodies, makes it highly probable that the antibody recognizes the antigen in question.

## Combined *In Situ* Hybridization and Immunoperoxidase

The cellular localization of IAPP in islets and the stomach has also been investigated with a combination of *in situ* hybridization and immunoperoxidase (52,53). In human islets, this technique demonstrates that islet cells containing IAPP mRNA also harbor IAPP and proinsulin, while lacking glucagon (Figure 3, A–C, see color plate in Addendum, p. A-16; unpublished results). In rat islets, IAPP mRNA is seen in centrally located cells; these cells also contain immunoreactive IAPP and proinsulin, whereas glucagon is absent (Figure 3D; 53). The architecture of human islets is slightly different from that of rat, in that the B cells occur in clusters rather than forming a central core. In the mouse antrum, combined *in situ* hybridization and immunoperoxidase demonstrates a moderate number of cells expressing IAPP and its mRNA (Figure 3E; 52). Additionally, IAPP mRNA can be localized to a subpopulation of somatostatin cells (Figure 3F).

Although immunocytochemistry under most conditions is a more sensitive approach to determine the neuroendocrine phenotype of cells, *in situ* hybridization and the technique of combined *in situ* hybridization and immunoperoxidase offer some additional benefits. First, since probes for *in situ* hybridization are designed, nonspecific binding can be avoided in a way that is

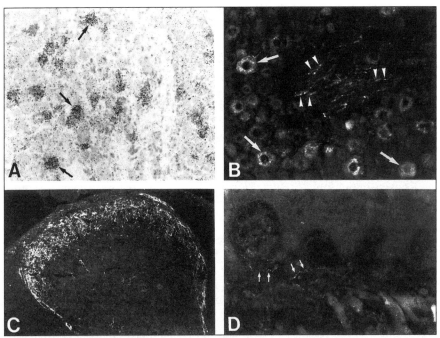

**Figure 2. Expression of IAPP in the sensory nervous system.** (A) *In situ* hybridization with a mixture of two IAPP oligoprobes in rat dorsal root ganglion demonstrates the occurrence of IAPP mRNA in a population of small to medium-sized nerve cell bodies (examplified by arrows). (B–D) In the rat, immunofluorescence with a monoclonal IAPP antibody reveals the presence of IAPP-like immunoreactivity in nerve cell bodies and nerve fibers in dorsal root ganglion (B; arrows and arrowheads, respectively), in nerve fibers in the superficial layers of the dorsal horn in spinal cord (C) and in a small population of subepithelial nerve fibers of the tongue (D, arrows).

not possible with immunization to obtain antibodies. Also, the conditions for hybridization can be changed to optimize target binding. Second, if the identity of an endocrine cell can be demonstrated by two independent techniques, optimally in the same section, it will add weight to the findings obtained. Also, cells that very actively synthesize protein or cells which are less differentiated (e.g., in certain neoplasias), may under certain conditions not be readily detected by immunocytochemical techniques due to a low cellular content of the secretory product (degranulated cells), whereas *in situ* hybridization will readily demonstrate the content of mRNA, thus revealing the neuroendocrine phenotype of the cell (67).

## QUANTIFICATION OF *IN SITU* HYBRIDIZATION

Conventional *in situ* hybridization is a good tool for the demonstration of gene expression and occurrence of the transcripts in neuroendocrine cells. However, for a number of scientific questions, there is a need for quantification of the transcript formation, in order to more deeply study the regulation of gene expression. To allow such studies, quantitative *in situ* hybridization has evolved.

Quantitative *in situ* hybridization can either be performed as a densitometric analysis of an x-ray film, which has been apposed to slides containing sections subjected to *in situ* hybridization, or by computerized image analysis of such slides dipped in autoradiographic emulsion. Using the latter approach, the cellular resolution of the quantification is maintained.

### Computerized Image Analysis

Computerized image analysis is a powerful tool when used in conjunction with cytochemical techniques. Several different commercially available systems are in use today. These systems basically operate in the same way: a video camera is connected to the microscope; the analog video signal is converted to a digitized image by a frame grabber in the computer and this image is then processed by computer software, in that distances and areas can be measured and, in the digitized image, the grey value of each picture element (pixel) can be determined. Then, depending on the sophistication of the system, additional features allow further analysis. Autoradiography is well suited for quantification by computerized image analysis. Two approaches are commonly used: either the grey value of a structure is determined (often converted via calibration to optical density) or, by thresholding, a binary image is created and enhanced, after which the number or area of grains is calculated. The data from such determinations can then be used for assessment of relative changes in gene expression, since the number/density of grains correlates with the amount of probe hybridized to a section and, thus, its content of the complementary mRNA. In our studies, we have chosen to measure the area of grains, in an effort to circumvent the problem of grain clustering.

## Quantification of IAPP and Insulin Gene Expression in Experimental Models of Diabetes

Since it was previously demonstrated that the ratio of IAPP and insulin mRNA levels is increased in pancreatic extracts after dexamethasone-treatment in the rat (11,60), we were curious to see if these results could be reproduced by quantitative *in situ* hybridization and if additional information would be yielded. Theoretically, the dissociation of IAPP and insulin gene expression could be a true dissociation at the cellular level or an up-regulation of IAPP in D cells, which have also been shown to express IAPP (53). Pancreatic tissues from rats treated with dexamethasone and from control rats were processed for *in situ* hybridization using oligonucleotide probes for IAPP and insulin mRNA, respectively. Darkfield images of the sections were digitized; the polarity of the image was reversed and a threshold set at a fixed grey level. The outline of individual islets was interactively defined and the area of the grains covering an islet and the actual area of the islet were determined. All sections were hybridized to the respective probe in the same experiment and analyzed in duplicate, determining grain area in several randomly selected islets. To avoid saturation of the emulsion and loss of its linear response to the radiolabeled probes, an extra set of slides was processed identically and used for assessment of the correct exposure time. The fact that the grain areas were all of the same magnitude ensured that the respective mRNA levels were quantitated in the same range of the response curve. The grain areas of the islets from the treated animals and their controls were compared using a two-tailed one-way ANOVA (Kruskal-Wallis' test) followed by Dunn's multiple comparisons test *post hoc*.

Results from the studies on IAPP and insulin expression in the dexamethasone-treated rat are summarized in Figure 4 and darkfield images of islets from dexamethasone-treated rats hybridized with the IAPP probe, to which quantitative *in situ* hybridization has been applied, are shown in Figure 6, A and B. Dexamethasone increases the islet levels of both IAPP and insulin mRNA. However, when the dexamethasone-induced increase of islet size is taken into account, we find that the insulin mRNA level per unit area is actually decreased, whereas that of IAPP is increased. Thus, this technique allows the conclusion that dexamethasone differentially regulates IAPP vs. insulin gene expression in islets (49).

In a previous report from Bretherton-Watt and collegues (11), using densitometric quantification of Northern blots, a 16-fold increase of IAPP mRNA and a 4-fold increase of insulin mRNA were reported in the same experimental model that we have used. Although the changes in gene expression that we observe are not as great, they are of the same magnitude and, more importantly, the ratio of IAPP and insulin mRNA levels is approximately the same (2.5 vs. 4). However, using the cytochemical approach, we can conclude that insulin gene expression is actually decreased at the cellular level following dexamethasone-treatment; the increase in total insulin gene expression, assessed in both reports, is due to hyperplasia/hypertrophy of B cells. This finding may be relevant to diabetes pathogenesis since it may represent selective fatigue of the B cells in response to peripheral insulin resistance. Conversely, the increased expression of IAPP under experimental conditions characterized by insulin resistance may be a useful marker and may possibly have pathogenetic implications, since IAPP at pharmacological doses can cause insulin resistance (see above).

In the same study by Bretherton-Watt and collegues (11), the effects of streptozotocin-treatment on IAPP and insulin gene expression in rats were assessed; they found that streptozotocin reduces the level of IAPP mRNA in pancreatic extracts 6-fold less than that of insulin mRNA. Using quantitative *in situ* hybridization as described above, we investigated the effect of a high dose of streptozotocin on islet gene expression of IAPP and insulin in the rat. The results from these studies are summarized in Figure 5, and darkfield images of islets from rats rendered diabetic with streptozotocin and hybridized with the IAPP probe are exemplified in Figure 6, C and D. In accordance with the findings of Bretherton-Watt and collegues (11), we find that streptozotocin-treatment differentially reduces the levels of IAPP and insulin mRNA, to 24% and 15% of controls, respectively (50). Bretherton-Watt and collegues demonstrate a reduction in the IAPP and insulin mRNA levels to 30% and 5% of controls, respectively (11). Again, these two studies show that similar results can be obtained when using two different methods, i.e., Northern hybridization and *in situ* hybridization. However, the cytochemical approach can yield additional information, because when *in situ* hybridization with the IAPP probe is combined with immoperoxidase for insulin and somatostatin (as described above), we can also demonstrate that the differential expression of IAPP and insulin in this experimental model of diabetes is determined by expression of IAPP in B cells, since IAPP mRNA predominantly occurs in cells immunostained for insulin, i.e., in B cells, and not in cells stained for somatostatin (50). Thus, it is unlikely that the minor population of IAPP/somatostatin cells is responsible for the differential expression of IAPP and insulin in experimental diabetes, although it may contribute.

## Methodological Aspects of Quantitative *In Situ* Hybridization

As illustrated above, quantitative *in situ* hybridization can in many experimental settings yield additional information that is not obtainable when quantifying gene expression in tissue or cell extracts. However, the monitoring of gene expression in extracts, in particular using nuclease protection assays, offers some advantages compared to quantitative *in situ* hybridization. For instance, standard curves allow determinations of mRNA concentrations, whereas quantitative *in situ* hybridization provides assessment of relative changes and, thus, to some extent is semiquantitative. On a Northern blot, the hybridization signal of the target mRNA can be normalized to that of a probe hybridized with β-actin mRNA, the rationale being that β-actin is expressed identically in similar specimens, thus enabling comparisons of hybridization signals between specimens.

When comparing relative changes in the levels of different mRNAs it is assumed that the probes hybridize with their target mRNAs with the same efficacy. This applies to all forms of nucleic acid hybridization. However, the kinetics of hybridization in tissue sections cannot be determined, while those of hybridization on filters or in solution can be described. Therefore, in an attempt to standardize hybridization conditions, the probes used should be designed to hybridize with approximately the same melting temperature ($T_m$; temperature at which half the probe molecules are hybridized with their target mRNA). This means the probes should be of the same length and display the same ratio of guanosin+cytidine/adenosine+thymidine; these factors determine the $T_m$ of the probe, in that an increased length of the probe or a greater proportion of guanosin+cytidine increases the $T_m$. Furthermore, the stringency at which hybridization occurs should be kept identical, i.e., the concentration of salt and organic solvent in the hybridization buffer; lowering of salt

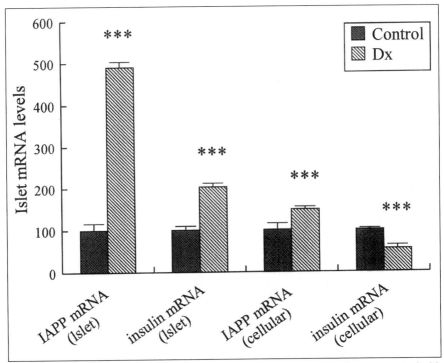

**Figure 4. Quantitative *in situ* hybridization performed on pancreatic sections from rats treated with dexamethasone (Dx), 2.0 mg/kg daily for 12 days, using radiolabeled probes for IAPP and insulin mRNA.** Results (mean ± SEM) are given as percent of controls and displayed as either the total area of autoradiographic grains in islets (islet level) or as a ratio of grain area/islet area (cellular level). \*\*\*$P$ <0.001 vs. controls (for details, see Reference 49).

concentration or increasing organic solvent concentration increases the stringency of hybridization. Finally, hybridization should take place at the same temperature; a raised temperature increases the stringency of hybridization.

Great care must be taken not to over- or underexpose the sections to the emulsion; this will result in a loss of linearity of the autoradiographic emulsion to the radiolabeled probe and, thus, a nonlinear hybridization signal. As pointed out above and emphasized by Gerfen and collegues (28), when relative changes in hybrization with different probes are compared it is crucial that the quantification is performed in the same range of the response curve of the autoradiographic emulsion; in practice, this means that the baseline values of the control groups should be similar. Some of these events can be controlled by inclusion of radioactive standards, to which the hybridization signal can be normalized. In this respect, $^{14}$C-standards are particularly suitable because the energy of the emitted radiation is approximately the same as for $^{35}$S, while its half-life is vastly longer. These standards are commercially available and can be recycled. Finally, standardization of the experimental procedures in quantitative *in situ* hybridization is instrumental. For instance, we have found that both the fixative used and fixation time can affect the hybridization signal. Therefore, it is recommended that specimens are collected at the same time and processed identically up to the actual hybridization.

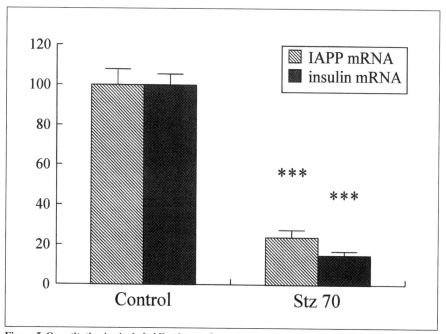

**Figure 5. Quantitative *in situ* hybridization performed on pancreatic sections from rats treated with streptozotocin (Stz), 70 mg/kg, using radiolabeled probes for IAPP and insulin mRNA.** Results (mean ± SEM) are given as percent of controls and displayed as the total area of autoradiographic grains in islets. ***$P$ <0.001 vs. controls (for details, see Reference 50).

Also, only slides hybridized in the same experiment should be used for quantification. Further, the thickness of the emulsion covering the slides must be kept identical in each experiment, since β radiation emitted from $^{35}$S penetrates the emulsion, resulting in a higher signal with a thicker emulsion layer.

## PROTOCOLS FOR *IN SITU* HYBRIDIZATION

Three protocols for *in situ* hybridization are given below. In our hands, these protocols are sensitive and reproducible tools for localizing and quantifying neurohormonal peptide gene expression in various tissues, using oligodeoxyribonucleotide probes ("oligos"). These protocols, which have been adopted from several workers and modified/improved (17,21,31,39), will be discussed from a practical standpoint. Probes and their labeling will also be discussed.

### Probes for *In Situ* Hybridization

We have used oligos for our *in situ* hybridization studies. There are several reasons for this. Oligos are designed by the investigator, based on the cloned cDNA sequence. Thus, a sequence can be chosen with the appropriate length

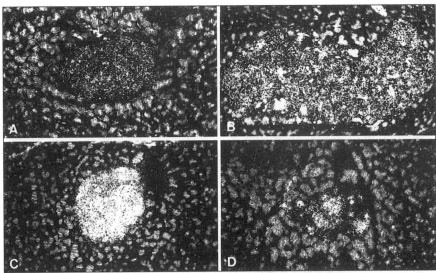

**Figure 6. *In situ* hybridization with an IAPP probe in islets from rats treated with dexamethasone (A-control; B-Dx) and streptozotocin (C-control; D-Stz).** In B, the typical enlargement of islets due to Dx-treatment is seen; in addition, note the increased density of grains, indicating an increased mRNA content per cell. The Stz-treated islet displays a disrupted architechture with scattered cells labeled by the probe. These darkfield images are digitized and analyzed by computerized image analysis. Note that the signal strength of the two control sections (A and C) is different. This is necessary in order to compare them with the treated animals, which, in the case of Dx (B), is an up-regulation of gene expression, and, thus, an increased signal, whereas streptozotocin (D) induces a down-regulation of gene expression, and, thus, a decreased signal.

and nucleotide composition. Currently, software is available to aid in the design of oligos for hybridization experiments; for instance, hairpin loops can be predicted, and the $T_m$ and nucleotide composition can be controlled. The sequence of the oligo should be checked for homologies to other nucleic acid sequences by a computerized search through a nucleic acid database, thereby reducing the likelihood of hybridization occurring with sequences other than the desired target mRNA. The oligos can either be aquired from any of several biotech companies offering such services or be directly synthesized in the laboratory, when such facilities are available. Since oligos are composed of deoxyribonucleotides (DNA), they are fairly stable in a standard laboratory environment, as opposed to ribonucleic acid probes (cRNA or "riboprobes"), which are very sensitive to degradation by ribonucleases (RNase). The relatively small size of oligos enables them to readily penetrate the target tissue. Oligos are labeled or endtailed by an enzyme called terminal transferase; this enzyme adds a number of nucleotides of choice to the 3′-end. We have used $^{35}$S-labeled dATP because the sensitivity of autoradiographic detection has not yet been surpassed. $^{35}$S-dATP emits radiation at an intermediate energy level; thus, exposure times to emulsion are reasonable for most probes used (2 days to 4 weeks) and the resolution of the autoradiographic signal is good. It is also possible to 3′-endtail oligos with enzyme (alkaline phosphatase)- or hapten (digoxigenin or biotin)-modified nucleotides. An additional benefit of isotope-labeling of probes is that autoradiography can be satisfactorily quantified. The tailing reaction is very straightforward; either an available protocol can be adopted or kits from several manufacturers can be employed. Finally, if the sensitivity needs to be enhanced, several oligos complementary to different parts of the target transcript can be used simultaneously as a "probe cocktail".

The sensitivity of riboprobes is greater than that of oligos; the $T_m$ of an RNA-RNA hybrid is higher than that of a DNA-RNA hybrid and, thus, hybridization can be performed at a higher temperature. This may be important when a very low level of expression is being detected. However, the drawback is that riboprobes are very unstable and that a certain degree of molecular biological expertise is needed. Riboprobes are generated by *in vitro* transcription; hence, there will be a need for knowledge of how to culture bacteria, handle vectors, use restriction enzymes, subclone cDNA and handle RNA polymerases. Riboprobes are sometimes hampered by "stickiness", meaning that they may bind to tissues or tissue components in a nonspecific manner.

## Protocol 1: Paraffin-Embedded Tissue

1. Specimens are immersed overnight in 4% paraformaldehyde in phosphate buffer, pH 7.4.
2. Specimens are rinsed in 70% ethanol and then dehydrated through graded concentrations of ethanol, cleared in xylene (Merck, Darm-

stadt, Germany) followed by embedding in paraffin.

3. Sections are cut (4 µm) in a microtome and mounted on chrome-alum-coated slides.

4. Section are deparaffinized by xylene (Merck), 5 min.

5. Rehydration: 100% ethanol, $2 \times 4$ min; 90% ethanol, 4 min; 70% ethanol, 4 min and finally rinsed in phosphate-buffered saline (PBS), 2 min.

6. Permeabilization: 0.01% Triton® X-100 in PBS (Sigma Chemical, St. Louis, MO, USA), 2 min.

7. Rinse in PBS, 5 min.

8. Enzyme-digestion: proteinase K (10 µg/mL in 20 mM Tris-HCl/5 mM EDTA, Sigma Chemical; 30 min at 37°C).

9. Post-fixation: 4% paraformaldehyde in PBS, 15 min followed by rinsing in PBS, 5 min.

10. Acetylation: 0.25% acetic anhydride (Sigma Chemical) in 0.1 M tri-ethanolamine/saline (0.9%; Merck), 10 min.

11. Dehydration: 70% ethanol, 3 min; 90% ethanol, 3 min; 100% ethanol, 5 min followed by air-drying of the sections.

12. Hybridization overnight at 37°C. Hybridization buffer (50 µL), containing probe to a concentration of 1 pmol/mL hybridization buffer, is applied to each slide, which typically contains two sections. The sections are covered with a plastic film to prevent evaporation. The slides are kept in a sealed moisturizing chamber [paper towels are wetted by $2\times$ SSC ($1\times$ SSC = 0.15 NaCl and 0.015 M sodium citrate)].

13. Removal of Parafilm slips: slips are gently removed in $1\times$ SSC at room temperature.

14. Post-hybridization washes: 0.5–$1\times$ SSC, $4 \times 15$ min, 55°C followed by once in $1\times$ SSC at room temperature, 30 min.

15. Blocking of endogenous peroxidase activity: 3% $H_2O_2$ (Merck), 5 min.

16. Permeabilization: 0.25% Triton X-100 in PBS, 15 min.

17. Incubation with primary antibody in moist chambers overnight at 4°C.

18. Rinsing: 0.25% Triton X-100 in PBS, $2 \times 15$ min.

19. Incubation with unlabeled secondary antibody (1:80) for 30 min at room temperature.

20. Rinsing: 0.25% Triton X-100 in PBS, $2 \times 15$ min.

21. Incubation with peroxidase anti-peroxidase complex (1:160; DAKO, Copenhagen, Denmark) for 30 min.

22. Rinsing: 0.25% Triton X-100 in PBS, 15 min followed by 0.05 M Tris (pH 7.6) for 15 min.

23. Incubation in diaminobenzidine tetrahydrochloride for 7 min.

24. Rinsing in distilled water for 10 min.

25. Dehydration: 70% ethanol in 0.3 M $NH_4Ac$, 3 min; 90% ethanol in 0.3 M $NH_4Ac$, 3 min; 100% ethanol, 5 min.

26. Sections are air-dried.

27. Emulsion-coating: Ilford K-5 emulsion is placed in a water bath, 42°C. When the emulsion has melted it is gently stirred (bubbles must be avoided) and is diluted 1:1 with distilled water and aliquoted. Dip the slides in a melted aliquot of emulsion with a steady and reproducible movement (even coating!) and place them vertically in a rack to dry. The dried slides are placed in light-sealed boxes containing a dehumidifying agent. Store the boxes at 4°C until development.
28. Development: Develop the slides in Kodak D-19, 5 min (Eastman Kodak, Rochester, NY, USA). Briefly rinse them with tap water and transfer them to Kodak Polymax, 10 min. Rinse the slides for 10 min in running tap water; air-dry.
29. Optional: Background staining, for example hematoxylin and eosin (H&E). Transfer the slides from the running tap water to the preferred stain.
30. Mounting: Air-dried sections are mounted with a water soluble medium (Kaiser's glycerin-gelatin; Merck)

Steps 14–23 are only included when *in situ* hybridization and immunocytochemistry are performed in combination.

### Protocol 2: Cryostat Sections

1. Specimens are frozen in OCT Compound (Miles, Elkhart, IN, USA) on dry ice and stored at -80°C until hybridization.
2. Sections are cut (10 μm) in a cryostat and thaw-mounted onto chromalum-coated slides.
3. Fixation: 4% paraformaldehyde in PBS, pH 7.4, 15 min.
4. Rinsing: $2 \times 5$ min in PBS.
5. Acetylation: 0.25% acetic anhydride in 0.1 M triethanolamine/saline (0.9%; Merck), 10 min.
6. Sections are dehydrated through 70% ethanol, 3 min; 90% ethanol, 3 min; 100% ethanol, 3 min, chloroform, 5 min, and finally 100% ethanol, 5 min.
7. Hybridization: Post-hybridization washes and emulsion-coating performed as in Protocol 1.

### Protocol 3: Cryostat Sections

Perform steps 1 and 2 as in Protocol 2.
3. Hybridization: Post-hybridization washes and emulsion-coating performed as in Protocol 1.

Hybridization buffer: 50% formamide (Merck) and 4× SSC, 1× Denhardt's solution (0.02% polyvinylpyrrolidone, 0.02% Ficoll®, 0.02% bovine serum albumin), 10% dextran sulfate, 0.24 μg/μL yeast tRNA, 0.5 μg/μL salmon sperm DNA and 1% sarkosyl. 0.2 M dithiothreitol (all from Sigma Chemical) was added to the hybridization buffer prior to hybridization.

## COMMENTS

Protocol 1 is suitable for parenchymatous organs, such as the gastrointestinal tract, pancreas or testis, because the fixation and paraffin embedding act as a glue for the tissue, yielding the best morphology of the protocols described. Protocol 1 is particularly suitable when studying gene expression in tissues rich with RNase, such as the pancreas, since RNases are inactivated by the paraformaldehyde. However, the sensitivity of *in situ* hybridization of Protocol 1 is lower than of Protocols 2 and 3. Protocol 2 is applicable when studying gene expression in neuronal tissues, yielding a good morphology and a high sensitivity of the *in situ* hybridization, but can also be used for other tissues, such as the gastrointestinal tract or adrenal gland. In neuronal tissues, such as the brain, spinal cord and various ganglia, Protocol 3 offers the highest sensitivity (approximately twice that of Protocol 2); a prerequisite for successful hybridization with this protocol, though, is a low abundance of RNase, since no measures in the tissue handling or in the protocol inhibit RNase activity.

The permeabilization with Triton enhances the penetration of the probes into the tissue. Similarly, the enzymatic degradation by proteinase K increases the access of the probes to the target mRNA. This step should be optimized for each individual probe, target mRNA and tissue, by trying different concentrations of enzyme and time of incubation. Also, the activity of the enzyme may vary between suppliers. Alternatively, pronase, pepsin or HCl can also be used in this protocol. The post-fixation with paraformaldehyde inhibits the enzymatic degradation; the time of fixation should also be optimized. Acetylation has been introduced on the basis that positively charged tissue components will electrostatically attract the negatively charged probe molecules. Instead, these sites are blocked by acetic anhydride prior to hybridization. The hybridization temperature is of crucial importance. If the experiments are hampered by unspecific background, it can be reduced by increasing the hybridization temperature to 42°C; however, this may also result in a decrease of specific hybridization signal. The concentration of formamide in the hybridization buffer also affects the stringency; for example, it can be reduced to 30% and the hybridization signal will increase. However, the expense may be an increased unspecific background. The stringency of the post-hybridization washes can also be increased (reduce SSC or increase temperature) to reduce unspecific background, although in our experience this is less effective than increasing the hybridization temperature.

Degradation of nucleic acids by RNases is a problem when working with *in situ* hybridization but it should not be overemphasized. It is beneficial to keep RNA work separate from other laboratory work, the handling of animals in particular. The investigator should wear gloves and a clean robe. All glassware used up until hybridization should be RNase-free. We no longer use RNase-free solutions, meaning solutions that are treated with diethylpyrocarbonate, which is in turn heat-inactivated. Instead, we have found that double-distilled water that has been autoclaved works equally well. It is imperative that all work with the autoradiographic emulsion be performed under safelight conditions; this must be checked regularly, since any light exposure will increase background and possibly even ruin the experiments if it is massive. It is a good idea to run experiments in duplicate or triplicate so that one set of slides can be developed after a short time to evaluate the optimal exposure time required for the experiment.

## CONCLUDING REMARKS

Cytochemical techniques are extremely valuable for studies on gene expression in neuroendocrine cells. Immunocytochemistry, when applied as single, double or triple immunofluorescence, reveals the chemical coding of neuroendocrine cells, by use of multiple antibodies detecting neuroendocrine messengers in cells simultaneously. Our results from the islets emphasize the importance of double or triple immunostaining, in that the occurrence of IAPP in somatostatin cells would not have been recognized otherwise. *In situ* hybridization localizes mRNA in tissue sections and is a complement to immunocytochemical techniques. *In situ* hybridization can be performed in combination with immunoperoxidase and, thus, simultaneously reveal the presence of mRNA and translated peptide/protein in the labeled endocrine cells. Combined *in situ* hybridization and immunocytochemistry will aid in securing the authenticity of both the immunoreactive and hybridization signal. Moreover, while immunocytochemistry is a predominantly qualitative technique, *in situ* hybridization can be quantified by computerized image analysis, yielding assessments of changes in gene expression at the level of individual cells. Thus, the morphologist is provided with a tool to monitor discrete up- or down-regulations of mRNA levels. These techniques allow a more thorough understanding of the function of neuroendocrine cells at the molecular level. Since the techniques may be used in research focused on the function of neuroendocrine cells under various conditions, they offer important tools for establishing the physiology and pathophysiology of the neuroendocrine system. We expect these techniques to be central to integrative research in the future.

## ACKNOWLEDGMENTS

The studies on which this review was based were supported by the Swedish Medical Research Council (Project No. 12X-4499, No. 14X-6834 and No. 12X-712), the Swedish Diabetes Association, the Novo Nordisk, Albert Påhlsson, Wiberg, Ernhold Lundström and Craford Foundations and by the Faculty of Medicine, University of Lund. The authors thank Dr. Per Westermark, Linköping, Sweden, for providing a polyclonal rat IAPP antibody and Dr. Amy Pearce, Dr. Mark Fineman and Dr. Joy Koda (Amylin Pharmaceuticals, San Diego, CA, USA) for providing the monoclonal human IAPP/ amylin antibody. Also, we thank Dr. Jan Holst, Malmö, Sweden, for providing specimens from human pancreas. These studies could not have been performed without the technical assistance of Lilian Bengtsson, Charlotte Daun, Lena Kvist, Ann-Christin Lindh, Doris Persson and Iréne Reimertz.

## REFERENCES

1. **Ahrén, B. and G. Sundkvist.** 1995. Long-term effects of alloxan in mice. Int. J. Pancreatol. *17*:197-201.
2. **Ahrén, B., G. Sundkvist, H. Mulder and F. Sundler.** 1996. Blockade of muscarinic transmission increases the frequency of diabetes after low-dose alloxan challenge in the mouse. Diabetologia *39*:388-390.
3. **Ahrén, B. and F. Sundler.** 1992. Localization of calcitonin gene-related peptide and islet amyloid polypeptide in the rat and mouse pancreas. Cell Tissue Res. *269*:315-322.
4. **Ar'Rajab, A. and B. Ahrén.** 1991. Effects of amidated rat islet amyloid polypeptide on glucose-stimulated insulin secretion in vivo and in vitro in rats. Eur. J. Pharmacol. *192*:443-445.
5. **Ar'Rajab, A. and B. Ahrén.** 1993. Long-term diabetogenic effect of streptozotocin in rats. Pancreas *8*:50-57.
6. **Beaumont, K., M.A. Kenney, A.A. Young and T.J. Rink.** 1993. High affinity amylin binding sites in rat brain. Mol. Pharmacol. *44*:493-497.
7. **Bennet, W.M., C.S. Beis, M.A. Ghatei, P.G.H. Byfield and S.R. Bloom.** 1993. Amylin tonally regulates arginine-stimulated insulin secretion in rats. Diabetologia *37*:436-438.
8. **Betsholtz, C., L. Christmanson, U. Engström, F. Rorsman, K. Jordan, T.D. O'Brien, M. Murtaugh, K.H. Johnson and P. Westermark.** 1990. Structure of cat islet amyloid polypeptide and identification of amino acid residues of potential significance for islet amyloid formation. Diabetes *39*:118-122.
9. **Betsholtz, C., L. Christmanson, U. Engström, F. Rorsman, V. Svensson, K.H. Johnson and P. Westermark.** 1989. Sequence divergence in a specific region of islet amyloid polypeptide (IAPP) explains differences in islet amyloid formation between species. FEBS Lett. *251*:261-264.
10. **Bhogal, R., D.M. Smith and S.R. Bloom.** 1992. Investigation and characterization of binding sites for islet amyloid polypeptide in rat membranes. Endocrinology *130*:906-913.
11. **Bretherton-Watt, D., M.A. Ghatei, S.R. Bloom, H. Jamal, G.J. Ferrier, S.I. Girgis and S. Legon.** 1989. Altered islet amyloid polypeptide (amylin) gene expression in rat models of diabetes. Diabetologia *32*:881-883.
12. **Chen, L., I. Komiya, L. Inman, K. McCorkle, T. Alam and R.H. Unger.** 1989. Molecular and cellular responses of islets during perturbations of glucose homeostasis determined by in situ hybridization histochemistry. Proc. Natl. Acad. Sci. USA *86*:1367-1371.
13. **Chen, L., I. Komiya, J. O'Neil, M. Appel, T. Alam and R.H. Unger.** 1989. Effects of hypoglycemia and prolonged fasting on insulin and glucagon gene expression: studies with in situ hybridization. J. Clin. Invest. *84*:711-714.
14. **Christmanson, L., F. Rorsman, G. Stenman, P. Westermark and C. Betsholtz.** 1990. The human islet amyloid polypeptide (IAPP) gene. Organization, chromosomal localization and functional identification of a promoter region. FEBS Lett. *267*:160-166.
15. **Cooper, G.J.S.** 1994. Amylin compared with calcitonin gene-related peptide: structure, biology, and relevance to metabolic disease. Endocr. Rev. *15*:163-201.
16. **Cooper, G.J., A.C. Willis, A. Clark, R.C. Turner, R.B. Sim and K.B. Reid.** 1987. Purification and characterization of a peptide from amyloid-rich pancreases of type 2 diabetic patients. Proc. Natl. Acad. Sci. USA *84*:8628-8632.
17. **Dagerlind, Å., K. Friberg, A.J. Bean and T. Hökfelt.** 1992. Sensitive mRNA detection using unfixed tissue: combined radioactive and non-radioactive in situ hybridization histochemistry. Histochemistry *98*:39-49.
18. **De Koning, E.J.P., E.R. Morris, F.M.A. Hofhuis, G. Posthuma, J.W.M. Höppener, J.F. Morris, P.J.A. Capel, A. Clark and J.S. Verbeek.** 1994. Intra- and extracellular amyloid fibrils are formed in cultured pancreatic islets of transgenic mice expressing human islet amyloid polypeptide. Proc. Natl. Acad. Sci. USA *91*:8467-8471.

19. **De Vroede, M., A. Foriers, M. Van de Winkel, O. Madsen and D. Pipeleers.** 1992. Presence of islet amyloid polypeptide in rat islet B and D cells determines parallelism and dissociation between rat pancreatic islet amyloid polypeptide and insulin content. Biochem. Biophys. Res. Commun. *182*:886-893.

20. **DeFronzo, R.A., R.C. Bonadonna and E. Ferrannini.** 1992. Pathogenesis of NIDDM. A balanced overview. Diabetes Care *15*:318-368.

21. **Denijn, M., R.A. De Weger, A.D. Van Mansfeld, J.A. van Unnik and C.J. Lips.** 1992. Islet amyloid polypeptide (IAPP) is synthesized in the islets of Langerhans. Detection of IAPP polypeptide and IAPP mRNA by combined in situ hybridization and immunohistochemistry in rat pancreas. Histochemistry *97*:33-37.

22. **Fan, L., G. Westermark, S.J. Chan and D.F. Steiner.** 1994. Altered gene structure and tissue expression of islet amyloid polypeptide in the chicken. Mol. Endocrinol. 8:713-721.

23. **Ferrier, G.J., A.M. Pierson, P.M. Jones, S.R. Bloom, S.I. Girgis and S. Legon.** 1989. Expression of the rat amylin (IAPP/DAP) gene. J. Mol. Endocrinol. *3*:R1-R4.

24. **Fox, N., J. Schrementi, M. Nishi, S. Ohagi, S.J. Chan, J.A. Heisserman, G.T. Westermark, A. Leckström, P. Westermark and D.F. Steiner.** 1993. Human islet amyloid polypeptide transgenic mice as a model of non-insulin-dependent diabetes mellitus (NIDDM). FEBS Lett. *323*:40-44.

25. **Frontoni, S., S.B. Choi, D. Banduch and L. Rossetti.** 1991. In vivo insulin resistance induced by amylin primarily through inhibition of insulin-stimulated glycogen synthesis in skeletal muscle. Diabetes *40*:568-573.

26. **Fürnsinn, C., H. Leuvenink, M. Roden, P. Nowotny, B. Schneider, M. Rohac, T. Pieber, M. Clodi and W. Waldhäusl.** 1994. Islet amyloid polypeptide inhibits insulin secretion in conscious rats. Am. J. Physiol. *267*:E300-E305.

27. **Gall, G. and M.L. Pardue.** 1969. Formation and detection of RNA-DNA hybrid molecules in cytological preparations. Proc. Natl. Acad. Sci. USA *63*:378-381.

28. **Gerfen, G.R., J.F. McGinty and W.S. Young III.** 1991. Dopamine differentially regulates dynorphin, substance P, and enkephalin expression in striatal neurons: in situ hybridization histochemical analysis. J. Neurosci. *11*:1016-1031.

29. **German, M.S., L.G. Moss, J. Wang and W.J. Rutter.** 1992. The insulin and islet amyloid polypeptide genes contain similar cell-specific promoter elements that bind identical beta-cell nuclear complexes. Mol. Cell. Biol. *12*:1777-1788.

30. **Hiramatsu, S., K. Inoue, Y. Sako, F. Umeda and H. Nawata.** 1994. Insulin treatment improves relative hypersecretion of amylin to insulin in rats with non-insulin-dependent diabetes mellitus induced by neonatal streptozocin injection. Metabolism *43*:766-770.

31. **Höfler, H., H. Childers, M.R. Montminy, R.M. Lechan, R.H. Goodman and H.J. Wolfe.** 1986. In situ hybridization methods for the detection of somatostatin mRNA in tissue sections using anti-sense RNA probes. Histochem. J. *18*:597-602.

32. **Höppener, J.W., J.S. Verbeek, E.J.P. de Koning, C. Oosterwijk, K.L. van Hulst, H.J. Visser Vernooy, F.M.A. Hofhuis, S. van Gaalen, M.J.H. Berends, W.H.L. Hackeng, H.S. Jansz, J.F. Morris and A. Clark.** 1993. Chronic overproduction of islet amyloid polypeptide/amylin in transgenic mice: lysosomal localization of human islet amyloid polypeptide and lack of marked hyperglycaemia or hyperinsulinaemia. Diabetologia *36*:1258-1265.

33. **Inoue, K., A. Hisatomi, F. Umeda and H. Nawata.** 1992. Relative hypersecretion of amylin to insulin from rat pancreas after neonatal STZ treatment. Diabetes *41*:723-727.

34. **Ishida-Yamamoto, A. and M. Tohyama.** 1989. Calcitonin gene-related peptide in the nervous system. Prog. Neurobiol. *33*:335-386.

35. **Johnson, K.H., T.D. O'Brien, D.W. Hayden, K. Jordan, H.K. Ghobrial, W.C. Mahoney and P. Westermark.** 1988. Immunolocalization of islet amyloid polypeptide (IAPP) in pancreatic beta cells by means of peroxidase-antiperoxidase (PAP) and protein A-gold techniques. Am. J. Pathol. *130*:1-8.

36. **Jordan, K., M.P. Murtaugh, T.D. O'Brien, P. Westermark, C. Betsholtz and K.H. Johnson.** 1990. Canine IAPP cDNA sequence provides important clues regarding diabetogenesis and amyloidogenesis in type 2 diabetes. Biochem. Biophys. Res. Commun. *169*:502-508.

37. **Kahn, S.E., W.Y. Fujimoto, D.A. D'Alessio, J.W. Ensinck and D. Porte, Jr.** 1991. Glucose stimulates and potentiates islet amyloid polypeptide secretion by the B-cell. Horm. Metab. Res. *23*:577-580.

38. **Kitamura, K., K. Kangawa, M. Kawamoto, Y. Ichiki, S. Nakamura, H. Matsuo and T. Eto.** 1993. Adrenomedullin: a novel hypotensive peptide isolated from human pheochromocytoma. Biochem. Biophys. Res. Commun. *192*:553-560.

39. **Kiyama, H. and P.C. Emson.** 1990. Distribution of somatostatin mRNA in the rat nervous sytem as visualized by a novel non-radioactive in situ hybridization histochemistry procedure. Neuroscience *38*:223-244.

40. **Larsson, H. and B. Ahrén.** 1995. Effects of arginine on the secretion of insulin and islet amyloid

135

polypeptide in humans. Pancreas *11*:201-205.

41. **Lawrence Jr, J.C. and J.N. Zhang.** 1994. Control of glycogen synthase and phosphorylase by amylin in rat skeletal muscle. Hormonal effects on the phosphorylation of phosphorylase and on the distribution of phosphate in the synthase subunit. J. Biol. Chem. *269*:11595-11600.

42. **Leffert, J.D., C.B. Newgard, H. Okamoto, J.L. Milburn and K.L. Luskey.** 1989. Rat amylin: cloning and tissue-specific expression in pancreatic islets. Proc. Natl. Acad. Sci. USA *86*:3127-3130.

43. **Leighton, B. and G.J. Cooper.** 1988. Pancreatic amylin and calcitonin gene-related peptide cause resistance to insulin in skeletal muscle in vitro. Nature *335*:632-635.

44. **Lukinius, A., E. Wilander, G.T. Westermark, U. Engström and P. Westermark.** 1989. Co-localization of islet amyloid polypeptide and insulin in the B cell secretory granules of the human pancreatic islets. Diabetologia *32*:240-244.

45. **McCabe, J.T., J.I. Morrell and D.W. Pfaff.** 1986. In situ hybridization as a quantitative autoradiographic method: vasopressin and oxytocin gene transcription in the Brattleboro rat. p. 73-96. *In* G.R. Uhl (Ed.), In Situ Hybridization in the Brain. Plenum Press, New York.

46. **Miyazato, M., M. Nakazato, K. Shiomi, J. Aburaya, H. Toshimori, K. Kangawa, H. Matsuo and S. Matsukura.** 1991. Identification and characterization of islet amyloid polypeptide in mammalian gastrointestinal tract. Biochem. Biophys. Res. Commun. *181*:293-300.

47. **Mosselman, S., J.W. Höppener, C.J. Lips and H.S. Jansz.** 1989. The complete islet amyloid polypeptide precursor is encoded by two exons. FEBS Lett. *247*:154-158.

48. **Mosselman, S., J.W. Höppener, J. Zandberg, A.D. van Mansfeld, A.H. Geurts van Kessel, C.J. Lips and H.S. Jansz.** 1988. Islet amyloid polypeptide: identification and chromosomal localization of the human gene. FEBS Lett. *239*:227-232.

49. **Mulder, H., B. Ahrén, M. Stridsberg and F. Sundler.** 1995. Non-parallelism of islet amyloid polypeptide (amylin) and insulin gene expression in rat islets following dexamethasone treatment. Diabetologia *38*:395-402.

50. **Mulder, H., B. Ahrén and F. Sundler.** 1995. Differential expression of islet amyloid polypeptide (amylin) and insulin in experimental diabetes in rodents. Mol. Cell. Endocrinol. *114*:101-109.

51. **Mulder, H., A. Leckström, R. Uddman, P. Westermark and F. Sundler.** 1995. Islet amyloid polypeptide (amylin) is expressed in sensory neurons. J. Neurosci. *15*:7626-7632.

52. **Mulder, H., A.C. Lindh, E. Ekblad, P. Westermark and F. Sundler.** 1994. Islet amyloid polypeptide is expressed in endocrine cells of the gastric mucosa in the rat and mouse. Gastroenterology *107*:712-719.

53. **Mulder, H., A.C. Lindh and F. Sundler.** 1993. Islet amyloid polypeptide gene expression in the endocrine pancreas of the rat: a combined in situ hybridization and immunocytochemical study. Cell Tissue Res. *274*:467-474.

54. **Nicholl, C.G., J.M. Bhatavdekar, J. Mak, S.I. Girgis and S. Legon.** 1992. Extra-pancreatic expression of the rat islet amyloid polypeptide (amylin) gene. J. Mol. Endocrinol. *9*:157-163.

55. **Nishi, M., S.J. Chan, S. Nagamatsu, G.I. Bell and D.F. Steiner.** 1989. Conservation of the sequence of islet amyloid polypeptide in five mammals is consistent with its putative role as an islet hormone. Proc. Natl. Acad. Sci. USA *86*:5738-5742.

56. **Nishi, M., T. Sanke, S. Seino, R.L. Eddy, Y.S. Fan, M.G. Byers, T.B. Shows, G.I. Bell and D.F. Steiner.** 1989. Human islet amyloid polypeptide gene: complete nucleotide sequence, chromosomal localization, and evolutionary history. Mol. Endocrinol. *3*:1775-1781.

57. **Novials, A., Y. Sarri, R. Casamitjana, F. Rivera and R. Gomis.** 1993. Regulation of islet amyloid polypeptide in human pancreatic islets. Diabetes *42*:1514-1519.

58. **O'Brien, T.D., P. Westermark and K.H. Johnson.** 1991. Islet amyloid polypeptide and insulin secretion from isolated perfused pancreas of fed, fasted, glucose-treated, and dexamethasone-treated rats. Diabetes *40*:1701-1706.

59. **Opie, E.L.** 1901. On the relation of chronic interstitial pancreatitis to the islands of Langerhans and to diabetes mellitus. J. Exp. Med. *5*:397-428.

60. **Pieber, T.R., D.T. Stein, A. Ogawa, T. Alam, M. Ohneda, K. McCorkle, L. Chen, J.D. McGarry and R.H. Unger.** 1993. Amylin-insulin relationships in insulin resistance with and without diabetic hyperglycemia. Am. J. Physiol. *265*:E446-E453.

61. **Porte Jr, D.** 1991. B cells in type 2 diabetes mellitus. Diabetes *40*:166-180.

62. **Sexton, P.M., J.S. McKenzie and F.A.O. Mendelsohn.** 1988. Evidence for a new subclass of calcitonin/calcitonin gene-related peptide binding site in rat brain. Neurochem. Int. *12*:323-335.

63. **Sexton, P.M., G. Paxinos, M.A. Kenney, P.J. Wookey and K. Beaumont.** 1994. In vitro autoradiographic localization of amylin binding sites in rat brain. Neuroscience *62*:553-567.

64. **Stojanovska, I., R. Gennaro and J. Proietto.** 1990. Evolution of dexamethasone-induced insulin resistance in rats. Am. J. Physiol. *258*:E748-E756.

65. **Stridsberg, M., S. Sandler and E. Wilander.** 1993. Cosecretion of islet amyloid polypeptide (IAPP) and insulin from isolated rat pancreatic islets following stimulation or inhibition of beta-cell function. Regul. Pept. *45*:363-370.

66. **Toshimori, H., R. Narita, M. Nakazato, J. Asai, T. Mitsukawa, K. Kangawa, H. Matsuo and S. Matsukura.** 1990. Islet amyloid polypeptide (IAPP) in the gastrointestinal tract and pancreas of man and rat. Cell Tissue Res. *262*:401-406.

67. **Van Gompel, J., T. Mahler, M. De Paepe and G. Klöppel.** 1993. Comparisons of in situ hybridization and immunocytochemistry for the detection of residual beta cells in the pancreas of streptozotocin-treated rats. Acta Diabetol. *30*:118-122.

68. **Van Mansfeld, A.D., S. Mosselman, J.W. Höppener, J. Zandberg, H.A. van Teeffelen, P.D. Baas, C.J. Lips and H.S. Jansz.** 1990. Islet amyloid polypeptide: structure and upstream sequences of the IAPP gene in rat and man. Biochim. Biophys. Acta. *1087*:235-240.

69. **Van Rossum, D., D.P. Menard, A. Fournier, S. St-Pierre and R. Quirion.** 1994. Autoradiographic distribution and receptor binding profile of [I-125]Bolton Hunter-rat amylin binding sites in the rat brain. J. Pharmacol. Exp. Ther. *270*:779-787.

70. **Verchere, C.B., D.A. Dalessio, R.D. Palmiter and S.E. Kahn.** 1994. Transgenic mice overproducing islet amyloid polypeptide have increased insulin storage and secretion in vitro. Diabetologia *37*:725-728.

71. **Wang, Z.L., W.M. Bennet, M.A. Ghatei, P.G. Byfield, D.M. Smith and S.R. Bloom.** 1993. Influence of islet amyloid polypeptide and the 8-37 fragment of islet amyloid polypeptide on insulin release from perifused rat islets. Diabetes *42*:330-335.

72. **Westermark, P., D.L. Eizirik, D.G. Pipeleers, C. Hellerström and A. Andersson.** 1995. Rapid deposition of amyloid in human islets transplanted into nude mice. Diabetologia *38*:543-549.

73. **Westermark, P., U. Engström, K.H. Johnson, G.T. Westermark and C. Betsholtz.** 1990. Islet amyloid polypeptide: pinpointing amino acid residues linked to amyloid fibril formation. Proc. Natl. Acad. Sci. USA 87:5036-5040.

74. **Westermark, P. and L. Grimelius.** 1973. The pancreatic islet cell in insular amyloidosis in human diabetic and non-diabetic adults. Acta Pathol. Microbiol. Scand. [A] *81*:291-300.

75. **Westermark, P., K.H. Johnson, T.D. O'Brien and C. Betsholtz.** 1992. Islet amyloid polypeptide—a novel controversy in diabetes research. Diabetologia *35*:297-303.

76. **Westermark, P., C. Wernstedt, T.D. O'Brien, D.W. Hayden and K.H. Johnson.** 1987. Islet amyloid in type 2 human diabetes mellitus and adult diabetic cats contains a novel putative polypeptide hormone. Am. J. Pathol. *127*:414-417.

77. **Westermark, P., C. Wernstedt, E. Wilander, D.W. Hayden, T.D. O'Brien and K.H. Johnson.** 1987. Amyloid fibrils in human insulinoma and islets of Langerhans of the diabetic cat are derived from a neuropeptide-like protein also present in normal islet cells. Proc. Natl. Acad. Sci. USA *84*:3881-3885.

78. **Westermark, P., C. Wernstedt, E. Wilander and K. Sletten.** 1986. A novel peptide in the calcitonin gene related peptide family as an amyloid fibril protein in the endocrine pancreas. Biochem. Biophys. Res. Commun. *140*:827-831.

79. **Yamamoto, H., Y. Uchigata and H. Okamoto.** 1981. Streptozotocin and alloxan induce DNA strand breaks and poly(ADP-ribose) synthetase in pancreatic islets. Nature *294*:284-285.

80. **Young, A.A.** 1994. Amylin regulation of fuel metabolism. J. Cell. Biochem. *55*:12-18.

81. **Young, A.A., P. Carlo, T.J. Rink and M.W. Wang.** 1992. 8-37hCGRP, an amylin receptor antagonist, enhances the insulin response and perturbs the glucose response to infused arginine in anesthetized rats. Mol. Cell. Endocrinol. *84*:R1-R5.

82. **Young, A.A., B. Gedulin, L.S.L. Gaeta, K.S. Prickett, K. Beaumont, E. Larson and T.J. Rink.** 1994. Selective amylin antagonist suppresses rise in plasma lactate after intravenous glucose in the rat - Evidence for a metabolic role of endogenous amylin. FEBS Lett. *343*:237-241.

83. **Young, A.A., B. Gedulin, D. Wolfe Lopez, H.E. Greene, T.J. Rink and G.J. Cooper.** 1992. Amylin and insulin in rat soleus muscle: dose responses for cosecreted noncompetitive antagonists. Am. J. Physiol. *263*:E274-E281.

# CHAPTER 7

# Fluorescence *In Situ* Hybridization

**Sunny Luke[1,3], Victoria Belogolovkin[1,2], Jerry A. Varkey[1,3] and Charles T. Ladoulis[1]**

[1]Department of Pathology, Maimonides Medical Center, Brooklyn, NY; [2]School of Medicine, SUNY Health Science Center at Brooklyn, Brooklyn, NY; [3]Department of Biology, Adelphi University, Garden City, NY, USA

## SUMMARY

*During the last decade, pathology has progressed remarkably with the incorporation of molecular techniques into the arena of diagnostics. Fluorescence* in situ *hybridization (FISH) is a molecular histopathological technique that can be readily used to identify DNA or RNA abnormalities at the cellular level with the use of an epifluorescence microscope. FISH technology is applicable to most forms of tissue preparation including peripheral blood, bone marrow, solid tumors, body fluids, amniotic fluids and products of conception. This technology enables the identification of gene, chromosomal and genomic abnormalities in histologically processed material, exfoliated cells, smears and monolayer cultures with appropriate fluorescently labeled DNA probes. The following chapter provides a broad overview and extensive technical details of FISH technology, thereby providing the reader with a complete picture of the field. In light of the applicability of this technology to interphase cells, FISH can serve to provide the pathologist with a genotypic analysis to accompany the phenotypic morphology of the cells for the purpose of diagnosis and prognosis. The utility of such a correlation defines FISH technology as a revolutionary technique in the field of research and diagnostics, particularly in areas of cancer, genetic abnormalities and infectious diseases.*

## INTRODUCTION

Ever since a correlation was found between the pathogenesis of diseases and genomic alterations, molecular cytogenetic techniques have found a place in molecular medicine. These techniques are used in tracing gene and genomic abnormalities that underlie the development of cancer and genetic diseases (55,82,126). Fluorescence *in situ* hybridization (FISH) has emerged as one of the most important molecular techniques for visualizing specific DNA or RNA sequences in chromosomal, nuclear and cytoplasmic compartments of a cell (25,58,106). FISH is based on the principle that target nucleic acid sequences can hybridize with labeled probes of complimentary sequences whereby the resulting hybridization sites can be observed using a fluorescent microscope. As a result, this technology is routinely used for detecting chromosomal and gene abnormalities without growing cells in culture, a field known as "interphase cytogenetics" (4,44,116). The high sensitivity and specificity of this technology has made it possible to localize cloned genes to specific chromosomal loci by physical mapping as well as in identifying micro-deletions/micro-duplications of chromosomes that are the cause of a number of contiguous gene syndromes (28). FISH is further exploited to assess the amplification or deletion of oncogenes and tumor suppressor genes together with their subsequent expression by mRNA hybridization (12,37,45). In infectious diseases, the technique of FISH is used to identify the presence of viral nucleic acids in individual cell populations of the host.

In 1969 Gall and Pardue introduced a technique known as *in situ* hybridization (ISH), to localize specific nucleic acids in individual cells (26). In those early years, the capabilities of ISH were limited to highly repetitive DNA sequences using radioactively labeled probes that were subsequently visualized by autoradiography. Initially nucleic acid probes were labeled with $^{32}P$, $^{35}S$, $^{3}H$ and $^{125}I$. However, there were clear disadvantages in using radioisotopically labeled probes, such as the exposure to radioactive materials, the turnaround time that took several weeks and the presence of high background requiring extraneous statistical analysis. In the ensuing years, efforts were taken to replace the drawbacks associated with radiolabeled probes with other label modalities and detection systems. These include various haptens, such as biotin, digoxigenin, rhodamine, bromodeoxy uridine, acetylaminofluorene, sulfonate and mercury trinitrophenol with a corresponding detection system (43,85). The described haptens are incorporated into the DNA or RNA by nick translation or, in some cases, by simple chemical reaction for probe antigenicity. The tagged probes are then detected with the corresponding antibody against the specific tag or, as in the case of biotin, with a fluorescently labeled avidin molecule. The use of such fluorimetric or colorimetric systems served to visualize the location and improve the spatial resolution of the DNA sequences of interest.

While FISH technology was first utilized in the early 80s, the last eight years have witnessed tremendous modifications. The innovations include: (1) chromosomal *in situ* suppression (CISS) hybridization for painting human chromosomes using flow-sorted chromosome libraries (22,59); (2) detection of hybridization signals using flow cytometry in a suspension of cells called FISHES (fluorescence *in situ* hybridization en suspension) (57); (3) forward and reverse chromosome painting (41); (4) comparative genomic hybridization (CGH) for genome-wide screening of DNA sequence copy number aberrations in tumors (40,44); (5) *in situ* PCR for detecting and amplifying viral DNA or RNA sequences in infected cells (58); (6) mitochondrial *in situ* hybridization (115); (7) primed *in situ* DNA synthesis (PRINS) as an alternative to *in situ* hybridization (32); and finally, (8) multicolor spectral karyotyping (SKY), which allows the simultaneous visualization of all 23 pairs of human chromosomes (96).

The evolution of recombinant DNA technology along with the development of fluorochromes, PCR, flow cytometry, fluorescent reagents, multi-parameter fluorescence imaging microscopy, spectroscopy and charged coupled device (CCD) imaging has laid the foundation for the emergence of FISH as an excellent tool for the identification of molecular markers that have both clinical and research applications (25,29). To introduce the reader to the wide applicability of this technique, technological details are included in this chapter.

## TECHNICAL ASPECTS

### Probes

The requirements for successful ISH begin with the selection of appropriate DNA or RNA probes. Human cells contain 6 billion nucleotides, including an abundance of RNA transcripts representing those portions of DNA that are actively being expressed. The advent of cloning, somatic cell hybridization, microdissection, micro-cloning, PCR and flow cytometry has made it possible to generate DNA probes from specific genes, DNA sequence families (localized and dispersed), chromosomal bands, and whole human chromosomes (74). The size of the DNA fragment that can be cloned ranges from 1 kilobase to 1000 kilobases. As a result, the size of the fragment will be the basis for determining the type of vector used in cloning; plasmids (insert limit of approximately 10 kilobases), cosmids (insert limit of approximately 40 kilobases), cosmid contigs (insert limit of approximately 60 kilobases), PI artificial chromosomes (insert limit of approximately 150 kilobases) and finally yeast artificial chromosomes (YAC) (insert limit of approximately 200–1000 kilobases) (35).

FISH probes can be grouped into eight categories depending on their genomic origin: (1) Centromere-specific satellite probes, which include alpha, beta and classical satellite probes that are complimentary to localized repetitive DNA sequence families at the centromeric and pericentromeric regions of human chromosomes. Alpha satellite (alphoid DNA) probes are the more common probes used in FISH because their monomers are diverged enough to provide centromeric specificity for the chromosome in question. These centromeric probes are applicable to both metaphase and interphase nuclei, allowing the rapid identification of chromosomal aneuploidies (121) (Figure 2, A and B). (2) Gene probes, which can be applied to both metaphase and interphase cells, are useful in assessing gene abnormalities (61,110). (3) Telomeric probes are a class of FISH probes containing repeating units of the six base pair sequence TTAGGG. These probes are used in FISH analysis for detecting abnormalities localized to the telomeric regions of chromosomes (64,92). Most recently, FISH probes representing the subtelomeric regions located approximately 100–300 kilobases from the ends of each chromosome containing unique sequences for each pair of chromosomal arms (short arm, p, and long arm, q) were discovered. They were cloned using half-YAC cloning technology. Consequently, these new FISH probes can be used effectively to identify cryptic translocations characteristic of several congenital diseases and the breakage and healing of cancer-specific chromosomes during clonal evolution (127). (4) Whole chromosome painting probes (WCPP) consist of a series of probes complimentary to unique sequences located at specific sites along the length of a single chromosome. When WCPP are used in a single hybridization, the length of the entire chromosome is "painted". WCPP's are particularly useful in identifying translocations of clinical significance when applied to cancer genetics for the purposes of screening chromosomal structural alterations of unknown origins. Furthermore, WCPPs are applicable to pre- and postnatal testing for the identification of chromosomal translocation (48,59,67) (Figure 2D). (5) Interspersed repetitive sequences (IRS) probes: the presence of such sequences in the human genome, Alu or LI repetitive families in particular, can be cloned or PCR-generated to produce Alu or LI FISH probes. Upon hybridization, the Alu probe produces R-bands, while LI probes create G-bands on human metaphase chromosomes (54). As a result, a combination of IRS probes with a gene probe using double-color hybridization enables the simultaneous localization of a particular gene to a specific chromosomal band (6). (6) Locus-specific probes for the critical regions of human chromosomes are established by creating regional-specific libraries from microdissected chromosomal bands. These probes are useful in detecting micro-deletions and micro-duplications characteristic of contiguous gene syndromes (47,66) (Figure 2, C and F). (7) Total genomic probes are produced by labeling isolated total human DNA. This is used for screening human chromosomes in somatic cell hybrids, in evolutionary studies, and in identifying non-spermic cells in semen (68). (8) Riboprobes (RNA probes) are single-stranded antisense or sense RNA probes. Antisense RNA probes are used for the detection and evaluation of gene expression by mRNA ISH. RNA probes with sequences identical to its target mRNA are also available as sense RNA probes (21,107). Single-stranded DNA cloned in the M13 vector, oligonucleotide (oligoprobes) and PCR-generated probes are also available for use in FISH (78).

**Probe Labeling**

Various nonradioactive labeling kits are commercially available for the efficient labeling of as little as 10 ng to as much as 3 µg of DNA per standard assay. Probe DNA can be labeled by nick translation, random priming of oligonucleotides, and/or photo activation (43). However, the majority of probe labeling for FISH technology is accomplished through nick translation. Pinkel et al. developed FISH in 1986 by incorporating biotin (Vitamin H) directly into the DNA (85). This was followed by the incorporation of another chemical, digoxigenin, which is a steroid derived from the plant *digitalis*. Both biotin and digoxigenin are important haptens ("reporter molecules") used to label DNA and are presently used in FISH technology; they provide distinct signals with a characteristically low background (high signal-to-noise ratio). However, directly conjugated fluorescently labeled nucleotides like FITC-dUTP (FITC = fluorescein isothiocyonate) and Texas-Red -dUTP (Texas-Red = Sulphorhodamine 101) are also used as direct labels. The major drawback associated with the use of a directly labeled fluorochrome probe is a weak hybridization signal. Oligonucleotides are labeled by 3' end tailing with terminal deoxynucleotidyl transferase (43,49).

After the target nucleotides have been labeled, the probe DNA can be separated by gel filtration using a spin column. One of the most important factors for successful *in situ* hybridization is sufficient incorporation of labeled nucleotide into the probe DNA as well as proper probe fragment size. Ideally probe fragments should be between 200–600 nucleotides in length. Probe fragmentation can be adjusted by varying the DNase concentration during the nick translation reaction where the resulting fragment size can subsequently be estimated by agarose gel electrophoresis (43,49,92).

## Specimen Preparation

FISH is applicable to practically all forms of tissue preparation including single-cell suspensions obtained from fresh solid tumors; frozen sections; formalin-fixed, paraffin-embedded blocks of preserved solid tumors (95,116); chromosomes obtained from growing cells derived from solid tissue (116,122); and PHA-stimulated peripheral blood (70). FISH can also be performed on cytospin preparations of body fluids, fine needle aspirates (FNA) (123), bone marrow smears, blood smears, human sperms, blastomeres (30) and amniotic fluid cells (89), as well as chorionic villus cells and products of conception (POCs) (113).

**Preparation and pretreatment of single cells.** Single cells obtained from a variety of sources need to be fixed in suitable fixative for optimal nucleic acid preservation. Enzymatic disaggregation of fresh solid tumors can yield malignant cells in suspension. Similarly, amniotic fluids, cytological preparations, imprints, peripheral blood smears and bone marrow aspirates provide single cells for FISH analysis. Fixation should be performed prior to the placement of the cells onto the pretreated slides. Some of the cell fixatives are mehanol:acetic acid (3:1), 70% ethanol, 4% formaldehyde in phosphate-buffered saline (PBS) or a combination of methanol and acetone. In the case of smears and imprints, however, fixation can be carried out by simply dipping the slides in fixative. The type of fixative will vary from one specimen type to another. Following fixation, the cells are spotted onto the pre-cleaned slides. This is followed by enzymatic digestion to improve probe penetration, which will improve the quality of hybridization signals. Moreover, protein digestion may also decrease the background that often results from autofluorescence. Slides with fixed nuclei (fixation removes cytoplasm and some nuclear proteins) should be subjected to a 5-min equilibration at 37°C in PBS followed by incubation in 1–2 mg of pepsin in 100 mL 0.01 *N* HCl for 10 min at 37°C. Following protein digestion, further washing in PBS is necessary before proceeding to hybridization. Care should be taken to optimize pepsin concentration and incubation time (70,87,89,113).

**Preparation and pretreatment of paraffin-embedded tissue sections.** It is a standard practice in most pathology laboratories to routinely fix solid tumors in formalin and embed them in paraffin. FISH procedure can be carried out on such preserved tissue sections. Generally paraffin sections 3- to 4-μm-thick are mounted on poly-L-lysine-coated slides. Deparaffinization can be carried out in xylene followed by washing in 100% methanol. After deparaffinization, pretreatment of tissue can be carried out in 1 M sodium thiocyanate at 45°C for 10 min. This procedure will result in the dissociation of histones from DNA. This step is followed by proteolytic enzyme digestion using proteinase K for 20 min at 45°C. After rinsing slides in double-distilled $H_2O$, dehydration of specimens should be carried out in ascending order of ethanol followed by air-drying (87,95).

**Preparation and pretreatment of chromosomes.** For metaphase FISH, the specimen preparation includes growing cells in T25 culture flasks for fibroblasts as well as tumor cells to attain monolayer according to established tissue culture protocols. On the other hand, lymphocytes in peripheral blood can be grown in suspension after stimulating with phytohaemagglutinin (PHA). Monolayers may be harvested when 80% confluency is attained. However, for PHA-stimulated blood, 72 h is appropriate. After metaphase arrest using colcemid treatment, cells can be harvested for chromosomes by standard hypotonic treatment and fixation with several changes of methanol/acetic acid (3:1 vol/vol). The harvested cells can be dropped onto wet slides for metaphase chromosomes. Slides are air-dried and can be stored before proceeding to FISH. If stored, the slides need to be washed in PBS and dehydrated in ascending order of methanol (75%, 85%, 95%, 100% for 5 min each) before going onto FISH. These slides can be stored in sealed boxes at 4°C. However, old slides stored at room temperature for more than a few days show poor hybridization results (49,73,85).

## Denaturation, ISH and Post-Hybridization Wash

Denaturation is the process by which double-stranded DNA of the probe as well as the target cells are separated into single strands. After the corresponding cell, chromosome and tissue preparations have been properly fixed, they are now ready for denaturation. Prior to denaturation, the probe (10–16 ng DNA probe per slide) is mixed with a hybridization buffer. The majority of hybridization buffers contain 50% (vol/vol) formamide, 10% (wt/vol) dextran sulfate, 2× SSC (standard saline citrate, which is 0.15 M sodium chloride and 0.15 M sodium citrate), 0.1 mM EDTA, 0.5 mM Tris-HCl (pH 7.6), and 100 μg of blocking DNA (unlabeled Cot-1, sheared salmon sperm DNA, or placental DNA). It is important to use blocking DNA when repetitive DNA probes are used. Denaturation of probe DNA is carried out in a probe mixture for 5 min at 71°–95°C. At the same time, slides containing the target DNA in interphase nuclei or chromosome preparations are also denatured in denaturation solution containing 70% formamide, 10% 2× SSC, 20% distilled $H_2O$ in a water bath heated to 71°C. The probe in the hybridization mixture is applied to the corresponding slide, coverslipped and sealed. However, in the case of formalin-fixed, paraffin-embedded tissue specimens, denaturation of probe DNA and tissue DNA is performed simultaneously by applying probe mixture onto the specimen in an oven at 90°C for 10 min. Hybridization is performed overnight in a humidified chamber at 37°C; with this reaction, single-stranded cellular DNA or RNA sequences and complementary sequences of the molecular probe anneal to form double-stranded hybrid molecules. Duration of hybridization varies from probe to probe. For repetitive DNA probes, the hybridization time can be as little as a few hours, while for cosmid gene probes a range of 24 to 36 h are recommended. In general, 12 h of hybridization is sufficient for most probes.

After hybridization, coverslips are removed by agitating in 2× SSC at room temperature. The next step of post-hybridization wash is critical and is referred to as the "stringency wash". Stringency washes serve to remove unbound and mismatched probe sequences from nonspecific targets to avoid cross-hybridization and false-positive results. Under conditions of high stringency only probes with perfect or near perfect homology to target sequences remain stable as hybrids after post-hybridization wash. This strategy is exploited for gene probes as well as for chromosome-specific alphoid and telomeric probes. High-stringency wash conditions are generally achieved with an elevated formamide concentration (65%), an increase in temperature (45°C) and a low salt concentration. A low-stringency wash is preferred when hybridization signals are expected to bind to sequences showing 65%–85% homology as in the centromeric and telomeric probes for all human chromosomes. Post-hybridization washes are generally performed for 20 min with periodic agitation under conditions of high, medium or low stringencies. Detailed studies concerning probe concentration, hybridization buffer, stringency washes, hybridization time and temperatures have been reviewed (49,85,95).

**Probe Detection**

Both direct and indirect systems of probe detection can be employed depending upon the labeling strategy. In directly labeled probes, counterstaining the nuclei with appropriate fluorochromes like propidium iodide (PI) or 6-diaminophenylindole dihydrochloride (DAPI) is sufficient to visualize the site of hybridization using a fluorescent microscope. However, this approach gives limited sensitivity and, as a result, high-quality fluorescent microscope enhancement or, in some cases, computer-based image enhancement is required for signal visualization. On the other hand, indirect detection methods that are similar to indirect immunocytochemical procedures are widely used. In this method, appropriate antibodies raised against the corresponding haptens (biotin or digoxigenin) are employed, yielding signals with a higher quality of resolution. Maximal signal-to-noise ratios can be obtained by incubating cells in blocking reagent to prevent nonspecific staining. Counterstaining of cells in the indirect method of detection is done with PI or DAPI with a suitable anti-fade solution. A procedural flow chart for the detection of biotinylated and digoxigenin-labeled probes are given below.

①*Biotinylated Probe*

➥Blocking reagent➔FITC-conjugated avidin➔biotinylated goat-anti-avidin antibody➔FITC-conjugated avidin➔counterstain with PI or DAPI with anti-fade solution➔fluorescence microscopy

②*Digoxigenin-labeled probe*

➥Blocking reagent➔mouse antibody against digoxigenin➔FITC-conjugated anti-mouse antibody➔counterstain with PI or DAPI with anti-fade solution➔fluorescence microscopy

The protein-fluorophore complex approach as shown for biotin- and digoxigenin-labeled probes has gained wide acceptance because these methods can be carried out in 24 h rather than weeks as required by autoradiography. Additionally, this method is also compatible with nonfluorescent detection systems. If needed, the cycle of biotinylated antibodies and avidin FITC can be repeated to increase the efficiency of hybridization signals. This "sandwich" method of biotin-avidin-fluorochrome was initially reported by Pinkel et al. for avidin-based detection systems and is now applied to single copy genes. Recently, further improvements replacing indirect methods have been reported where direct coupling of fluorophores to probe DNA and detection via anti-fluorochrome antibodies have been reported. Wiegant et al. employed such a system and reported the visualization of probes as small as 1 kilobase. In the case of haptens other than biotin, signal amplification is accomplished by the addition of corresponding primary antibodies like anti-digoxigenin, anti-thymine dimer, anti-dinitrophenol, anti-sulfonate and anti-acetylaminofluorene also yield signal qualities similar to biotin. A significant drawback of fluorescence detection is that signals tend to fade rapidly. The incorporation of an anti-fade solution into the counterstain will serve to reduce fading and allow for prolonged exposure time and microscopic analysis.

A number of reports have shown that CISS hybridization using whole chromosome libraries cloned in blue scribe plasmids is an important diagnostic tool in analyzing structural abnormalities of metaphase chromosomes obtained from malignant as well as other cells grown in culture. The FISH procedure paints the whole chromosome in question. WCCPs directly labeled with fluorochromes are commercially available. Some popular fluorochromes include Spectrum Orange and Spectrum Green. Such probes are

**Table 1. Spectroscopic Properties for Commonly Used Fluorochromes in *In Situ* Hybridization Detection**

| Fluorochrome | Excitation Maxima (nm) | Emission Maxima (nm) |
|---|---|---|
| FITC | 490 | 520 |
| FITC-antibody | 490 | 525 |
| TRITC | 560 | 580 |
| Texas Red (TR) | 590 | 615 |
| CY5 | 646 | 663 |
| Propidium Iodide (PI) | 520 | 610 |
| Quinacrine | 420 | 500 |
| DAPI | 365 | 420 |
| AMCA | 350 | 430 |
| Spectrum Orange | 510 | 590 |
| Bodipy | 505 | 520 |
| EITC | 522 | 540 |

Note: FITC = fluorescein isothiocyanate; TRITC = tetra methyl rhodamine isothiocyanate; CY5 = pentamethene cyanine dye isothiocyante; DAPI = 6-diaminophenylindole dihydrochloride; AMCA = 7-amino-4-methyl eoumarin-3 acetic acid; Bodipy = bora dipyrro methene difluoride; EITC = eosin isothiocyanate.

always mixed with human Cot-1 blocking DNA. Since the target area of hybridization is large, the "painted" chromosome can be visualized approximately one hour after post-hybridization wash. A combination of differently labeled WCPP probes provide the luxury of simultaneous visualization of chromosomal abnormalities between two chromosomes such as balanced and/or unbalanced translocations (1,9,49,85,95).

**Microscopic Visualization and Photography**

FISH signals are analyzed with high-quality fluorescence microscope (Zeiss, Nikon, Olympus, Leitz, etc.). The important components are (1) a high-power HBO 100-W mercury lamp; (2) an appropriate filter set consisting of an excitation filter, dichroic mirror (beam splitter) and an emission filter; (3) objective lenses optimized for fluorescence microscopy; and (4) an image processing system. Fluorochromes are characterized by the capacity to be excited by the light of short wavelengths and then emit light of a higher wavelength. This phenomenon is called "Stoke's shift". As a result, the epifluorescence microscope should be able to take the advantage of Stoke's shift using a dichroic mirror, excitation filter and emission filter using a fluorochrome's excitation maximum and emission maximum. Table 1 gives the details of excitation maximum and emission maximum for different fluorochromes and DNA counterstains generally used in probe labeling and FISH protocols. Figure 1 gives the optics in an epifluorescence microscope directing the light path. A mercury lamp emits light containing different wavelengths. However, upon reaching the excitation filter, only the selected wavelength (shorter wavelength) will be allowed to pass through the specimen by the reflection of the dichroic mirror. Once the specimen is irradiated with shorter wavelengths (excitation radiation), after absorbing some photons, the emitted light (emission radiation) reaches longer wavelengths. This emitted light is allowed to pass through the emission filter to reach the eyepiece as fluorescence emission, which is used for visualization. Hence, it is critical to select appropriate filter sets that excite the fluorophores and can select the resultant emitted signal for visualization. In addition to the proper selection of filters, it is also important to obtain high-quality objective lenses that have high transmission and appropriate immersion oil designed for use with a flu-

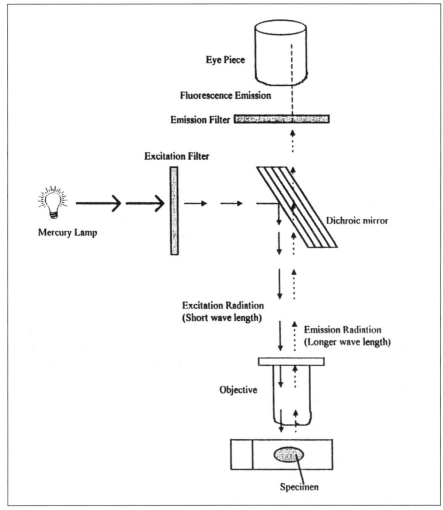

Figure 1. Light path in an epifluorescence microscope.

orescence microscope. The common pairs of fluorochromes generally used in FISH are FITC as the conjugated fluorochrome with PI or DAPI as the counterstain, as well as coumarines (e.g., AMCA=7-amino-4-methyl coumarin-3-acetic acid) with Hoechst-33258, Spectrum Green with quinacrine, TRITC (tetramethyl rhodamin isothiocyanate) counterstained with DAPI, or Cy5 with YOYO-3. A double band-pass filter allows two fluorochromes differing in spectral properties (e.g., FITC and PI) to be viewed simultaneously, which is essential for single-color FISH. (PI gives red interphase nuclei or chromosomes while FITC gives yellow hybridization signals). It is important to consult your manufacturer for specific filters appropriate for the fluorophore in your experiment. More recently, triple band-pass filter for double-color FISH and quadruple band-pass filter for triple-color FISH have become commercially available. As the latest innovation, Applied Spectral Imaging (Carlsbad, CA, USA) released Spectral Cube SD 200 for six-color FISH. In this system, probes directly labeled with Spectrum Orange, Spectrum Green and Aqua, as well as indirectly labeled biotin and digoxigenin with Texas Red, FITC and Texas Red/FITC, can be simultaneously viewed with their filter sets. Filters needed for FISH can be mounted in specific filter holders attached to the fluorescent microscope (1,7,9).

Fluorescence photography depends on the quality of the microscope, selection of proper films and finding correct exposure times. Kodak Ektachrome 400 slide film or higher speed films are preferred and excellent for FISH photography (Eastman Kodak, Rochester, NY, USA). The length of the exposure should be determined empirically or automatically using exposure control systems. For more sophisticated evaluation like three dimensional imaging, multiple-color FISH, visualization of weak gene signals and exploring nuclear architecture in context of gene or chromosomal location, specialized camera systems like CCD or confocal laser scanning microscopy can be used. For readers interested in digitilized imaging please refer to the appropriate literature (7,19,25,31,35, 88).

**Evaluation and Interpretation of FISH Signals**

The availability of unique repetitive DNA probes for the centromere of each homologous pair of human chromosomes now provide the opportunity to analyze the entire chromosome complement in interphase cells. As a result, interphase cytogenetic analysis offers direct connection between morphology and genetic abnormalities. But care should be taken in the proper interpretation of hybridization signals. As in other analytical techniques, the efficiency of hybridization should be determined by running appropriate controls. Both positive and negative controls are necessary in every experiment to confirm hybridization efficiency. Positive control experiments reveal the expected interphase signals. Generally, this can be confirmed for each probe under study by running lymphocyte ISH or on normal diploid cells obtained from the same block of tumor tissue. In normal lymphocytes, the percentage

**Table 2. Criteria for Evaluating FISH Signals**

1. Artifactual "nullisomy" (negative nuclei) should not exceed 25%. If it exceeds, optimal pretreatment steps should be established.
2. Establish a base line frequency for monosomic population to control "truncation" artifacts resulting from cut nuclei during microtomy.
3. Autofluorescence and background fluorescence reducing the sensitivity of signals and can be controlled by optimizing proteolytic digestion.
4. Data should be scored from areas showing uniform fluorescence intensity.
5. Avoid overlapping nuclei as it overestimates chromosome copy number.
6. Split signals should be considered as one hybridization site.
7. Signals that can appear as a single spot due to somatic pairing of chromosomes should be avoided.
8. Care should be taken to consider alphoid DNA polymorphism resulting in heterogeneity of signals.

of diploid signals is expected to exceed 95% while in archival specimens 85% can be considered as optimal. For assessing gene abnormalities, tumor cell lines showing the expected gene abnormalities can be purchased from commercial sources and FISH procedure can yield an accurate measure of gene copy numbers. Since hybridization signals are smaller for genes compared to centromeric probes, accurate hybridization efficiency is difficult and may not exceed above 70%. Negative experimental control should also be included by probe omission during the procedure or pretreatment of tissue sections by nuclease digestion (DNase or RNase) to prove that hybridization depends on DNA or RNA and not on nonspecific reactions during the detection system. FISH signals are generally evaluated by counting 200 nuclei. The criteria summarized in Table 2 allows the proper estimation of FISH signals in tissue sections, while some are also applicable to isolated tumor cells from different sources (1,49,87,95).

## APPLICATIONS OF FISH TECHNOLOGY

The broad applicability of FISH to most forms of tissue preparation has made this technology essential to both researchers and clinicians alike. Although FISH technology has limitations, its applicability in the arenas of research and clinical diagnosis are unbounded.

### Cancer Diagnostics

The primary applicability of FISH in cancer research revolves around the usefulness of the technique in identifying chromosomal and gene markers of disease. The last twenty years have witnessed a conceptual breakthrough in our understanding of cancer development and progression by analyzing chromosomes, genes and gene products (97). The aberrations that characterize carcinogenesis range from gene deletions or amplifications to chromosomal copy number changes (aneuploidy) and structural aberrations. Cytogenetic analysis is a standard technique for the identification of chromosomal changes occurring in cells capable of mitotic division *in vivo* and *in vitro*. The usefulness of chromosomal analysis is dependent on the ability to obtain sufficient metaphase spreads from the cells of interest. It is the identification of recurring chromosomal and gene abnormalities and their correlation with the different subtypes of leukemia and lymphomas that have distinct clinical, morphologic and immunophenotypic features is the key in understanding carcinogenesis. Such information is vital for diagnosis and prognosis as well as monitoring patients for the presence of residual disease and relapse. Indeed, cytogenetics has enabled the identification of genomic changes associated with leukemia and lymphomas since hematopoietic progenitor cells obtained from bone marrow aspirates as well as leukocytes obtained from peripheral blood provide metaphase spreads for karyotypic analysis. However, cancer cells obtained from solid tumors are difficult to culture and obtain metaphase spreads. This is the principal problem associated with cytogenetic analysis of solid tumor cells. It is this very problem that has made the identification of genetic and chromosomal abnormalities associated with solid tumors extremely difficult. FISH eliminates the need for metaphase spreads enabling the identification of gene and chromosomal anomalies on interphase cells (116). Secondly, results can be obtained within 24 h, whereas conventional cytogenetics may require one to two weeks.

**Hematological malignancies.** In conjunction with immunophenotyping, FISH can be used (1) to determine the clonal origin of leukemia and lymphomas (76,119); (2) monitor the efficacy of therapy for the treatment of leukemia and lymphomas (3,52); (3) detect the presence of residual disease following therapy (52); (4) identify early relapse (52); and (5) detect numerical chromosomal anomalies and gene amplifications/deletions in micrometastatic cell populations found in the bone marrow aspirates of cancer patients (36). Chemotherapy diminishes the number of aberrant cells. Arkesteijn et al. reported the use of FISH to monitor patients following chemotherapy (3). FISH is particularly useful for follow-up analysis as the decreasing number of aberrant cells makes it increasingly difficult to obtain metaphase spreads for cytogenetic analysis. Of note, FISH findings were in agreement in 30 of 38 cases prior to and following therapy, which indicates the possible presence of subclonal populations or distinct chromosomal changes occurring as a result of clonal relapse. Mehrotra et al. performed FISH along with immunophenotyping to trace the clonal origin of acute myeloid leukemia (AML) and identified the presence of aberrant cells in CD34-positive and CD34-negative pluripotential stem cell compartments (76).

It has been determined that chronic myelogenous leukemia (CML) is characterized by a balanced translocation involving chromosomes 9 and 22, resulting in what is known as the Philadelphia chromosome (Ph). Molecular analysis further specified this translocation as involving the transfer of the c-*abl* oncogene located on 9q34 to a region on chromosome 22 known as bcr (breakpoint cluster, 22q12) (109). Ph is not a marker of CML alone as it is found in association with other forms of leukemia including acute nonlymphocytic leukemia (ANLL) and acute lymphocytic leukemia (ALL) (23). Presently, FISH is used to detect the *bcr-abl* translocation in CML and ALL (5,10,23). This is accomplished via dual-color FISH where cosmid probes specific for portions of both *abl* and *bcr* are each labeled with different fluorochromes and applied to metaphase or interphase nuclei. If both signals are found on the same chromosome, the test is positive for the translocation. Application to interphase nuclei is equally feasible when two distinct hybridization signals appear fused or in close approximation to one another. Of note, approximately 5%–10% of CML patients are Ph-negative on standard karyotypic analysis but present with symptoms characteristic of CML. In this case, double color FISH is essential for the identification of the *bcr-abl* translocation for definitive genotypic diagnosis (23) (Figure 3E). Acute promyelocytic leukemia (APL) is associated with a translocation involving chromosomes 17 and 15 but, unlike Ph, this translocation is unique to APL. FISH technology is currently used to detect t (15;17) characteristic of APL(101). The presence of the Ph chromosome in CML yields a good prognosis while its presence in ANLL is a poor prognostic indicator. Subsets of acute myeloid leukemia are characterized by t (8;21) or inversion of chromosome 16 (52). Chronic and acute myeloid leukemia have long been known to present with trisomy 8. Specifically, trisomy 8 is the most common chromosomal anomaly karyotypically identified in APL and has been successfully identified by Kwong et al. using FISH technology (50,129). Similarly, trisomy 12 has been consistently associated with chronic lymphocytic leukemia (CLL) with poor prognosis (14,77,91,104). B cell CLL is the most common adult leukemia and is commonly found to have trisomy 12 along with structural aberrations involving the region of 13q14. Band q14 in chromosome 13 is the site for the tumor suppressor gene RB1 (63). In patients having CLL, FISH using biotin-labeled RB1 probe has demonstrated monoallelic deletion of the RB1 gene indicating a possible pathogenic link between CLL and monoallelic deletion of RB1. Conventional cytogenetic analysis on bone marrow aspirates obtained from CLL patients presents with particular difficulty because neoplastic B cells fail to grow well in culture. Hence, conventional cytogenetics performed on patients suspected of having CLL would be unable to confirm the presence of trisomy 12. FISH analysis can readily identify trisomy 12 as well as RB1 deletion on interphase nuclei making it a more sensitive diagnostic tool for the diagnosis of CLL (104,105). In fact, Reining et al. found FISH to be a more sensitive diagnostic technique for identifying trisomy 12 as compared to PCR (91). The identification of such tumor-specific markers is essential for diagnosis and prognosis (3,4,36,52,101,119, 128,129).

Traditionally the diagnosis of such specific chromosomal aberrations as discussed above has been done with conventional cytogenetic technology. The functionality of FISH as an adjunctive or possibly as a replacement technique to conventional cytogenetic analysis has been under investigation. The usefulness of FISH in the identification of such abnormalities was put to the test in a study reported by Jenkins et al. who sought to look at the sensitivity and specificity of FISH using chromosome 8 as a marker on bone marrow aspirates positive for myeloid malignancy and other hematological disorders. Sensitivity was defined as the percentage of abnormal cases diagnosed as abnormal, and specificity is defined as the percentage of cases determined to be normal and diagnosed as such. Results obtained yielded an overall sensitivity of 90.2% and an overall specificity of 98.6% (52). A similar blinded study conducted by the Mayo Clinic (Rochester, MN, USA) and the University of Chicago (Chicago, IL, USA) sought to determine the ploidy status of sample cells with respect to chromosome 8 of normal and pathological bone marrow samples. Results obtained were similar with an optimal sensitivity of 95% and an optimal specificity of 100%, which lead researchers to conclude that FISH is an effective technique in the diagnosis of trisomy 8 and can substitute for conventional cytogenetic analysis. Esendier et al. found FISH technology was 2.6 times more effective in identifying trisomy 12 than conventional cytogenetic technology performed on metaphase spreads (25). Such findings reflect positively on the usefulness of FISH as a diagnostic tool for the identification of tumor-specific markers associated with leukemia.

A recent study conducted by Huegel et al. (37) examined the applicability of FISH on uncultured bone marrow smears using centromeric specific probes for chromosomes 18, X and Y heterochromatin-specific probe. Results yielded the expected number of signals in 87%–97% of cells scored for chromosomes 18 and X and 95%–97% of cells scored for chromosome Y, leading the investigators to conclude that FISH technology can be employed on uncultured bone marrow smears (Figure 3D).

**Solid Tumors.** Unlike leukemia, specific tumor markers associated with the onset and progression of most solid tumors have not been fully identified. This is largely attributable to the inability of solid tumor cells to grow *in vitro*. FISH enables one to determine tumor-specific and stage-specific markers associated with most solid tumors because it can be performed on interphase nuclei (Figure 3A). This is a clear advantage over traditional cytogenetics. The identification of such markers is significant not only from a diagnostic standpoint but also from a prognostic one as well. Conventional cytogenetics along with interphase cytogenetics have identified many tumor- and stage-specific markers of clinical significance. FISH technology can serve as an adjunctive tool in the diagnosis and prognosis of solid tumors by providing the pathologist with a DNA-based analysis of the tumor of interest to correlate with the phenotypic morphology of the cells (124). Genotypic analysis of tumor cells provides a specific diagnosis and prognosis (Figure 3C).

A recent study conducted by Alcaraz et al. performed FISH on paraffin-embedded specimens of high-grade prostate tumors using a variety of chromosome-specific probes with interesting results (2). Specifically, the researchers found a correlation with poor survivorship and the presence of trisomy 7, identifying a possible tumor-specific chromosome marker of prognostic significance. Ovarian cancer has been consistently associated with trisomy 12 as a principal change that is not only found to exist in carcinoma but in phenotypically benign neoplasms such as fibromas, adenomas and thecomas involving the ovary (108,125). Likewise, bladder cancers have been consistently identified with losses of parts of chromosome 9 as an early alteration in tumorigenesis. Furthermore, bladder cancers have also been characterized by complete losses of chromosomes 9, 11, 16, 17 and 18 as well as gains of chromosomes 7 and parts of chromosome 8, along with mutation of the p53 gene (17p13.1) (62,105). Mutation of the p53 gene is common event in nearly all human tumors (8). Sauter et al. utilized FISH technology on paraffin blocks of bladder tumors and found there to be a higher incidence of p53 deletion in higher grade bladder tumors (94). Similarly, Sauter et al. found c-*myc* (8q24) amplification and over-expression as frequent events in bladder tumors with the use of FISH (93).

Similarly, von Hippel-Lindau (VHL) disease is an inherited autosomal dominant disease involving mutations in the VHL tumor suppressor gene (located in the chromosomal band of 3p25-26). Typically, VHL tumors include renal cell carcinomas, angiomatosis retinae, cerebellar and spinal hemangioblastomas, pheochromocytomas and cysts localized to parenchymal organs. Decker et al. reported on the use of FISH in identifying the homozygous loss of the VHL gene using YAC-cloned probes representing the regions of 3p and 3qter in a patient suspected of having VHL disease (19). Additionally, cancer cells have long been identified with the presence of double minute chromosomes that are believed to be sites of gene amplification. Giollant et al. utilized a chromosome-specific library that represented chromosome 9 as a probe for FISH analysis to identify the origin of double minutes in a patient presenting with astrocytoma. Approximately 50% of astrocytomas display double minutes. In the latter case, nearly the entire double minutes were painted by the chromosome-9-specific WCPP identifying their origin (31). Harris and Swain reported a study where it was concluded that c-*erb* B2 (17q11.2-q12) oncogenes over-expressing breast cancers appear to be resistant to alkylator-based chemotherapy. The latter study also points to previous findings that correlate the expression of mutant p53 and *bcl*-2 genes with resistance to chemotherapy (38).

Current clinical tests being conducted on solid tissue tumors involve the identification of chromosomal aneuploidies including the identification of trisomy 12 in ovarian cancer as well as +7 and -9 in bladder tumor along with the identification of gene amplifications for c-*erb* B2 in breast cancer (Figure

2E), N-*myc* amplification in neuroblastoma (Figure 3F) and small cell carcinoma of the lung, as well as deletion of p53 in a variety of solid tumors (8). The identification of such tumor-specific chromosomal and gene markers allows genotypic characterization of tumors that would not only serve to facilitate diagnosis but also aid in deciding treatment modalities more suitable for such molecularly categorized tumors (12,45,86,94,125).

Hydatidiform moles may subsequently result in gestational trophoblastic tumors. Hydatidiform moles are of two types: (1) complete hydatidiform moles (CHM) and (2) partial hydatidiform moles (PHM), which are cytogenetically different. CHMs generally have a diploid karyotype, usually 46 XX while PHMs generally present with a triploid karyotype, usually 69 XXX or 69 XXY. The difficulty in distinguishing between CHM and PHM arises when there is a twin pregnancy in which CHM coexists with a fetus. PHMs are characterized by the presence of a fetus unlike CHMs, which do not present with a fetus. CHMs are five times more likely than PHM to result in gestational trophoblastic tumors (choriocarcinomas). A still higher incidence of choriocarcinoma is expected if the CHM is associated with a fetus. FISH technology can be applied in such a situation to differentiate between the two forms of hydatidiform moles. Choi-Hong et al. utilized X and Y alpha satellite probes for FISH analysis to confirm the presence of dizygotic twinning in cases of CHM associated with a normal fetus (16).

Although mitochondrial DNA constitutes less than 1% of the total human genome, the association of this extranuclear DNA with several human diseases have been reported. An increase in the number of mitochondria is a diagnostic marker in certain tumors such as oncocytoma and non-neoplastic conditions like Hashimoto's thyroiditis (18,120). Verma et al. established that mitochondrial FISH can serve as an important tool in assessing high copy numbers of mDNA. Primers were designed to amplify the 154 base pair target region of the ND4 coding region in the mitochondrial genome. After labeling with digoxigenin, these mitochondrial probes were used for hybridizing mitochondria, in formalin-fixed, paraffin-embedded oncocytic lesions. Oncocyte's exhibited strong signals when hybridized with mitochondrial probe proving that mitochondrial ISH works effectively for a positive diagnosis (115). In future, this technology may be employed in identifying other mitochondrial diseases characterized by DNA deletions.

**Cytology.** Cytological samples include fine needle aspirates, body fluid samples, touch preparations (imprints) and smears prepared from surgically resected tissue samples. Cytology samples are ideally suited for FISH analysis because the cells are dispersed and generally arranged as a monolayer of cells on the slide (15). The application of FISH on body fluids enables a genotypic characterization of tumor cells by the identification of the presence or absence of specific chromosomal and gene anomalies as characteristic markers. FISH has been routinely used in the analysis of fine needle aspirates. Wolman et al. reports the utilization of FISH on touch preparations of breast tissues from biopsies, using centromere-specific probes for chromosomes 1, 16, 17, 18 and X as well as cosmid probe for c-*erb* B2 and identified genetic abnormalities of clinical relevance (123). Similarly, Sneige and colleagues have reported the use of FISH on cytological smears of mammographically detected breast lesions using directly labeled pericentromere-specific probes for chromosomes 7 to 12 as well as 17, 18 and X. This study successfully identified the presence of aneuploidy on cytology smears of histologically defined areas permitting an effective correlation between the genomic anomalies and the cell morphology (98). Soini et al. describes the utilization of ISH on fine needle aspirates obtained from thirteen breast tissue samples for the purpose of measuring c-*erb* B2 and *myc* gene expression (100). All carcinomas were found to express *myc* while only eight of the tumors studied demonstrated c-*erb* B2 gene expression. Elevated levels of gene expression were not dependent on gene amplification. However, tumors that showed *myc* and c-*erb* B2 gene amplification had a corresponding increase in the number of mRNA transcripts representing those genes.

ISH can be used effectively to evaluate and monitor gene expression by evaluating the mRNA content of the genes of interest. Unlike immunohistochemistry, ISH is directly dependent on the cell content of the mRNA transcript representing the gene of interest. Hence, ISH techniques for the evaluation of gene expression are more specific for the expression of a gene because they measure mRNA content rather than protein product which has undergone post-transnational modification and potential storage within the cell (21).

## Congenital Disorders

**Prenatal testing.** The applicability of FISH in prenatal clinical diagnosis has only recently begun to be explored and applied. Prenatal diagnosis can be applied to amniotic as well as to chorionic villus samples where conventional cytogenetic technology would require those cells to be cultured, which often requires anywhere from one to two weeks. Common human aneuploides of autosomal and sex chromosomal origin that permit maturation of the fetus to full term include trisomies for chromosome 21 (Down's Syndrome), 18 and 13, Turner's syndrome (45 X), Klinefelter's syndrome (47, XXY) and 47, XYY (73). FISH technology can be performed on uncultured amniotic and chorionic villus cells as well as on peripheral fetal lymphocytes to identify the the above-described aneuploidies. Results can be obtained within 24 h, which greatly improves the turnaround time for prenatal diagnosis; this is clearly advantageous for both the clinician and prospective parent (89).

In a study conducted by Ward and colleagues using 4500 non-cultured amniocytic samples, the usefulness of FISH as a viable technique was evaluated. Chromosome-specific centromeric probes for chromosomes X, Y, 13,18 and 21 were utilized. The detection rate for chromosomal aneuploidies was 73.3% with an accuracy rate of 93.9%. Ward and colleagues concluded that FISH can be a rapid and effective evaluative technique for the identification of aneuploidies prenatally (117). Several studies have looked at the applicability of FISH in prenatal diagnosis not only as a adjunctive technique to cytogenetics but also as a primary diagnostic tool in the evaluation of prenatal anomalies. Down's syndrome is the most common cause of mental retardation in living populations (generally live born trisomies for 18 and 13 do not live beyond 3 months, except in mosaics) (73). The most frequently asked question by a pregnant women above the age of 35 is "Does my amniotic fluid sample present with Down's syndrome?" The accuracy of FISH technology in detecting Down's syndrome is limited by the specificity of the technology. The currently available alphoid probes for chromosome 21 is not 100% effective. Primarily, chromosome 21 alphoid probe shares sequence homology with the centromeric regions of chromosome 13. As a result, on hybridization both chromosomes are identified simultaneously. Secondly, the alphoid sub-family of chromosome 13/21 exhibit "DNA heteromorphism" resulting in the inability to detect trisomy 21 at the interphase level. A population screening showed that close to 2% may not have sufficient repeat size for registering hybridization signals at interphase level and would show false-negative diagnostic results (114). The pitfalls associated with screening Down's syndrome using alphoid probe have been overcome by introducing a locus-specific chromosome 21 probe (Figure 2C). This probe is specific for the Down's syndrome critical region residing on the long arm of chromomsome 21 (17,65,81). Several published reports attest that this single cosmid clone is effective in the detection of Down's syndrome at interphase because it does not possess DNA heteromorphism or share sequence homology with other chromosomal regions (17,51,81,92,118).

Most recently, Lapidot-Lifson et al. have adapted a technical protocol to detect anomalies involving chromosomes 13, 18, 21, X and Y in a period of 8 h for the purpose of rapidly evaluating chromosomal anomalies in fetuses determined to be at high risk (51). Hence, FISH technology with apropriate probes for chromosomes 21, 18, 13, X and Y can be effectively employed to detect the level of mosaicism in growing fetuses as well as in live-born individuals. Mosaic phenotypes generally exhibit dilutions of the phenotypic characters hence deceiving health professionals (73).

Prenatal genetic diagnosis currently favors the post-implantation period (9th–16th week of gestation) for testing. However, if genetic technology is employed on preimplantation embryos, couples predisposed to genetic abnormalities can be assured of having normal babies. FISH combined with polymerase chain reaction (PCR) and *in vitro* fertilization (IVF) can identify specific chromosomal or gene abnormalities in embryos during the preimplantation period. During this period, the human embryo is amenable to cell biopsy, cell culture and even cryopreservation. Centers for reproductive medicine use FISH technology in sexing human embryos and in identifying genetically aberrant embryos using probes for X, Y, 21, 18 and 13. Future prospects may employ ISH in combination with PCR as a routine dignostic

157

test for the presence of genetic disease (30). FISH is also used to study sperm function and mammalian fertilization (13,80).

It is estimated that 50% to 60% of all early spontaneous abortions are due to chromosomal abnormalities; of those, 50% may have autosomal trisomies, 25% are polyploids and 20% present with monosomy X (45, X) (73). However, growing the tissue samples from POCs for standard karyotypic analysis is hampered by higher rates of culture failures due to the absence of viable cells, contamination of tissue samples and slow growth of some fetal cells. In such cases, the chromosomal cause of spontaneous abortion can be determined by performing FISH with chromosome-specific probes (Figure 3B). These samples are obtained by isolating cells from fetal parts or by using formalin-fixed, paraffin-embedded specimens (113). Double-color FISH using alphoid probe for chromosomes 16 and X can easily identify trisomy 16 or monosomy X as well as triploidy in interphase cells. Trisomy 16, monosomy X and triploidy are the most lethal conditions and constitute 30% of all early fetal losses as substantiated by our own FISH study (unpublished data).

**Postnatal diagnosis of contiguous gene syndromes.** Contiguous gene syndromes, also known as micro-deletion/micro-duplication syndromes, are characterized by minute deletions and/or duplications. Some common micro-deletion syndromes include DiGeorge syndrome [del (22q11)], Miller-Dieker syndrome [del (17p13.3)], Prader-Willi syndrome [del (15q11q13)], Angelman syndrome [del (15q 11q13)] Cri-du-chat (5p-), Langer-Giedion [del(8q24.11q24.13)], Wolf-Hirschhorn [del(4p16.1)], WARG (Wilm's tumor, aniridia, gonadoblastoma, retardation) [del(11p13)], myotonic dystrophy [del(19q13)] and Tuberous selerosis type 1 [del(9q34)] or type II [del(11q23)] (20,28,34,47,66). These disorders are characterized by distinct clinical manifestations in adulthood that are not as readily identifiable in infancy and early childhood, which suggests the need for postnatal testing of those suspected to be afflicted with any of the latter syndromes. Conventional cytogenetics requires the use of high-resolution banding, at least 750 bands, to detect such minute deletions/duplications. However, high-resolution banding is unable to detect micro-deletions/micro-duplications that are as small as 2 million base pairs; this makes the detection of syndromes such as Prader-Willi and Angelman difficult because they are characterized by deletions that are approximately 2 million base pairs (11,28). FISH technology has demonstrated a greater sensitivity than high-resolution banding in the diagnosis of micro-deletion/micro-duplication syndromes primarily because FISH analysis can detect anomalies that are less than 1 million base pairs. A probe specific for the micro-deletion/micro-duplication is used in conjunction with a marker probe for the same chromosome to ensure proper hybridization. Hence, the presence of two signals on each chromosome would indicate no deletion or duplication (Figure 2F). Charocot-Marie-Tooth disease (type A) is a condition that results from DNA duplication. Lupski et al. showed the use of a locus-specific, single-cosmid probe to identify this disease on interphase nuclei (72). Although it appears that in many cases of suspected contiguous gene syndromes, there is no apparent micro-deletion or duplication, the ability of FISH to detect anomalies as small as 1 million base pair makes FISH the ideal diagnostic tool for postnatal evaluation of suspected micro-deletion/micro-duplication syndromes.

FISH is a powerful tool in identifying constitutive marker chromosomes and chromosomal translocations including Robertsonian types inherited and *de novo*. Published reports using cases from pre- and postnatal samples demonstrated the significance of this technology for the careful evaluation of the proband to predict genetic and clinical consequences (48,67,70,83,84).

**Research Applications**

Recent technological advances in forward and reverse genetics, somatic cell genetics, chromosome-mediated gene transfer (CMGT) and irradiation and fusion gene transfer (IFGT) have been exploited to clone genes of interest. It is estimated that the human genome has close to one hundred thousand genes and only about five thousand have been cloned to date. ISH has been used to map genes to their respective chromosomal location. In the early 80's, direct visualization of single copy genes was achieved by isotopic ISH. In 1987 Garson et al. first used non-isotopic gene probes to map the N-*myc* and B-*NGF* genes of approximately 1 kilobase in length to their respective chromosomal bands (27). The hybridization signals they received were not strong enough for routine mapping experiments. Later, Lichter expanded this technique by incorporating repetitive sequence families such as Alu or Kpn into cosmids and YACs to increase the hybridization targets for single copy DNA sequences (60). These gene probes were then prehybridized with unlabeled competitor DNA before applying to metaphase chromosomes, a procedure known as chromosome *in situ* suppression hybridization (CISS) (60). Biotin- and digoxigenin-labeled probes are now heavily employed in CISS technology for "gene mapping" and "gene ordering" in both metaphase and interphase cells. Today, FISH is used to map genes as small as 0.8 to 1 kilobase as well as ordering cosmid clones along a particular chromosome. Lichter could order 50 cosmid clones along chromosome 11 at a distance of 1 megabase while Trask used his technique in interphase gene ordering for several genes using a multicolor FISH (60,110,111,112). Recently, using extended chromatin fiber FISH and DNA fiber FISH, constructing a high-resolution human gene map is a top priority for the human genome project (60,88, 111,112,118). The FISH probes are further introduced as phylogenetic markers for "evolutionary walks" through the genomes of higher primates for tracking human evolution and identifying appropriate animal models for research (65,68,69,71).

A combination of different gene probes with a particular whole chromosome library probe via multicolor FISH in a single hybridization experiment gives the appearance of "chromosome bar coding" (56). Chromosome bar coding, a modified FISH procedure, can be effectively used in the study of interphase nuclear architecture, cell cycle dynamics and tumor-associated chromosomal and gene aberrations. In tumor nuclei, coincidence of signals from different probes on a particular chromosomal domain represent normal genomic organization while separation of probe signals indicate structural chromosomal aberration at interphase.

Other research applications of FISH include arrangements and transport of mRNAs, biological dosimetry (79), detection of viral integration sites in chromosomes, detection of amplified gene localization sites in cancer chromosomes (27), DNA replication studies (24,99), characterization of somatic cell hybrids, molecular analysis of the human genome (103,112) and developmental studies (13).

## TECHNOLOGICAL ADVANCES

### PRINS

PRINS is a variation of FISH in which the principles of PCR technology are incorporated with partial FISH procedure. In this alternative form of FISH, the sequence-specific detection of nucleic acid *in situ* is achieved through epifluorescence microscopy using unlabeled probe as opposed to labeled probe used in FISH. Briefly, the unlabeled probe is mixed with deoxyribonucleotide triphosphates where one of the nucleotides is hapten-labeled. The mixture is then applied in the presence of DNA polymerase (*Taq* 1 polymerase) to metaphase or interphase cells on a slide. The unlabeled probe functions as a primer for the chain elongation and initiates the synthesis of labeled DNA at the target site by extending the 3' end using labeled nucleotides. After sequence-specific annealing of the probe DNA to the template DNA (target sequences) on the metaphase chromosome or interphase nuclei, direct visualization of the hybridized area is possible. The primary advantage of this technique over traditional FISH technology is the turnaround time. Labeling, hybridization and visualization can be accomplished in the span of an hour. The technique can be used very efficiently in gene mapping. Pellestor et al. report the utilization of dual-color PRINS to detect aneuploidies in human spermatozoa in less than two hours. PRINS has been used successfully to diagnose trisomy 21 in amniotic fluid samples. This technology, in principle, has the same applicability as FISH (25,32,33,78,81).

### CGH

The primary limitation of FISH technology is that it can only identify a known region of the human genome using an appropriate probe. However, CGH can analyze abnormalities in unidentified regions of the human genome without the use of cloned probes or sequence knowledge. Using this technique, the entire human genome can be examined at one time for the loss or gain of genetic material. CGH is used on a variety of specimens, including fresh, frozen or paraffin-embedded tissue (44,46,90,102).

Briefly, a normal reference DNA in the normal human genome is labeled with tetramethylrhodamine isothiocynate (TRITC) while the target DNA from a tumor genome is labeled with fluorescein isothiocynate (FITC). The two are then hybridized in the presence of excess unlabeled human Cot1 DNA onto a normal metaphase spread. Using three separate band-pass filters and an imaging system, the intensity ratio of FITC and TRITC are then used to display the over- or under-representation of the target DNA.

CGH can identify various genetic abnormalities such as gains or losses of genetic material that are of at least 2 megabases in size. Joos et al. identified the gain of chromosome arms 7q, 8q, 9q and 16p and the entire chromosome 20 and 22 as well as the losses of chromosome 16q and 18q in prostate carcinoma using CGH (42). Similarly, Iwabuchi et al. have used CGH to identify the following genomic abnormalities in benign, low-grade and high-grade ovarian carcinomas; gains of copy 3q25-26, 8q24 and 20q13, along with losses on 16q and 17qter-q21. It observed that gains at 8q24 occurred primarily in high-grade tumors, while gains of 3q25-26 and 20q13 were found most often in low-grade tumors (39).

However, CGH cannot be used to identify translocations, inversions or other small genetic changes such as point mutations, intragenic rearrangements, and changes that occur in only a small subpopulation of cells. The use of Cot1 DNA results in the exclusion of the pericentromeric and telomeric regions because they are highly repetitive regions. Consequently abnormalities confined to these regions are undetectable by CGH. As with molecular cytogenetic technologies, CGH has limitations that include: (1) turnaround time that takes 3–4 days because of the hybridization; (2) identified genomic abnormalities cannot be analyzed in conjunction with tissue and cell morphology; and (3) the high cost of equipment.

## SKY

Technical advancements such as CGH and the latest technology, SKY, enable one to visualize and examine the entire human genome at once. This is not plausible with FISH. SKY technology permits the differential labeling of all 24 different chromosomes (75,96). Briefly, PCR is used to generate a chromosome library that is then labeled directly with nucleotides conjugated with CY2, CY3, CY5, Spectrum Green and Texas Red, in combinations. A probe consisting of these nucleotides is subsequently hybridized onto a metaphase spread with the use of Cot-1 DNA. Cot-1 DNA prevents the hybridization of large repetitive sequences like those located at the telomeric and centromeric regions. This is then analyzed with a combination of Fourier spectroscopy, CCD imaging and optical microscopy with the capability to measure visible and near-infrared wavelengths. SKY can detect genomic anomalies that range from 500–1500 kilobase pairs.

SKY has been used effectively to identify translocations. Furthermore, SKY technology is more sensitive than standard G-banding for detecting translocation such as t(1:11)(q44;p153). This technology can be used most effectively to study cancer genetics involving translocations and amplifications onto different chromosomes. Similarly, this technology will be equally valuable in evolutionary biology. The primary limitations of using SKY technology is the inability to detect amplifications, deletions, micro-deletions, micro-duplications or inversions that occur within the same chromosome.

## PROTOCOL

FISH has proven to be an exemplary tool for both research and clinical diagnostics. The widespread use of FISH in these fields has led to the development of various easy-to-use protocols (1,9,49,95). However, several variations in protocols do exist, due to the diversity of haptens used; subsequently their detection methods differ, which might present some difficulty to those unaccustomed to ISH. Presented here are standard protocols that are used in various laboratories, which are variations from the influential publication of Pinkel et al. (85). However, standardization is required for each lab depending on the source of the specimens and the type of haptens used. The flow chart (Figure 4) presents an overview of the FISH procedure for the various specimens that are routinely used in our lab.

### Materials

* Proteinase K (Sigma Chemical, St. Louis, MO, USA)
* Xylene +
* Methanol +
* 1 M sodium thiocyanate (NaSCN) +
* Poly-L-lysine-coated slides +
* PBS +
* DAPI stain (Sigma Chemical)
*Biotinylated DNA probes +
*Formamide (Sigma Chemical)
*20× SSC +
*Herring sperm DNA (Sigma Chemical)
*Placental DNA (Sigma Chemical)
*PN Buffer +
*Modified PN buffer +
*Nonidet® P-40 (Sigma Chemical)
*Nonfat dry milk +
*Sodium azide (Sigma Chemical)
*Fluorescein-labeled avidin (Oncor, Gaithersburg, MD, USA)
*Anti-Avidin antibody (Oncor)
*Antifade Mounting Medium (Oncor)

*Propidium Iodide (Oncor)
*Plastic cover slips +
*Rubber cement +
*Coplin jars with lids +
*Dextran sulfate (Sigma Chemical)
*Sodium Chloride (Sigma Chemical)
*Sodium Citrate (Sigma Chemical)
*Water bath +
*Oven +
*Slide warmer +
*distilled $H_2O$ +
*Microcentrifuge tube +
*Fluorescence microscope equipped with appropriate filters (Oncor)

Note: (+) indicates products are available from various commercial vendors.

## Reagent Preparations

1. **Probe DNA.** Probe DNA can be labeled with biotin using nick translation. Biotinylated probes can be generated using Bio Nick kit from Life Technologies (Gaithersburg, MD, USA) according to the manufacturer's instructions. Generally labeling kits contain nucleotide mixture (dNTP mixture), bio-11-dUTP, enzyme mixture containing DNase1 and DNA polymerase 1 and EDTA. Check the size of the probe DNA by running in 1% agarose gel. Purify the labeled probe using Nick Column (Pharmacia Biotech, Piscataway, NJ, USA).
2. **20× SSC.** 17.53 g Na Cl and 8.82 g trisodium citrate in 100 mL distilled $H_2O$ and store at room temperature.
3. **2× SSC.** 1 part of 20× SSC in 9 parts of sterile water and adjust to pH 7.0 using HCl.
4. **Hybridization mixture.** 5 mL formamide, 2 mL of 50% (wt/vol) dextran sulphate and 1 mL of 20× SSC. Adjust to pH 7.0. This solution can be stored at 4°C for several weeks.
5. **Carrier DNA.** Dissolve herring sperm DNA in distilled $H_2O$ to obtain the final cone of 1 mg/mL. Sonicate to 400 base pairs in length.
6. **Blocking DNA.** Use placental DNA at a concentration of 1 mg/mL.
7. **Probe mixture.** Make probe mixture in a microcentrifuge tube by adding the following:

   | | |
   |---|---|
   | *Hybridization mixture | 105 µL |
   | *Human placental DNA | 15 µL |
   | *Herring sperm DNA | 15 µL |
   | *Probe DNA | 15 µL |

   Vortex mix and store aliquots of 30 µL at -20°C
8. **Denaturation solution.**

   70% formamide solution:

163

Prepare denaturation solution by adding the following:

| | |
|---|---|
| *Formamide | 70 mL |
| *20× SSC | 10 mL |
| *Distilled water | 20 mL |

Mix and adjust to pH 7.0. This solution can be stored in 4°C for a week.

9. **PN buffer.** 25.2 g $Na_2HPO_4$. 7 $H_2O$ and 0.83 g Na $H_2PO_4$ $H_2O$ in 1000 mL of distilled $H_2O$. Add 0.6 mL of Nonidet P-40 (NP-40). Mix and adjust to pH 8.0. This buffer can be kept at room temperature for a year.

10. **Modified PN buffer (blocking buffer).** Add 5% nonfat dry milk (Carnation) and 0.02% sodium azide to PN buffer. This solution can be kept for several months at 4°C.

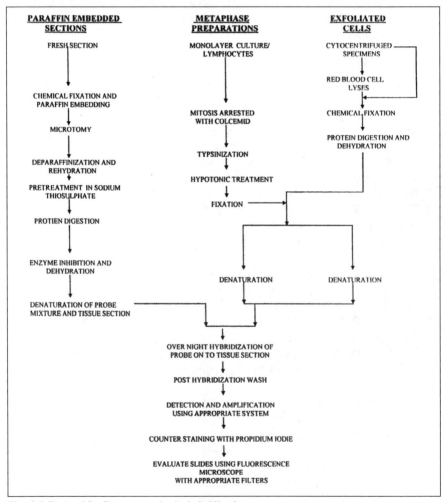

Figure 4. Protocol for fluorescence *in situ* hybridization.

11. **Post-hybridization wash solution I.**
    *Formamide        25 mL (50%)
    *Distilled $H_2O$    20 mL (40%)
    *20× SSC          5 mL (10%)
12. **Post-hybridization wash solution II.**
    *Formamide        32.5 mL (65%)
    *Distilled $H_2O$    12.5 mL (25%)
    *20× SSC          4.0 mL (10%)

## FISH on Paraffin-Embedded Sections

### Preparation of formalin-fixed, paraffin-embedded tissue sections:

1. Obtain 4-$\mu$m-thick paraffin-embedded tissue sections using a standard rotary microtome.
2. Transfer the section onto poly-L-lysine-coated slides and incubate overnight in a 55°C oven.
3. Remove paraffin from the sections using two changes of xylene for 10 min each.
4. Dehydrate the tissue by passing through two changes of 100% ethanol.
5. Air-dry the slides and then proceed to incubate in 1 M Na SCN at 60°C for 15 min.
6. Wash slides in two changes of distilled $H_2O$.
7. Dilute lyophilized proteinase K in PBS to obtain 1 g/mL concentration. Incubate slides in this protein solution for 15–20 min at 45°C.
8. Wash the slides in distilled $H_2O$.
9. Dehydrate in 70%, 80% and 95% ethanol.
10. Visualize optimal protein digestion using a fluorescent microscope and DAPI as a DNA-specific stain.
11. Wash slides in distilled $H_2O$ and dehydrate in alcohol series.
12. Slides can be stored in room temperature or used for FISH.

### Denaturation and hydridization:

1. Prewarm 30 $\mu$L of probe mixture at 37°C for 5 min. This probe mixture contains 10 ng/$\mu$L of biotin-labeled alphoid DNA probe. Vortex mix briefly.
2. Pipet 30 $\mu$L of the probe solution onto each slide containing appropriate tissue section. Care should be taken to avoid disturbing tissue.
3. Coverslip and carefully seal all four sides with rubber cement.
4. Place slides in an oven at 92°C for 15 min. This will lead into the denaturation of both probe DNA as well as tissue DNA.
5. Hybridize slides in a humidified chamber for 8 to 24 h.

### Post-hybridization wash:

6. Prewarm 50 mL of post-hybridization wash solution 1 in a Coplin jar to 37°C in a water bath.
7. Carefully remove rubber cement using forceps. Allow glass slide to fall

off by soaking the slides in 2× SSC at room temperature for 3 min. Care should be taken to avoid sliding the coverslip as it may remove the tissue from the slide.

8. Transfer the slide to post-hybridization wash solution 1 for 15 min. Agitate slightly at 5-min intervals.

9. Wash the slides further in 40 mL of 0.1× SSC at 37°C for 30 min. Post-hybridization wash allows for the unbound probe to fall off from the tissue.

10. Transfer slides to 40 mL of PN buffer. From this point on, the slide should not be allowed to dry. If need be, store the slide in 4°C for up to 14 days before proceeding with detection.

**Immunocytochemistry for detection and amplification:**

11. Remove slides from PN buffer and add 100 µL of modified PN buffer (blocking reagent) containing 5% nonfat dry milk and 0.02% $NaN_3$. Cover with plastic coverslips and incubate at room temperature in a humidified chamber for 10 min.

12. Wash twice in PN buffer at 5 min intervals.

13. Add 100 µL of fluorescein-labeled avidin (avidin-FITC) to each slide. Cover with plastic coverslip and incubate at 37°C for 20 min in a humidified chamber.

14. Carefully remove the coverslip and wash three times for 5 min each in PN buffer at room temperature.

15. The intensity of biotin-linked fluorescence can be amplified by adding anti-avidin antibody to each slide and then incubate at 37°C in a humidified chamber for 30 min.

16. Wash the slide as in step 14.

17. Apply another layer of fluorescein-avidin as in step 13.

18. Wash as in step 14.

19. Wash for 5 min in PBS at room temperature.

20. Stain tissue sections in 15 µL of anti-fade medium containing PI (0.5 µg/mL) or DAPI (0.5 µg/mL).

21. Cover with glass coverslip.

22. View slides using an epifluorescent microscope with appropriate filters.

23. For photography, use ektachrome ASA 400 color slide film. Generally a 30- to 80-s exposure is needed. Obtain the data.

**FISH on Metaphase Preparations**

Fibroblast, solid tumor, peripheral blood, aminocyte, bone marrow and chorionic villi cultures can be established according to standard cytogenetic methods using monolayer, coverslip or suspension culture techniques. Chromosomes can be prepared from respective cultures after treatment with colcemid. Metaphase cells followed by hypotonic treatment can be fixed in methanol:acetic acid (3:1 vol/vol). They are now ready to be dropped onto microscopic slides. After the following pretreatments, these slides can be used for metaphase FISH.

1. Dehydrate slides through a series of three ice-cold ethanol washes for 2 min each (60%, 70% and 95%) alcohol.
2. Air-dry slides and proceed for denturation or store at -70°C for long-term storage. (See section C [steps 11–15] for denaturation, hybridization, post-hybridization wash and immunocytochemistry).

## FISH on Exfoliated Cells

1. Cytocentrifuge specimens onto precleaned super frost slides, at 1500 rpm for 5 min.
2. Treat bloody specimens for 5 min in 0.8% ammonium chloride before cytocentrifugation if needed to clear red blood cells.
3. Fix slides in ice cold Carnoy's solution (3:1, methanol:acetic acid) for 20 min.
4. Fixed slides can be stored in -15°C or proceed for FISH.
5. Stain slides with DAPI to assess the cellularity and protein content using an epifluorescence microscope.
6. Wash slides in PBS to remove the cover slip and residual DAPI.
7. Treat slides in 1 M sodium thiocyanate for 5 min at 37°C.
8. Wash slides in two changes of distilled water.
9. Dilute proteinase K in PBS to obtain 1 μg/mL concentration. Incubate slides in this protein solution at 37°C until the protein background is clear.
10. Wash slides in 2× SSC, followed by dehydration in ice cold 70%, 80%, and 95% ethanol solutions.

## Denaturation and hybridization:

11. Slides carrying metaphase spreads and exfoliated interphase cells can be denatured in 70% formamide solution. The denaturation solution as prepared in the reagent preparation should be prewarmed to 70°C in a water bath. Denature cellular DNA as well as chromosomal DNA by immersing the slides in denaturation solution for 2 min at 70°C. This should be followed by denturation in a 4°C ethanol series: 80%, 90% and 100%. Air-dry slides.
12. Add approximately 1.5 μL of biotin-labeled alpha satellite probe in 40 μL of probe mixture (See Reagent Preparation) in a microcentrifuge tube and denature at 70°C for 5 min in a water bath. Chill quickly in ice bath.
13. Add probe mixture to slides and cover slip. Seal coverslip with rubber cement and incubate slides in a humidified chamber at 37°C overnight for hybridization.

## Post-hybridization wash

14. Remove rubber cement and soak slides in 2× SSC in order to remove

coverslip. Carry out post-hybridization wash at 45°C for 20 min in post-hybridization wash solution 1, i.e., 50% formamide concentration for exfoliated cells while at 42°C for chromosome preparations in post-hybridization wash solution 11 (65% formamide).

15. Transfer the slides to 2× SSC and wash for 8 min. Place the slides in PN buffer and proceed for immunocytochemical detection and amplification from section A (steps 11 to 24).

## COMMENTS

YAC, cosmid, WCP and other human DNA probes of superior quality are available from various commercial sources (Oncor, Life Technologies, Vector Laboratories). They have a good collection of pathologically important probes giving excellent results. These probes avoid the need for labeling. Some of them are biotin labeled while others are digoxigenin labeled. Directly congugated with Spectrum Orange, Spectrum Green and other fluorochrome probes are also available.

## ACKNOWLEDGMENTS

The authors wish to acknowledge the financial support from Maimonides Research and Development Foundation, Brooklyn, NY. We would also like to thank Ms. Dawn McHugh (Oncor, Gaithersburg, MD) for providing the following slides: *bcr/abl* translocation, N-*myc* gene amplification, Prader Willi/Angelman deletion, and double-color FISH on bone marrow smear), which are also included in this publication.

## REFERENCES

1. **Abati, A., J.S. Sanfor, P. Fetseh, F.W. Marincola and S.R. Wolman.** 1995. Fluorescence *in situ* hybridization (FISH): a user's guide to optimal preparation of cytologic specimens. Diagn. Cytopathol. *13*:486-492.

2. **Alcaraz, A., S. Takahashi, J.A. Brown, J.P. Herath, E.J. Bergstrath and J.J. Larson-Keller.** 1994. Aneuploidy and aneusomy of chromosome 7 detected by fluorescence *in situ* hybridization are markers of poor prognosis in prostate cancer. Cancer Res. *54*:3998-4002.

3. **Arkesteijn, G.J., S.L. Erpelinck, A.C. Martens and A. Hagemeijer.** 1996. The use of FISH with chromosome specific repetitive DNA probes for the follow up of leukemia patients. Correlations and discrepancies with bone marrow cytology. Cancer Genet.Cytogenet. *88*:69-75.

4. **Anatasi, J.** 1991. Interphase cytogenetic analysis in the diagnosis and study of neoplastic disorders. Am. J. Clin. Pathol. *95*:522-528.

5. **Arnoldus, E.P.J., J. Wiegant, I.A. Noordmeer, J.W. Wessels, G.C. Beverstock, G.C. Grosveld, M. Vander Ploeg and A.K. Raap.** 1990. Detection of the Philadelphia chromosome in interphase nuclei. Cytogenet. Cell. Genet. *54*:108-111.

6. **Baldini, A. and D.C. Ward.** 1991. *In situ* hybridization of human chromosomes with Alu PCR products: a simultaneous karyotype for gene mapping studies. Genomics *9*:770-774.

7. **Ballard, S.G. and D.C. Ward.** 1993. Fluorescence *in situ* hybridization using digital imaging microscopy. J. Histochem. Cytochem. *41*:1755-1759.

8. **Bartek, J., J. Bartkova, B. Vojtesek, Z. Staskova, J. Lukas, A. Rejthar, J. Kovarik, C.A. Midgley, J.V. Cannon and D.P. Lane.** 1994. Aberrant expression of the p53 oncoprotein is a common feature of a wide spectrum of human malignancies. Oncogene *6*:1699-1703.

9. **Bartsch, O. and E. Schwinger.** 1993. A simplified protocol for fluorescence *in situ* hybridization with repetitive DNA probes and its use in clinal cytogenetics. Clin. Genet. *40*:47-56.

10. Bentz, M., G. Cabot, M. Moos, M.R. Speicher, A. Ganser, P. Lichter and H. Dohner. 1994. Detection of chimeric BCR-ABL genes on bone marrow samples and blood smears in chronic myeloid and acute lymphoblastic leukemia by *in situ* hybridization. Blood *83*:1922-1928.

11. Bettio, D., N. Rizzi, D. Giardino, G. Grungi, V. Briscioli, A. Selicorni, F. Carnevale and L. Larizza. 1995. FISH analysis in Prader-Willi and Angelman syndrome patients. Am. J. Med. Genet. *56*:224-228.

12. Borg, A., B. Baldotrop, M. Ferno, H. Olsson and H. Sigurdsson. 1992. c-myc amplification is an independent prognostic factor in postmenopausal breast cancer. Int. J. Cancer *51*:687-691.

13. Brandriff, B.F. and L.A. Gordon. 1992. Spatial distribution of sperm derived chromatin in zygotes determined by fluorescence *in situ* hybridization. Mutat. Res. *296*:33-42.

14. Brynes, R.K., A. McCourty, N.C. Sun and C.H. Koo. 1995. Trisomy 12 in Richter's transformation of chronic lymphocytic leukemia. Am. J. Clin. Pathol. *104*:199-203.

15. Cajulis, R.S. and M. Bittner. 1994. Interphase cytogenetics in cytology. Cytopathol. Annu. 1-16.

16. Choi-Hong, S.R., D.R. Genest, C.P. Crum, R. Berkowitz, D.P. Goldstein and D.E. Schofield. 1995. Twin pregnancies with complete hydatidiform mole and coexisting fetus: use of fluorescent *in situ* hybridization to evaluate placental X- and Y-chromosomal content. Hum. Pathol. *26*:1175-1180.

17. Conte, R.A., S. Luke and R.S. Verma. 1995. Characterization of a ring chromosome 21 by FISH-technique. Clin. Genet. *48*:188-191.

18. Cotton, D. 1990. Oncocytomas. Histopathology *16*:507.

19. Decker, H.J.H., C. Neuhaus, A. Jouch, M. Speicher, T. Ried, M. Bujard, H. Brauch, S. Storkel, M. Storkle, B. Seliger and C. Huber. 1996. Detection of germ line mutation and somatic homozygous loss of the von Hippel-Lindau tumor suppressor gene in a family with a de novo mutation. A combined genetic study, including cytogenetics, PCR/SCCP, FISH and CGH. Hum. Genet. 97:770-776.

20. Daw, S.C., C. Taylor, M. Kraman, K. Call, J. Mao, S. Schuffenhauer, T. Meitinger, T. Lipson, J. Goodship and P. Scambler. 1996. A common region of 10p deletion in DiGeorge and velocardiofacial syndromes. Nat. Gen. *13*:458-460.

21. Delellis, R.A. 1994. *In situ* hybridization technique for the analysis of gene expression: applications in tumor pathology. Hum. Pathol. *25*:580-585.

22. Deng, H.X., K.I. Yoshiura, R.W. Dirks, N. Harada, T. Hirota, K. Tsukamato, Y. Jinno and N. Niikawa. 1992. Chromosome band specific painting: chromosome *in situ* suppression hybridization using PCR products from a micro dissected chromosome band as a probe pool. Hum. Genet. *89*:13-17.

23. Dewald, G.W., C.R. Sehad, E.R. Christensen, A.L. Tiede, A.R. Zinsmeister, J.L. Spurbeck, S.N. Thibodeau and S.M. Jalal. 1993. The application of fluorescent *in situ* hybridization to detect bcr/abl fusion in variant Ph chromosomes in CML and ALL. Cancer Genet. Cytogenet. *71*:7-14.

24. Ferguson, M. and D.C. Ward. 1992. Cell cycle dependent chromosomal movement in pre-mitotic human T lymphocyte nuclei. Chromosome *101*:557-560.

25. Fox, J.L., M.S. Hsu, L.E. Legator and S.A. Morison. 1995. Fluorescence *in situ* hybridization: powerful molecular tool for cancer prognosis. Clin. Chem. *41*:1554-1559.

26. Gall, J.G. and M.L. Pardue. 1969. Formation and detection of RNA-DNA hybrid molecules in cytological preparations. Proc. Natl. Acad. Sci. USA *63*:378-383.

27. Garson, J.A., J. Vanden Berghe and J.T. Kemshead. 1987. Novel non-isotopic *in situ* hybridization technique detects small (1 kb) unique sequences in routinely g-banded human chromosomes: fine mapping of N-myc and B-NGF genes. Nucleic Acids Res. *15*:4761-4770.

28. Gopal, V.V., H. Roop and N.J. Carpenter. 1995. Diagnosis of micro deletion syndromes: high-resolution chromosome analysis versus fluorescence *in situ* hybridization. Am. J. Med. Sci. *309*:208-212.

29. Gray, J.W., D. Moore and J. Piper. 1994. Molecular cytogenetic approaches to the development of biomarkers. Biomarkers for worker. Health Monitoring 100-120.

30. Griffin, D.K., L.J. Wilton, A.H. Handyside, G.H.G. Attinson, R.M.L. Winston and J.D.A. Delhanty. 1993. Diagnosis of sex in preimplantation embryos by fluorescent *in situ* hybridization. Br. Med. J. *306*:1382.

31. Giollant, M., S. Bertrand, P. Verrelle, A. Tchirkov, S. du Manoir, T. Reid, F. Mornex, J.F. Dore, T. Cremer and P. Malet. 1996. Characterization of double minute chromosomes "DNA content" in a human high grade astrocytoma cell line by using comparative genomic hybridization and fluorescence *in situ* hybridization. Hum. Genet. *98*:265-270.

32. Gosden, J.R., D. Hanrtty, J. Starling, J. Fantes, A. Mitchell and D. Porteous. 1991. Oligonucleotide primed *in situ* DNA synthesis (PRINS): a method for chromosome mapping, banding and investigation of sequence organization. Cytogenet. Cell. Genet. *57*:100-104.

33. Gosden, J. and G. Scopes. 1996. Uncultured blood samples can be labeled by PRINS and ready for chromosome enumeration analysis 1H after collection. BioTechniques *21*:88-91.

34. Greulich, K.O. 1992. Chromosome microtechnology: micro dissection and micro cloning. Trends

169

Biotechnol. *10*:48-51.

35.**Hulspas, R. and J.G. Bauman.** 1992. The use of fluorescent *in situ* hybridization for the analysis of nuclear architecture visualized by confocal laser scaning microscopy. Cell Biol. Int. Rep. *16*:739-747.

36.**Han, K., W. Lee, C. Harris, W. Kim, S. Shim and L. Meisner.** 1994. Quantifying chromosome changes and lineage involvement in myelodysplastic syndrome (MDS) using fluorescent *in situ* hybridization (FISH). Leukemia 8:81-86.

37.**Huegel, A., L. Coyle, R. McNeil and A. Smith.** 1995. Evaluation of interphase fluorescence *in situ* hybridization on direct hematological bone marrow smears. Pathology *27*:86-90.

38.**Harris, L. and M. Swain.** 1996. The role of primary chemotherapy in early breast cancer. Semin. Oncol. *23*:31-42.

39.**Iwabuchi, H., M. Sakamoto, H. Sakunga and Y.Y. Ma.** 1995. Genetic analysis of benign, low-grade, and high-grade ovarian tumors. Cancer Res. *55*:6172-6180.

40.**Hermsen, M.A.J.A., G.A Meijer and J.P.A. Baak.** 1996. Comparative genomic hybridization: a new tool in cancer pathology. Hum. Pathol. *27*:342-349.

41.**Joos, S., H. Scherthan, M.R. Speicher, J. Schlegel, T. Eremer and P. Lichter.** 1993. Detection of amplified DNA sequences by reverse chromosomal painting using genomic tumor DNA as a probe. Hum. Genet. *90*:584-589.

42.**Joos, S., U.S. Bergerheim, Y. Pan, H. Matsuyama, M. Bentz, S. du Manoir and P. Lichter.** 1995. Mapping of chromosomal gains and losses in prostate cancer by comparative genomic hybridization. Genes Chromosom. Cancer *14*:267-276.

43.**Kassler, C.** 1992. Nonradioactive Labeling and Detection of Biomolecules. Springer-Verlag, Berlin.

44.**Kallionemi, A., O.P. Kallionemi, D. Sudar, D. Rutovitz, J.W. Gray, F. Waldman and D. Pinkel.** 1992. Comparative genomic hybridization for molecular cytogenetic analysis of solid tumors. Science *258*:818-21.

45.**Kallioniemi, A., O.P. Kallizoniemi and W. Kurisu.** 1992. Erb-B2 amplification in breast cancer analyzed by fluorescence *in situ* hybridization. Proc. Natl. Acad. Sci. USA *89*:5321-5325.

46.**Kallioniemi, A., T. Visakorpi, R. Karhu, D. Pinkel and O-P. Kallioniemi.** 1995. Gene copy number analysis by fluorescence *in situ* hybridization and comparative genomic hybridization. *In* M. Schwab (Ed.), Methods in Human Genome Analysis. Academic Press, San Diego.

47.**Kao, F.T.** 1993. Micro dissection and micro cloning of human chromosome regions in genome and genetic disease analysis. Bioessays *15*:141-146.

48.**Krakar, W.J., T.J. Borell, C.R. Sehsd, M.J. Pennington, P.S. Karnes, G.W. Dewald and R.B. Jenkins.** 1992. Fluorescence *in situ* hybridization: use of whole chromosome paint probes to identify unbalanced chromosome paint probes to identify unbalanced chromosome translocation. Mayo Clin. Proc. *67*:658-662.

49.**Knoll, J.H.M.** *In situ* hybridization to metaphase chromosmes and interphase nuclei. Current Protocols in Human Genetics. *In* N.C. Dracopoli, J.L. Korf, D.T. Moir, C.C. Morton, C.E. Seidman, J.G. Seidman and D.R. Smit (Eds.), John Wiley and Sons, New York.

50.**Kwong, Y.L., K.F. Wong and T.K. Chan.** 1995. Trisomy 8 in acute promyelocytic leukemia: an interphase study by fluorescence *in situ* hybridization. Brit. J. Haematol. *90*:697-700.

51.**Lapidot-Lifson, Y., R.V. Lebo, R.R. Flandermeyer, J.H. Chung and M.S. Golbus.** 1996. Rapid aneuploid diagnosis of high-risk fetuses by fluorescence *in situ* hybridization. Am J. Obstet. Gynecol. *174*: 886-890.

52.**Lebau, M.M.** 1993. Detecting genetic changes in human tumor cells: have scientists "gone fishing"? Blood *81*:1979-1983.

53.**Lee, W., K. Han, C. Harris, S. Shim, S. Kim and L. Meisner.** 1993. Use of FISH to detect chromosomal translocations and deletions: analysis of chromosomal rearrangements in synovial sarcoma cells from paraffin embedded specimens. Am. J. Pathol. *143*:15-19.

54.**Lemieus, N., B. Dutrillaux, E. Vieges-Peqnignot.** 1992. A simple method for simultaneous R or G-banding and fluorescence *in situ* hybridization of small single copy genes. Cytogenet. Cell. Genet. *59*:311-312.

55.**Levine, A.J.** 1995. The genetic origins of neoplasia. JAMA *273*:592.

56.**Lengaur, C., M.R. Speicher, S. Popp, A. Jauch, M. Taniwaki, R. Nagaraja, H.C. Riethman, H. Doniskaller, M. Duroso, D. Schlessinger and T. Cremer.** 1993. Chromosomal bar codes produced by multicolor fluorescence *in situ* hybridization with multiple YAC clones and whole chromosme painting probes. Hum. Mol. Genet. *2*:505-512.

57.**Li, B.D., E.A. Timm, M.C Riedy, S.P. Harlow and C.C. Stewart.** 1994. Molecular phenotyping by flow cytometry method. Cell Biol. *42*:95-130.

58.**Lichter, P. and D.C. Ward.** 1990. Is non-isotopic *in situ* hybridization finally coming of age? Nature *345*:93-95.

59. Lichter, P., T. Cremer, J. Borden, L. Manueldis and D.C. Ward. 1988. Delineation of individual human chromosomes in metaphase and interphase cells by *in situ* suppression hybridization using recombinant DNA libraries. Hum. Genet. *80*:224-234.

60. Lichter, P., C.J. Chang-Tang, K. Call, D. Hermanson and D.C. Ward. 1990. High resolution mapping of human chromosome 11 by *in situ* hybridization with cosmid clones. Science *247*:64-69.

61. Linehan, W.M., M.I. Lerman and B. Zbar. 1995. Identification of the von Hippel-Lindau (VHL) gene. JAMA *273*:564-570.

62. Liebert, M. and J. Seigne. 1996. Characterisation of invasive bladder cancer: histological and molecular markers. Semin. Urol. Oncol. *14*:62-72.

63. Liu, Y., M. Hermanson, D. Grander, M. Merup, X. Wu, O. Rasool, G. Juliusson, G. Gahrton, R. Detlofsson, N. Nikiforova, C. Buys, S. Soderhall, N. Yankovsky, E. Zabarovsky and S. Einhorn. 1995. 13q deletions in lymphoid malignancies. Blood *86*:1911-1915.

64. Luke, S., R. Birnbaum and R.S. Verma. 1994. Centromeric and telomeric repeats are stable in nonagenerians as revealed by double hybridization fluorescent *in situ* technique. Genet. Anal. Tech. Appl. *11*:77-80.

65. Luke, S., S. Gandhi and R.S. Verma. 1995. Conservation of the Down syndrome critical region in humans and apes. Gene *161*:283-285.

66. Luke, S., R.S. Verma, R. Giridharan, R.A. Cone and M.J. Macera. 1994. Two Prader-Willi/Angelman syndrome loci present in an isodicentric marker chromosome. Am. J. Med. Genet. *51*:232-233.

67. Luke, S., G. Aggarwal, D.G. Stetka and R.S. Verma. 1994. Alphoid DNA diversity of a so-called monocentric Robertsonian fusion. Chromosome Res. 2:73-75.

68. Luke, S. and R.S. Verma. 1993. The genomic synteny at DNA level between human and chimpanzee chromosome. Chromosome Res. *14*:215-219.

69. Luke, S. and R.S. Verma. 1993. Telomeric repeat (TTAGGG)n sequences of human chromosmes are conserved in chimpanzee (Pan troglodytes). Mol. Gen. Genet. *273*:460-462.

70. Luke, S., R.S. Verma, R.A. Conte and T. Mathews. 1992. Molecular characterization of the secondary constriction region (qh) of human chromosome 9 with pericentric inversion. J. Cell. Sci. *103*:919-923.

71. Luke, S. and R.S. Verma. 1995. The genomic sequence for Prader-Willi/Angelman syndromes' loci of human is apparently conserved in the great apes. J. Mol. Evol. *41*:250-252.

72. Lupski, J.R., R. Montes de Oca-Luna, S. Slaugenhaupt, I. Penatao, V. Guzzeta, B.J. Trask, O. Saucedo-Cardeas, D.F. Barkers, J.K. Killian, C.A. Garcia, A. Chakravarti and P.I. Patel. 1991. DNA duplication associated with Charcot-Marie tooth disease Type IA. Cell *66*:219-232.

73. Macgregor, H.C. 1993. An introduction to animal cytogenetics. Chapman and Hall, New York.

74. Manuelidis, L. 1990. A view of interphase chromosomes. Science *250*:1533-1540.

75. Marx, J. 1996. New method for expanding the chromosomal paint kit. Science *273*:430.

76. Mehorotra, B., T.I. George, K. Kavanau, H. Avet-Loiseau, D. Moore II, C.L. Willman, M.L. Slovak, S. Atwater, D.R. Head and M.G. Pallavicini. 1995. Cytogenetically aberrant cells in the stem cell compartment ( CD34 + lin⁻) in acute myeloid leukemia. Blood *86*:1139-1147.

77. Martin, M.L., M.L. Marques, A. Garcia, M.A. Montalban, R. Toscano, A. Moreno, M.J. Gomez and E. Barreiro. 1995. Cytogenetic alterations found by chromosome analysis and FISH technique in two patients with variant chronic lymphocytic leukemia. Sangre *40*:425-429.

78. Mogensen, J.K., S. Pedersen, H. Fischer, J. Hindkjaer, S. Kolvraa and L. Bolund. 1992. Fast-one-step procedure for the detection of nucleic acids *in situ* by primer-induced sequence-specific labeling with fluorescein-12-dUTP. Cytogenet. Cell. Genet. *60*:1-3.

79. Natarajan, A.T., R.C. Vyas, F. Darroudi and S. Vermeules. 1992. Frequencies of x-ray induced chromosome translocations in human peripheral B lymphocytes as detected by *in situ* hybridization using chromosome specific libraries. Int. J. Radiat. Biol. *61*:199-203.

80. Pellestor, F., A. Girardet, L. Cignet, B. Andreo, F. Pelleston, A. Girardet, L. Coignet, B. Andreo and J.P. Charlieu. 1996. Assessment of aneuploidy for chromosomes 8, 9, 13, 16 and 21 in human sperm by using primed *in situ* labeling techique. Am. J. Hum. Genet. *58*:797-802.

81. Pellestor, F., A. Girardet, G. Lefort, B. Andreo and J.P. Charlieu. 1995. Rapid *in situ* detection of chromosome 21 by PRINS technique. Am. J. Med. Genet. *56*:393-397.

82. Pienta, K.J., A.W. Partin and D.S. Coffey. 1989. Cancer as a disease of DNA organization and dynamic cell structure. Cancer Res. *49*:2525-2532.

83. Plattner, R., N.A. Hearema and P.A. Jacobs. 1992. A nonisotopic in situ hybridization study of the chromosomal origin in 15 supernumerary marker chromosomes in man. J. Med. Genet. 29:699-703.

84. Plattner, R., N.A. Hearema, Y.B. Yurov and C.G. Palmer. 1993. Efficient identification of marker chromosomes in twenty-seven patients by stepwise hybridization with alpha satellite DNA probes. Hum. Genet. 91:131-140.

85. **Pinkel, D., T. Straume and J. Gray.** 1986. Cytogenetic analysis using quantitative high sensitivity fluorescence hybridization. Proc. Natl. Acad. Sci. *83*:2934-2938.

86. **Poddighe, P.J., F.C.S. Ramaekers, S.W.G.B. Smeets, G.P. Vooijs and A.H.N. Hopman.** 1992. Structural chromosomal aberations in transitional cell carcinoma of the bladder-interphase cytogenetic combining a centromeric, telomeric, and library DNA probe. Cancer Res. *52*:3929-3934.

87. **Qian, J., G. Bostwick, S. Takahashi, T.J. Borell, J.A. Brown, M.M. Lieber and R.B. Jenkins.** 1996. Comparison of fluorescence *in situ* hybridization analysis of isolated nuclei and routine histoloical sections from paraffin-embedded prostatic adenocarcinoma specimens. Am. J. Pathol. *149*:1193-1199.

88. **Ried, T., A. Baldini, T.C. Rand and D.C. Ward.** 1992. Simultaneous visualization of seven different DNA probes by *in situ* hybridization using combinatorial fluorscence and digital imaging microscopy. Proc. Natl. Acad. Sci. USA *89*:1388-1392.

89. **Reid, T., G. Landes, W. Dackowski, K. Klinger and D.C. Ward.** 1992. Multicolor fluorescence *in situ* hybridization for the simultaneous detection of probe sets for chromosmes 13, 18, 21, X and Y in uncultured amniotic fluid cells. Hum. Mol. Genet. *1*:307-313.

90. **Ried, T., I. Peterson, H. Holtgreve-Grez, M.R. Speicher, E. Schrock, S. du Manoir and T. Cremer.** 1984. Mapping of multiple DNA gains and losses in primary small cell lung carcinoma by comparative genomic hybridization. Cancer Res. *54*:1801-1806.

91. **Reining, G., K. Clodi, M. Konig, K. Geissler, O.A. Haas and C. Mannhalter.** 1994. Detection of trisomy 12 in chronic lymphocytic leukemia: comparison of polymerase chain reaction based technique with fluorescence *in situ* hybridization. Br. J. Haematol. *87*:843-845.

92. **Rietman, H.C., R.K. Moyzis, J. Meyne, D.T. Burk and M.V. Olson.** 1989. Cloning human telomeric DNA fragments into Saccharomyces cervisia using a yeast artificial chromosome vector. Proc. Natl. Acad. Sci. USA *84*:6240-6244.

93. **Sauter, G., P. Caroll, H. Moch, A. Kallioniemi, R. Kerschmann, P. Narayan, M.J. Mihatsh and F.M. Waldmen.** 1995. c-myc copy number gains in bladder cancer detected by fluorescence *in situ* hybridization. Am. J. Pathol. *146*:1131-1139.

94. **Sauter, G., G. Deng, H. Moch, R. Kerschmann, K. Matsumura, S. De Vries, T. George and F.M. Waldman.** 1994. Physical deletion of the p53 gene in bladder cancer. Detection by fluorescence *in situ* hybridization. Am. J. Pathol. *144*:756-766.

95. **Schofield, D.E.** 1995. Determination of aneuploidy using paraffin embedded tissue. *In* N.C. Dracopoli, J.L Haines, B.R. Korf, D.T. Moir, C.C. Morton, C.E. Seidman, J.G. Seidman and D.R. Smith (Eds.), Current Protocols in Human Genetics. John Wiley and Sons, New York.

96. **Schrock, E., S. du Manoir, T. Veldman, B. Schoell, J. Wienberg, M.A. Ferguson-Smith, Y. Ning, D.H. Ledbetter, I. Bar-Am, D. Soenksen, Y. Garini and T. Reid.** 1996. Multicolor spectral karyotyping of human chromosomes. Science *273*:494-497.

97. **Shackney, S.E., C.A. Smith, B.W. Miller, D.R. Burhdt, K. Martha, H.R. Gilas, D.M. Ketterer and A.A. Police.** 1989. Model for the genetic evolution of human solid tumors. Cancer Res. *49*:3344-3354.

98. **Sneige, N., A. Sahin, M. Dinh and A. El-Naggar.** 1996. Interphase cytogenetics in mammographically detected breast lesion. Hum. Pathol. *27*:33-35.

99. **Speel, E.J.M., J. Herbergs, F.C. Ramackers and A.H. Hopman.** Combined immunocytochemistry and FISH for simultaneous tricolor detection of cell cycle genomic and pheotypic parameters of tumor cells. J. Histochem. Cytochem. *42*:961-966.

100. **Soini, Y., A. Mannermaa, R. Winqvist, E. Kamel, K. Poikonen, H. Kiviniemi and P. Paakko.** 1994. Applications of fine needle aspirates to the demonstration of Erb-B2 and myc expression by *in situ* hybridization in breast carcinoma. J. Histochem. Cytochem. *42*:795-803.

101. **Speicher, M.R, A. Jauch, A. Parr and R. Beecher.** 1993. Delineation of translocation t (15:17) in acute promyelocytic leukemic by chromosomal *in situ* suppression hybridization. Leuk. Res. *17*:359-364.

102. **Speicher, M.R., S.D. Manoir, E. Schrock, H. Holtgreve-Grez, B. Schoell, C. Lengauer, T. Cremer and T. Ried.** 1993. Molecular cytogenetic analysis of formalin-fixed, paraffin embedded solid tumors by comparative genomic hybridization after universal DNA amplification. Hum. Mol. Genet. *2*:1907-1914.

103. **Stewart, A.** 1990. The functional organzation of chromosomes and the nucleus. Trends Genet. *6*:377-379.

104. **Stilgenbauer, S., H. Dohner, M. Bugley-Morschel, S. Weitz, M. Bentz and P. Lichter.** 1993. High frequency of monoallelic retinoblastoma gene deletion in B-cell chronic lymphoid leukemia shown by interphase cytogenetics. Blood *81*:2118-2124.

105. **Stilgenbaur, S., E. Leupolt, S. Onl, G. Weib, M. Schroder, K. Fischer, M. Bentz, P. Lichter and H. Dohner.** 1995. Heterogeneity or deletions involving RB-1 and D13525 locus in B-Cell chronic

lymphocytic Leukemia revealed by fluorescence *in situ* hybridization. Cancer Res. *55*:3475-3477.

106. **Stotes, M.H.** 1993. *In situ* hybridization a research technique or routine diagnostic test. Arch. Pathol. Lab. Med. *117*:478-481.

107. **Szakacs, J.G. and S.K. Livingston.** 1994. mRNA *in situ* hybridization using biotinylated oligonucleotide probes. Implications for the diagnostic probes. Implications for the diagnostic laboratory. Ann. Clin. Lab. Sci. *24*:324-338.

108. **Thompson, F.H., J. Emerson, D. Alberts, Y. Liu, X.Y. Guan, A. Burgess, S. Fox, R. Taetle, R. Weinstein, B. Marker, D. Powell and J. Trent.** 1994. Clonal chromosome abnormalities in 54 cases of ovarian cancer. Cancer Genet. Cytogenet. *73*:33-45.

109. **Tkaachuk, D.C., C.A. Westbrook, F.M. Andreal, T.A. Donlon, M.L. Slearly, K. Suryanaryan, M. Homge, A. Redner, J. Gray and D. Pinkel.** 1990. Detection of bcr/abl fusion in chronic myelogenous leukemia by *in situ* hybridization. Science *250*:559-562.

110. **Trask, B.J.** 1991. Fluorescence *in situ* hybridization applications in cytogenetics and gene mapping. Trends Genet. *7*:149-154.

111. **Trask, B.J., H. Massa, S. Kenwrick and J. Gitschier.** 1991. Mapping of human chromosome Xq28 by two-color fluorescence *in situ* hybridization of DNA sequences to interphase cell nuclei. Am. J. Hum. Genet. *48*:1-15.

112. **Trask, B.J., D. Pinkel and G. Van den Engh.** 1990. The proximity of DNA sequences in interphase cell nuclei is correlated to genomic distance and permits ordering of cosmid spanning 250 kilobase pairs. Genomics *5*:710-717.

113. **Van Lijnschoten, G., J. Albreents and M. Vallinge.** 1994. Fluorescence *in situ* hybridization on paraffin-embedded abortion material a means of retrospective chromosome analysis. Hum. Genet. *94*:518-522.

114. **Verma, R.S. and S. Luke.** 1992. Varations in alphoid DNA sequences escape detection of aneuploidy at interphase by FISH technique. Genomics *14*:113-116.

115. **Verma, V.A., C.M. Cerjan, K.L. Abbot and S.B. Hunter.** 1994. Non-isotopic *in situ* hybridization for mitochondria in oncocytes. J. Histochem. Cytochem. *42*:273-6.

116. **Waldman, F., G. Sauter and J. Isola.** 1996. Molecular cytogenetics of solid tumor progression, p. 68-78. *In* J. Li. (Ed.), Hormonal Carcinogenesis. Springer-Verleg, New York.

117. **Ward, B.E., S.L. Gersen, M.P. Caricelli, N.M. Mcguire, W.R. Dackowski, M. Weinstein, C. Sandlin, R. Warren and K.W. Klinger.** 1993. Rapid prenatal diagnosis of chromosomal aneuploidies by fluoresence *in situ* hybridization. Am. J. Hum. Genet. *52*:854-865.

118. **Waye, J.S., G.M. Grieg and H.F. Willard.** 1987. Detection of novel centromere polymorphisms associated with alpha satellite DNA from human chromosome II. Hum. Genet. *77*:151-156.

119. **Weber-Matthiesen, K., J. Deerberg, A. Muller-Hermelink, B. Schlegelberger and W. Grote.** 1993. Rapid immunophenotypic characterization of chromsomally aberrant cells by the new fiction method. Cytogenet. Cell. Genet. *63*:123-125.

120. **Welter, C., G. Kovacs, G. Seitz and N. Blin.** 1989. Alteration of mitochondrial DNA in human oncocytomas. Genes Chromosom. Cancer *1*:79-82.

121. **Willard, H.F. and J.S. Waye.** 1987. Hierarchal order in chromosome specific human alpha satellite DNA. Trends in Genet. *3*:192-198.

122. **Wolman, S.R., F.M. Waldman and M. Balazs.** 1993. Complementarity of interphase and metaphase chromosome analysis in human renal tumors. Genes Chromosom. Cancer *6*:17-23.

123. **Wolman, S.R., J.S. Sanford, K. Flom, H. Feiner, A. Abati and C. Bedrossian.** 1995. Genetic probes in cytology. Diagn. Cytopathol. *13*:429-435.

124. **Wolman, S.R.** 1995. Applications of fluorescent *in situ* hybridization (FISH) to genetic analysis of human tumors. Pathol. Annu. 30 2:227-243.

125. **Wolf, N.G., F.W. Abdul-Karim, N.J. Schork and S. Sehwartx.** 1996. Origins of heterogeneous ovarian carcinomas. A molecular cytogenetic analysis of histologically benign, low malignant potential and fully malignant components. Am. J. Pathol. *149*:511-20.

126. **Yamamoto, T.** 1993. Molecular basis of cancer: oncogenes and tumor suppressor genes. Microbiol. Immunol. *37*:11-22.

127. **Yung, J.F.** 1996. New FISH probes-the end in sight. Nat. Genet. *14*:10-12.

128. **Zhao, L., H.M. Kantarjian, J. van Oort, A. Cork, J.M. Trujillo and J.C. Liang.** 1993. Detection of residual proliferating leukaemic cells by fluorescence *in situ* hybridization in CML patients in complete remission after interferon treatment. Leukemia *7*:168-171.

129. **Zhao, L., J. Van Oort, A. Cork and J.C. Liang.** 1993. Comparison between interphase and metaphase cytogenetics in detecting chromosome 7 defects in hematological neoplasias. Am. J. Hematol. *43*:205-211.

# CHAPTER 8

# An Optimized Protocol for *In Situ* Hybridization Using PCR-Generated $^{33}$P-Labeled Riboprobes

## Lucy H. Lu[1] and Nancy A. Gillett[2]

[1]PE Applied Biosystems, Foster City, CA; [2]Sierra Biomedical, Sparks, NV, USA

## SUMMARY

*This report describes in detail an* in situ *hybridization protocol that has been optimized to produce consistent, sensitive results across a wide range of probe constructs and tissue types. Specific advances incorporated into this protocol include (1) the use of PCR gene fragments to generate cRNA probes, (2) the use of $^{33}$P-labeled probes and (3) the use of ultrafiltration microconcentrators to purify the probe and eliminate background. The use of PCR to generate RNA probes and the use of $^{33}$P-labeled probes in* in situ *hybridization have been described previously; however, this protocol is unique because it combines these techniques in a simple, streamlined procedure that includes the use of microconcentrators to purify the probe. This protocol is particularly appropriate for high-volume* in situ *hybridization laboratories using a wide variety of probe constructs and tissues.*

## INTRODUCTION

*In situ* hybridization is a powerful tool for the detection of gene expression within individual cells and tissue. This technique has increased in popularity, with numerous applications in both clinical and research settings. As the methodology has evolved, a number of different protocols have been described in the literature (2,12,13,15,19,20,24). More recently, non-isotopic methods for probe labeling have become popular and in many cases have replaced isotopic *in situ* hybridization in a variety of applications, particularly in the clinical arena (4,9,22,23). Non-isotopic methods of signal detection are probably the method of choice where the probe construct is well characterized, the tissue sets are limited and well defined, and repeated experiments using the probe are envisioned. Non-isotopic labeling provides a very rapid result and eliminates the necessity of dealing with radioactivity, both of which are important considerations, particularly for clinical or diagnostic applications. However, there are also disadvantages associated with the non-isotopic detection systems. The optimal probe concentration and associated antibody dilution must be determined by a series of titrations for each new probe construct and potentially for different types of tissue. Thus there is a component of method development that occurs with each new probe construct.

In our laboratory, *in situ* hybridization is used as a standard technique to examine gene expression of multiple gene constructs on a wide variety of tissues for a number of different researchers. In general we are addressing questions about endogenous gene expression in normal tissue; hence a high sensitivity for detection is necessary, because often the level of gene expression is not high. The high-volume workload encountered in our laboratory has necessitated development of a standard *in situ* hybridization protocol that works routinely without modification for a wide variety of both probe constructs and tissue types. Previously we used a standard isotopic [35]S-labeled RNA probe protocol that has been described (3,26) and recently reviewed in depth (25). We have incorporated several basic changes in that protocol that have greatly improved the sensitivity and specificity of mRNA detection and reduced both the time necessary to conduct the experiment and the length of exposure required for autoradiography. These changes are: (1) use of PCR fragments to generate RNA probes, (2) use of [33]P-labeled probes and (3) usue of ultrafiltration microconcentrators to purify the probe and eliminate background. Although the use of polymerase chain reaction (PCR) to generate RNA probes and the use of [33]P-labeled probes in *in situ* hybridization experiments have been described in the recent literature, this protocol is unique because it combines these techniques and includes the use of microconcentrators to purify the probe. We have developed a simplified, streamlined protocol that has been used with more than 30 probe constructs and 50 different tissue types. Over 40 experiments averaging more than 100 slides each were performed with a high level of success. We report here a detailed description of our protocol.

## MATERIALS AND METHODS

### Template Generation

Our original protocol to generate cRNA probes relied on plasmid templates that were constructed by inserting the gene into commercial cloning vectors containing bacteriophage RNA polymerase promoters (e.g., pBluescript™; Stratagene, La Jolla, CA and pGEM®, Promega, Madison, WI, USA). After cloning, the plasmid required linearization with a restriction enzyme to produce 5′ protruding or blunt ends in order to avoid the formation of extraneous transcripts during the transcription reactions. The linearization step ensured the synthesis of "run off" transcripts without the contamination of any vector sequences. Specific sense or antisense cRNA were then selectively transcribed using the appropriate RNA polymerase.

Currently, the majority of our cRNA probes are generated from PCR-amplified gene fragments, similar to previously described procedures (11,21,

27). PCR is used to generate gene constructs having different bacteriophage RNA polymerase promoter sequences to flank each end and serve as the template for *in vitro* transcription. This procedure eliminates the need for subcloning the gene into a ribovector and allows great flexibility in probe design. PCR can also be used to amplify a gene fragment from a riboclone (i.e., plasmid containing RNA polymerase promoters flanking the gene fragment), using primers that are specific for the promoter sequences. When transcribed with the appropriate bacteriophage RNA polymerase, PCR-amplified gene fragments from either a riboclone or other plasmid will produce either sense or antisense cRNA probes (Figure 1).

We design PCR primers using the Oligo® 4.0 Primer Analysis Software program (National Biosciences, Plymouth, MN, USA). Typically an 18–25 mer is designed with sequence specificity for the gene of interest. In addition, a 23–26 mer promoter sequence, consisting of the entire promoter and 6 additional bases at the 5′ end are added, resulting in a PCR primer that is 41–51 bases, which consists of both the promoter and gene-specific sequence. Typically we use the RNA promoters T3 or T7; we have also used the SP6 promoter successfully. The software program is used to aid in the primer design and to avoid the formation of duplexes, hairpins, etc. In general, we design primers to generate a probe that is 200–1000 bp in length. We success-

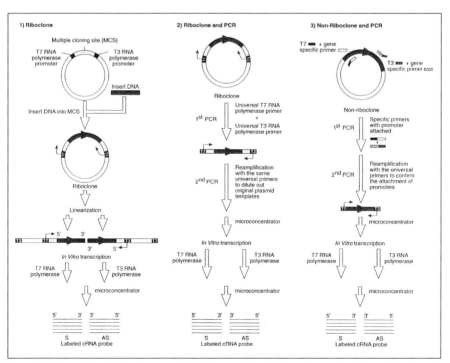

**Figure 1. Schematic outlining three methods of generating cRNA probes, using either a riboclone directly or PCR in conjunction with either a riboclone or non-riboclone.**

Table 1. Synthetic Primer Sequences Specific for Various RNA Polymerase Promoters Used in PCR

| Bacteriophage RNA Polymerase Promoter | Length | Primer Sequence[a] | # Of Additional Bases Beyond 5' End of Promoter | # Of Additional Bases From TSS[b] |
|---|---|---|---|---|
| T7 | 20-mer | 5'-TCT.<u>AAT.ACG.ACT.CAC.TAT</u>.AG-3' <br>(+1) | 3 | 1 |
| T7 | 23-mer | 5'-ACA.TCT.<u>AAT.ACG.ACT.CAC.TAT.AG</u>-3' | 6 | 1 |
| T7 | 27-mer | 5'-GGA.TTC.<u>TAA.TAC.GAC.TCA.CTA.TAG</u>.GGC-3' <br>(+1) | 7 | 4 |
| T3 | 20-mer | 5'-TGA.<u>ATT.AAC.CCT.CAC.TAA</u>.AG-3' <br>(+1) | 3 | 1 |
| T3 | 23-mer | 5'-ACG.TGA-<u>ATT.AAC.CCT.CAC.TAA</u>.AG-3' <br>(+1) | 6 | 1 |
| T3 | 27-mer | 5'CTA.TGA.<u>AAT.TAA.CCC.TCA.CTA.AAG</u>.GGA-3' <br>(+1) | 7 | 4 |
| SP6 | 26-mer | 5'GCA.TCA.GAT.<u>TTA.GGT.GAC.ACT.ATA</u>.GA-3' | 6 | 2 |
| SP6 | 30-mer | 5'-GCA.TAC.GAT.<u>TTA.GGT.GAC.ACT.ATA.GAA</u>.TAC-3'. <br>(+1) | 6 | 6 |

aUnderlined sequences contain bacteriophage RNA polymerase promoters.
bTTS: Transcription start site (+1).

fully generated probes up to 1.8 kb using PCR; however, use of the longer probes can increase the chance of problems during PCR or subsequent *in vitro* transcription.

We use the same PCR program for all of our applications—a "hot start" (7) and "touch-down" combination program in a PE Applied Biosystems Thermal Cycler 9600 (Foster City, CA, USA). The program can accommodate primer sequences with a wide range of annealing temperatures to amplify a variety of gene constructs empirically without relying on the calculated melting temperature ($T_m$) value. All PCRs were performed using a GeneAmp™ PCR kit purchased from PE Applied Biosystems. The reaction tube contained 10 µL of 10× PCR buffer II, 2 mM of $MgCl_2$, 200 µM each of dATP, dCTP, dGTP and dTTP, 1 µM of each primer and 0.1 µg of uncut DNA template. The Ampli-Taq® DNA polymerase (2 units/tube; PE Applied Biosystems) is withheld until the reaction mixture has been heated to 84°C for 5 min (hot start) and kept at 60°C. At the onset of the 60°C file, the *Taq* DNA polymerase is added and 10 cycles of the 3 temperature "touch-down" PCR begin with 30 s at 94°C, 30 s at 68°C, and 1 min at 72°C; the annealing temperature is programmed to drop 2°C in each following cycle (68°C decreasing to 50°C). At the completion of 10 cycles, another three-temperature PCR program with a fixed annealing temperature of 55°C is programmed to run for an additional 15 cycles; then a 7-min extension time at 72°C is done to complete the synthesis.

We routinely perform two rounds of PCR amplification. The first round is performed using the specific primers consisting of the gene-specific sequence plus promoter sequences. Aliquots of the PCR products are run on a 3% NuSieve® (3:1) agarose gel (FMC, Rockland, ME, USA) along with the size markers (AmpliSize™; Bio-Rad, Hercules, CA, USA) for size and concentration visualization. A PCR re-amplification is performed by using 2 µL of the PCR mixture and primer sets directed against only the T3 and T7 promoter. This confirms the linkage of the RNA polymerase promoter sequences. This step is also useful in diluting the plasmid template, particularly if riboclones have been used as template, and thus may also be active in the transcription reaction. We have used a variety of primer constructs of the T3 and T7 promoters, that include a variable number of additional bases on the 3′ and 5′ ends (Table 1).

The PCR product is then concentrated using ultrafiltration microconcentrators (Microcon-30™; Amicon, Beverly, MA, USA) to remove all primers. The concentrated PCR products are diluted to 0.5 µg/µL and run on a 3% Nusieve (3:1) agarose gel to confirm the size and the quality of the product. The amount of DNA is then calculated using a spectrophotometer (Gene-Quant™; Pharmacia, Piscataway, NJ, USA).

**Probe Synthesis: *In Vitro* Transcription Reactions**

*In vitro* transcription is performed to generate riboprobes using the Promega Riboprobe® System (Promega), with a modified protocol. In order

179

to synthesize cRNA probes with a very high specific activity, the total transcription reaction volume is reduced down to 10 µL to raise the effective concentration of the radionucleotide. Previously, all riboprobes in our laboratory were labeled with uridine 5'-($\alpha$-$^{35}$S) thiotriphosphate (>800 Ci/mmol; Amersham, Arlington Heights, IL, USA) for isotopic labeling. We are currently using uridine 5'-($\alpha$-$^{33}$P) triphosphate (>2000 Ci/mmol; Amersham). We typically use 125 µCi of $^{33}$P-UTP in each reaction. In addition, a small amount of nonradioactive UTP ($5 \times 10^{-5}$ mmol per reaction) is added to increase the final UTP concentration to approximately 10 µM. This ensures that the yield of full-length transcripts is not reduced due to a low concentration of the limiting UTP nucleotide and reduces the expense of generating the probe. The 10-µL reaction mixture contains 1× transcription buffer, 0.25 mM each ATP, CTP and GTP, 125 µCi ($\alpha$-$^{33}$P)-UTP, 0.5–1.0 µg of PCR-generated DNA fragments, 40 units of RNasin® (Promega), 10 mM dithiothreitol (DTT) and 15 units of the appropriate bacteriophage RNA polymerase. The reactions are carried out at 37°C for 1 h, then 1 unit of RQ DNase (Promega) is added and incubated for another 15 min at 37°C to destroy the template. At the end of the incubation, the reaction mixture is diluted with 90 µL of Tris EDTA (TE). One microliter of the labeling mixture is pipetted onto duplicate sets of DE81 filters. One set of filters is washed with phosphate buffer (17), and all the filters are counted in the scintillation counter with 6 mL of Bioflour™ II (DuPont-NEN; Boston, MA, USA). The percent incorporation and specific activity of each probe are calculated following established methods (17). Using the above protocol, we typically synthesize probes with specific activity as high as $1 \times 10^{10}$ cpm/µg. Several alternate methods to purify the probe have been tested in our laboratory, including phenol/chloroform extraction and ethanol precipitation, NucTrap® push columns (Stratagene), Microcon-30 and Microcon-100 microconcentrators (Amicon) and spin columns (Sephadex® G-50 Select®-D (RF); 5 Prime → 3 Prime, Boulder, CO, USA). One million cpm of each labeled cRNA transcripts are run on 6.5% mini urea-sequencing gel (Integrated Separation Systems, Natick, MA, USA). Following electrophoresis, the gel is wrapped with plastic wrap and exposed to Kodax XAR-2 film (Eastman Kodak, Rochester, NY, USA).

### Tissue Harvesting and Sectioning

In general, all tissues used in the *in situ* hybridization protocol are prepared according to techniques that have been previously described (19,25). The tissues are immersion-fixed in freshly prepared 4% paraformaldehyde at 4°C for 1 to 3 h. The 4% paraformaldehyde fixative solution is made by heating 1 liter of 0.1 M sodium phosphate buffer (pH 7.4) to 70°C and stirring in 40 grams of powder paraformaldehyde (EM grade; Polysciences, Warrington, PA, USA) until dissolved. The solution is filtered through the 0.45-µm Nalgene™ filter unit (Nalge, Rochester, NY, USA) and chilled to 4°C before use. Tissues are then rinsed and kept in filter-sterilized phosphate-buffered saline (PBS)/ 15% sucrose at 4°C for 2–18 h (i.e., can be stored overnight in sucrose). Tissues are embedded in cassette molds (Baxter, McGaw Park, IL, USA) in OCT (Baxter) frozen in isopentane chilled to -70°C (cryobath), wrapped with aluminum foil and stored at -80°C. Sections approximately 7 μm in thickness are cut on a Bright cryostat (model OTF/AS; Hacker Instruments, Fairfield, NJ, USA) and collected on Probe-on-Plus™ microslides (Fisher Scientific, Pittsburgh, PA, USA). It is important to wear gloves while sectioning the tissues to prevent RNase on the skin from contaminating the tissues. The sections are stored at -80°C in a slide box (VWR, Philadelphia, PA, USA) with a desiccant capsule (Indicator Humicap; United Desiccants Gates, Pennsauken, NJ). Using this protocol we have collected extensive tissue banks of rhesus monkey, rat and mouse. Both the tissue blocks and sections can be used successfully for *in situ* hybridization following long-term storage at -80°C. We have had success using tissues that have been stored as sections for more than five years.

## *In Situ* Hybridization

A typical *in situ* experiment in our laboratory is performed on 100–200 slides, using from 2–10 different riboprobes. All of the stock solutions, including the enzyme stocks, are made following the Molecular Cloning manual (17) unless otherwise specified. In general, we have not found it necessary to treat all the solutions and plasticwares with dimethyl pyrocarbonate (DEPC). However, a separate stock of chemicals and baked glasswares is always used to make the sterile solutions. The procedure is performed over 2 days, as outlined below.

**Day 1 of hybridization:** Plastic staining dishes (Baxter) are treated with chloroform to denature the ribonuclease before use. Typically, this treatment is only done once per month. Gloves are worn throughout the entire hybridization procedure in day 1.

1. The sections are removed from the freezer and air-dried in the hood for 10 min in aluminum trays, followed by 5 min incubation in the 55°C incubator to ensure good tissue adhesion to the slides.

2. Sections are then post-fixed in 4% paraformaldehyde for 10 min on ice, and rinsed in 0.5× standard saline citrate (SSC) for 10 min at room temperature.

3. Sections are immersed in 0.5 M NaCl, 10 mM Tris, pH 8.0 buffer with 0.5 μg/mL proteinase K (Sigma Chemical, St. Louis, MO, USA) for 15 min at 37°C.

4. The sections are rinsed in 0.5× SSC for 10 min and desiccated through graded EtOH, 70%, 90%, 100%, for 2 min each.

5. The dry slides are laid flat in utility boxes (Baxter) that are lined with buffer (50% formamide, 4× SSC)-saturated filter paper. Fifty microliters of hybridization buffer (10% dextran sulfate, 2× SSC and 50%

**Table 2. Primer Sequences Used for PCR for Human-Guanylin (HG) Construct From Riboclone and Non-Riboclone**

Riboclone template: pBlueScript HG

| | |
|---|---|
| T7/Sense primer[a] | 5'-TCT.<u>AAT.ACG.ACT.CAC.TAT.</u>AG-3' |
| T3/Antisense primer | 5'-TAC.<u>AAT.AAC.CCT.CAC.TTA.</u>AG-3' |

Non-riboclone template: pHEBO[b]HG

| | |
|---|---|
| T7/Sense primer[c] | 5'-CT.<u>AAT.ACG.ACT.CAC.TAT.</u>AGG.GCG.AAT.**GCC.TTC.CTG.CTC.TTC.GCA.CTG.TGC**-3' |
| T3/Antisense primer[d] | 5'TAC.<u>ATT.AAC.CCT.CAC.TAA.</u>AGG.GAA.CAA.**GCT.GGG.TGT.TGA.AGT.GGC.AGG.GAA**-3' |

[a]Underline sequences contain bacteriophage RNA polymerase promoters.
[b]See reference 5
[c]Sequence in bold corresponds to guanylin nucleotides 6–31.
[d]Sequence in bold corresponds to guanylin nucleotides 544–519.

formamide) is applied to each slide and incubated at 42°C for 1–3 h.

6. The labeled cRNA probes (1 × 10^6 cpm/slide) are denatured at 95°C in the presence of bakers yeast tRNA (15 µg/slide; Sigma Chemical) for 3 min, transferred on ice, and ice-cold hybridization buffer is added (enough to make the final volume of 50 µL per slide). This solution is mixed thoroughly before adding to the hybridization buffer already covering the slide. Hybridizations take place in a 55°C incubator overnight.

**Day 2 of Hybridization:** A second set of staining dishes is used to avoid cross-contamination from the ribonuclease solution used in step 1.

1. Slides are rinsed twice for 10 min in 2× SSC at room temperature and immersed in RNase buffer (0.5 M NaCl, 10 mM Tris, pH 8.0) with 20 µg/mL RNase A (Sigma Chemical) for 30 min at 37°C.

2. Slides are rinsed twice with 2× SSC at room temperature before subjecting to high-stringency wash in 4 L 0.1× SSC, 1 mM EDTA for every 100 slides at 55°C with gentle agitation for 2 h.

3. Slides are rinsed twice in 0.5× SSC at room temperature and dehydrated in graded EtOH, 70%, 80%, 90%, containing 0.3 M ammonium acetate for 2 min each.

4. Slides are dried in a vacuum desiccator for 2–3 h.

5. The dried slides are taped onto a piece of cardboard paper and overlaid with Hyperfilm™-βmax (Amersham) in the metal cassette for 15–60 h at room temperature.

6. The x-ray film is removed for development. The slides are then coated with Kodak NTB2 Nuclear Track Emulsion (International Biotechnologies, New Haven, NJ, USA); (diluted 1:1 with water) at 42°C, dried in the dark for 3 h and exposed at 4°C in light-tight boxes (VWR) that are sealed with tape and wrapped in aluminum foil. After 1–5 weeks of exposure, slides are developed in Kodak D-19 (diluted 1:1 with water) for 3 min, rinsed briefly in water and in Kodak GBX Fixer for 3 min. All three solutions are maintained at 15°C in an ice bath during use. Slides are counterstained with hematoxylin and eosin and mounted with Permount® (Fisher Scientific). Hybridization signals (silver grains) are visualized by epiluminescence microscopy.

## RESULTS AND DISCUSSION

Specific improvements to previous *in situ* hybridization protocols are threefold: (1) use of PCR gene fragments to generate RNA probes, (2) use of $^{33}$P-labeled probes and (3) the use of ultrafiltration concentrators to clean up the probe without gel purification.

Prior to the use of PCR to generate riboprobe templates, all of our riboprobes were produced from gene constructs that were cloned into commercial ribovectors. The advantages of using PCR to generate the riboprobe template are multiple. First, this has greatly increased our flexibility in custom-designing probes of a certain size, from a specific region, for the *in situ* hybridization experiments. It has eliminated the need for subcloning; typically the gene of interest would not have been inserted into a ribovector; hence subcloning was frequently the first step in an *in situ* experiment. Essentially any vector containing at least a portion of the gene of interest can be used as a template for PCR (Figure 1). The use of PCR to generate riboprobe templates for *in situ* hybridization has been previously described (11,21,27). Young et al. reported successful use of the RNA polymerase promoters T7, T3 and SP6. In contrast, Logel et al. found that T7 and T3 efficiently transcribed riboprobes from the small gene construct produced by the PCR amplification but were unable to successfully transcribe riboprobes using SP6. They suggested that SP6 may have more stringent requirements for additional DNA sequences to permit the polymerase to bind to the gene construct. We typically include the entire promoter sequence plus 3–7 additional bases at the 5′ end and 1–6 additional bases at the 3′ end from the transcription start site to our primer sequence (Table 1). This is in agreement with the literature that recommends additional sequences upstream of the core RNA polymerase sequence be included for maximal transcription efficiency from the promoter (8,14). Using this approach we have had successful transcription of riboprobes from all three polymerase promoters, SP6, T3 and T7, attached to the construct produced by PCR. We have not seen any major differences in transcription efficiency, using constructs with a variable number of bases between the promoter and the transcription start site (data not shown).

Our protocol typically includes a different promoter sequence in the upstream and downstream primer, such that the resulting PCR fragments can be used to generate sense and antisense probes. The most frequently used promoter sequences in our laboratory are the 23 mer T3 and T7 promoter pair. This approach does result in cRNA transcripts with the promoter sequence at the 3′ end. However, as described by others (11), we have not found that this approach decreases the specificity of our signal, yet it does reduce the number of primers and templates needed to generate both sense and antisense riboprobes. Using the human guanylin gene, comparative studies between riboprobes generated by ribovector versus those generated by PCR (Table 2) have not shown any difference in the *in situ* hybridization results obtained (Figure 2, see color plate in Addendum, p. A-19).

The second major improvement that we have incorporated into our protocol is the use of $^{33}$P as the isotopic label of choice instead of $^{35}$S. $^{33}$P has only recently been advocated as a radioisotopic label for *in situ* hybridization (1,13,16). Prior to the availability of $^{33}$P, $^{35}$S with an energy of 0.167 MeV was frequently considered to be the isotope label of choice for *in situ* hybridization. $^{35}$S-emitted β particles are more energetic than $^3$H (β energy = 0.018 MeV), thereby allowing shorter exposure times, yet less energetic than those of $^{32}$P (β energy = 1.71 MeV), thereby allowing better resolution at

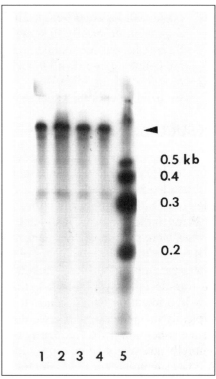

0.5 kb
0.4
0.3
0.2

1 2 3 4 5

**Figure 5. Analysis of cRNA probes purified by different methods.** α-$^{33}$P-UTP labeled cRNAs were transcribed from the PCR fragment of retGC kinase domain corresponding to nucleotides 1541 to 2374 of the full-length cDNA clone. The radio-labeled probes were transcribed by T3 RNA polymerase and subjected to different protocols to remove unincorporated ribonucleotides and the DNase degraded templates. One million cpm each of the cleaned probes were run on the mini sequencing gel and exposed to x-ray film. **Lane 1.** Phenol/chloroform extraction followed by ethanol precipitation. **Lane 2.** Sephadex G-50 spin column. **Lane 3.** NucTrap push column. **Lane 4.** Micron-30 microconcentrator. **Lane 5.** RNA marker (RNA Century; Ambion, Austin, TX, USA).

the cellular level. However, the [35]S probes are frequently associated with high nonspecific background hybridization. [33]P ($\beta$ energy = 0.249 MeV) emits more energetic $\beta$ particles than [35]S, thereby reducing the exposure duration (28). We have found a dramatic decrease in the length of exposure time necessary to see specific signal using [33]P-labeled probes. The primary benefit appears to be the decrease in background nonspecific binding with the [33]P label (Figure 3, see color plate in Addendum, p. A-20). The improvement in the sensitivity of the probe with the [33]P label is more dramatic with genes that have a low level of expression (i.e., retinal-specific guanylyl cyclase receptor; Figure 3) as compared to those with a moderate to high level of expression (i.e., Gly-CAM-1; Figure 4, see color plate in Addendum, p. A-21). The increased signal-to-noise ratio resulting from the decrease in background hybridization from nonspecific binding is probably the biggest advantage to using [33]P-labeled probes, as reported by others (1,13).

The final major change that we introduced into the protocol was the use of ultrafiltration microconcentrators at two steps in the protocol. These microconcentrators are used after the second PCR amplification to remove the RNA polymerase specific primers and short fragments. This step is important as the RNA polymerase specific primers could interfere with the subsequent transcription reaction. We have observed that the inclusion of this step has been instrumental in decreasing the background nonspecific hybridization observed on the final slide (data not shown). In addition, the ultrafiltration microconcentrators are used following the transcription reaction to remove unincorporated nucleotides as well as the degraded DNA template. We have eliminated phenol/chloroform extraction from the procedure without any noticeable difference in the results. We have compared ethanol precipitation, push columns, or spin columns and Microcon microconcentrators to purify probes and found that there is no major difference in terms of recovery yield and gel pattern (Figure 5). However, we have noticed that there is more variability in nonspecific background hybridization from the

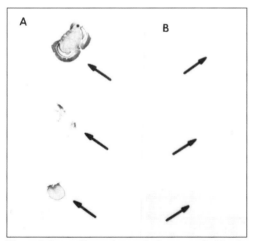

**Figure 6. Autoradiography results from a typical** *in situ* **hybridization experiment.** Photograph of Amersham Hyperfilm-$\beta$ max x-ray film following a 60 h exposure to slides subjected to *in situ* hybridization with $\alpha$-[33]-P-UTP-labeled human $\beta$-actin cRNA probes. **A.** Sections of rat brainstem, aorta and tongue probed with human $\beta$-actin cRNA antisense probe illustrating positive hybridization. **B.** Corresponding serial sections probed with $\beta$-actin cRNA sense probes do not show hybridization as expected for the negative control.

185

NucTrap push column cleaned probes as compared to probes purified by other means (data not shown). We prefer the Microcon microconcentrators because of their relative ease of use and their multiple applications (concentrate, desalt and removal of primers and labels). It is very important not to spin the microconcentrators at speeds that exceed the maximum *g* force recommended by the manufacturer, and it may be necessary to wash the membrane with a small amount of TE to improve the yield.

We now use the Hyperfilm-βmax to screen the results of the *in situ* hybridization experiment prior to dipping the slides in emulsion. Although the resolution of the film is usually not sufficient to give the definitive results of cellular localization, it can confirm the expression of the antisense probe in known positive tissues (positive control) and demonstrate the lack of expression with the sense probe (negative control) (Figure 6). Therefore we have found it a valuable addition in providing interim results and also gives indication for subsequent slide exposure time.

The protocol outlined above has been optimized for large-volume *in situ* hybridization studies where a variety of probe constructs and tissue types from numerous species are routinely used. Within a four-week period, our laboratory typically performs 4–8 *in situ* hybridization experiments, each using 2–10 different probe constructs, many of which are novel, and more than 50 different tissue types resulting in experiments comprising 100–200 slides each. Although this protocol may not be optimal for each individual probe within a specific tissue, we have found that it gives very consistent results across the spectrum of probes and tissues. This protocol is a simple, streamlined procedure that we have found easy to teach to a variety of individuals with different backgrounds. It is useful for laboratories where a large volume of *in situ* hybridization experiments are envisioned using different probe constructs or, conversely, in laboratories where *in situ* hybridization is an infrequently used procedure.

## RECIPES FOR MAKING THE REAGENTS

### Fixation of Cells/Tissues:

1. 4% Paraformaldehyde fixative solution: heat up 1 L 0.1 M sodium phosphate buffer (pH 7.4) to 70°C and stir in 40 g powder paraformaldehyde (EM grade; Polysciences) until dissolved. The solution is filtered through the Nalgene filter unit (0.45 μm) and chilled to 4°C before use.
2. 15% Sucrose solution; dissolve 15 g sucrose (molecular biology grade; Sigma Chemical) in 1 L 1× PBS (pH 7.4). The solution is filtered through the Nalgene filter unit (0.45 μm) and chilled to 4°C before use.

### *In Situ* Hybridization:

Hybridization buffer: 10% dextran sulfate, 2× SSC, and 50% formamide (Boehringer Mannheim, Indianapolis, IN, USA).

RNase buffer: 0.5 M NaCl, 10 mM Tris, pH 8.0.

Box buffer: 50% formamide, 4× SSC.

Wash buffer: 2×-SSC, 1 mM EDTA.

Stringency wash buffer: 0.1× SSC, 1 mM EDTA.

RNase stock solution (10 mg/mL): 10 mg RNase A (Sigma Chemical) in 1 mL of 10 mM Tris, pH 7.5, 15 mM NaCl. Heat to 100°C for 5 min. Allow to cool slowly to room temperature. Store in small aliquots at -20°C.

Proteinase K stock solution (10 mg/mL): 10 mg proteinase K (Boehringer Mannheim) in 1 mL of sterile water.

tRNA stock solution (50 mg/mL): 10 000 units of type X-SA bakers yeast (Sigma Chemical) in 8 mL of TE, pH 7.6, 0.1 M NaCl.

**Protocol 1: DNA Template Generation by PCR (GeneAmp PCR kit)**

For each 100-μL PCR reaction, prepare the following:

| | |
|---|---|
| 10.0 μL | 10× PCR GeneAmp buffer II |
| 6.0 μL | MgCl$_2$, 25 mM |
| 8.0 μL | dNTPs mixture (dA+dT+dC+dG), 10 mM |
| 1.0 μL | Upstream primer, 0.3 μg/μL |
| 1.0 μL | Downstream primer, 0.3 μg/μL |
| 1.0 μL | DNA template, 1 μg/μL |
| 72.3 μL | ddH$_2$O |

1. Withhold the *Taq* DNA polymerase and heat up the mixture to 84°C for 5 min, then keep at 60°C while adding 0.3 μL of *Taq* DNA polymerase (5 U/μL) to each tube.

2. Ten cycles of three-temperature "touch-down" PCR: begin with 30 s at 94°C, 30 s at 68°C, and 1 min at 72°C; the annealing temperature is programmed to drop 2°C in each following cycle (68°C decreasing to 50°C).

3. At the completion of 10 cycles, another three-temperature PCR program with a fixed annealing temp of 55°C is programmed to run for an additional 15 cycles; then a 7-min extension time at 72°C is done to complete the synthesis. Clean up and concentrate the PCR products by Microcon-30.

**Protocol 2: Probe Synthesis: *In Vitro* Transcription Reaction (Promega Riboprobe System)**

1. *In vitro* transcription reaction: each tube contains 125 μCi of dried [33]P-UTP and the following ingredients:

| | |
|---|---|
| 2.0 μL | 5× Transcription buffer |
| 1.0 μL | DTT, 100 mM |
| 1.0 μL | RNasin® |
| 1.0 μL | DNA template, 1 μg/μL |
| 2.0 μL | nucleotide working solution* |

2.0 µL   nuclease-free $H_2O$

Mix thoroughly by pipetting up and down and centrifuge. Add 1.0 µL of the appropriate RNA polymerase to each tube. Incubate 60 min at 37°C for reaction to occur.

* From the stock nucleotide solution (10 mM each of GTP, CTP and ATP) make a working solution of 2.5 mM dNTPs mixture (i.e., 10 µL GTP + 10 µL CTP + 10 µL ATP + 10 µL nuclease-free $H_2O$)

2. Add 1.0 µL of RQ1 DNase (Promega) to stop the reaction. Incubate 15 min at 37°C on heat block.

3. After the incubation, add 90 µL of TE to the mixture. Take 1.0 µL of the reaction mixture and dot onto duplicate sets of DE81 filter paper. One set of filters is washed with sodium phosphate buffer, and all filters are counted in scintillation counter with 6 mL of Bioflour II (DuPont-NEN) to monitor the percent incorporation.

4. Remove the unincorporated nucleotides from the remaining reaction mixture by Microcon-30. Store at -70°C until use.

### Protocol 3: *In Situ* Hybridization

#### Day 1 of hybridization:

Plastic staining dishes (Baxter) are treated with chloroform to denature the ribonuclease before use. Typically, this treatment is only done once per month. Gloves are worn throughout the entire hybridization procedure in day 1.

1. The sections are removed from the freezer and air-dried in the hood for 10 min in aluminum trays, followed by 5 min incubation in the 55°C incubator to ensure good tissue adhesion to the slides.

2. Sections are then post-fixed in 4% paraformaldehyde for 10 min on ice, and rinsed in 0.5× SSC for 10 min at room temperature.

3. Sections are immersed in 0.5 M NaCl, 10 mM Tris, pH 8.0, buffer with 0.5 µg/mL proteinase K (Sigma Chemical) for 15 min at 37°C.

4. The sections are rinsed in 0.5× SSC for 10 min and desiccated through graded EtOH, 70%, 90%, 100%, for 2 min each.

5. The dry slides are laid flat in utility boxes (Baxter) that are lined with buffer (50% formamide, 4× SSC)-saturated filter paper. Fifty microliters of hybridization buffer (10% dextran sulfate, 2× SSC and 50% formamide) are applied to each slide and incubated at 42°C for 1–3 h.

6. The labeled cRNA probes ($1 \times 10^6$ cpm/slide) are denatured at 95°C in the presence of bakers yeast tRNA (15 µg/slide; Sigma Chemical) for 3 min and transferred onto ice, and then ice-cold hybridization buffer is added (enough to make the final volume of 50 µL per slide). This solution is mixed thoroughly before adding to the hybridization buffer already covering the slide. Hybridizations take place in a 55°C incubator overnight.

## Day 2 of hybridization:

A second set of staining dishes is used to avoid cross contamination from the ribonuclease solution used in step 1.

1. Slides are rinsed twice for 10 min in 2× SSC at room temperature and immersed in RNase buffer (0.5 M NaCl, 10 mM Tris, pH 8.0) with 20 µg/mL RNase A (Sigma Chemical) for 30 min at 37°C.

2. Slides are rinsed twice with 2× SSC at room temperature before subjecting to high-stringency wash in 4 L 0.1× SSC, 1 mM EDTA for every 100 slides at 55°C with gentle agitation for 2 h.

3. Slides are rinsed twice in 0.5× SSC at room temperature and dehydrated in graded EtOH, 70%, 80%, 90%, containing 0.3 M ammonium acetate for 2 min each.

4. Slides are dried in a vacuum desiccator for 2–3 h.

5. The dried slides are taped onto a piece of cardboard paper and overlaid with Hyperfilm βmax (Amersham) in the metal cassette for 15–60 h at room temperature.

6. The x-ray film is removed for development. The slides are then coated with Kodak NTB2 Nuclear Track Emulsion (diluted 1:1 with water) at 42°C, dried in the dark for 3 h and exposed at 4°C in light-tight boxes (VWR) that are sealed with tape and wrapped in aluminum foil. After 1–5 weeks of exposure, slides are developed in Kodak D-19 (diluted 1:1 with water) for 3 min, rinsed briefly in water and in Kodak GBX Fixer (Eastman Kodak) for 3 min. All three solutions are maintained at 15°C in an ice bath during use. Slides are counterstained with hematoxylin and eosin and mounted with Permount (Fisher Scientific). Hybridization signals (silver grains) are visualized by epiluminescence microscopy.

## ACKNOWLEDGMENTS

We thank Allison Shaheen for some of the *in situ* hybridization data presented in this manuscript, Allison Bruce for drawing Figure 1 and Mark Vasser and Peter Ng for oligonucleotide synthesis. We also thank David Lowe, Allison Shaheen, Jim Hully and Curtis Chan for helpful discussions during the development of the protocol modifications. We thank David Lowe, Fred deSauvage and Larry Lasky for plasmid constructs used in this study.

## REFERENCES

1. **Baskin, D.G., A.J. Sipols, M.W. Schwartz and B.J. Goldstein.** 1993. *In situ* hybridization for insulin receptor mRNA in rat brain with ($^{35}$S) and ($^{33}$P) riboprobes. J. Histochem. Cytochem. *41*:1128.
2. **Bresser, J. and M.J. Evinger-Hodges.** 1987. Comparison and optimization of *in situ* hybridization procedures yielding rapid sensitive mRNA detections. Gene Anal. Tech. *4*:89-104.
3. **Breuss, J.M., N. Gillett, L. Lu, D. Sheppard and R. Pytela.** 1993. Restricted distribution of integrin β6 in primate epithelial tissues. J. Histochem. Cytochem. *41*:1521-1527.
4. **Bucana, C.D., R. Radinsky, Z. Dong, R. Sanchez, D.J. Brigati and I.J. Fidler.** 1993. A rapid colori-

metric *in situ* mRNA hybridization technique using hyperbiotinylated oligonucleotide probes for analysis of mdr1 in mouse colon carcinoma cells. J. Histochem. Cytochem. *41*:499-506.

5. **Cachianes, G., C. Ho, R.F. Weber, S.R. Williams, D.V. Goddel and D.W. Leung.** 1993. Epstein-Barr virus-derived vectors for transient and stable expression of recombinant proteins. BioTechniques *15*:255-259.

6. **de Sauvage, F.J., S. Keshav, W.J. Kuang, N. Gillett, W. Henzel and D.V. Goddel.** 1992. Precursor structure, expression, and tissue distribution of human guanylin. Proc. Natl. Acad. Sci. USA *89*:9089-9093.

7. **Erlich, H.A., D. Gelfand and J.J. Sninsky.** 1991. Recent advances in the polymerase chain reaction. Science *252*:1643-1651.

8. **Kain, K.C., P.A. Orlandi and D.E. Lanar.** 1991. Universal promoter for gene expression without cloning: expression-PCR. BioTechniques *10*:366-373.

9. **Kiayama, H., P.C. Emson and M. Yohyama.** 1990. Recent progress in the use of the technique of non-radioactive *in situ* hybridization histochemistry: new tools for molecular neurobiology. Neurosci. Res. *9*:1-21.

10. **Lasky, L.A., M.S. Singer, D. Dowbgenko, Y. Imai, W.J. Henzel, C. Grimley, C. Fennie, N. Gillett, S.R. Watson and S.D. Rosen.** 1992. An endothelial ligand for L-selectin is a novel mucin-like molecule. Cell *69*:927-938.

11. **Logel, J., D. Dill and S. Leonard.** 1992. Synthesis of cRNA probes from PCR-generated DNA. BioTechniques *13*:604-610.

12. **McCafferty, J., L. Cresswell, C. Alldus, G. Terenghi and R. Fallon.** 1989. A shortened protocol for *in situ* hybridization to mRNA using radiolabeled RNA probes. Technique - A Journal of Methods in Cell and Molecular Biology *1*:171-182.

13. **McLaughlin, S.K. and R.F. Margolskee.** 1993. [33]P is preferable to [35]S for labeling probes used in *in situ* hybridization. BioTechniques *15*:506-511.

14. **Milligan, J.F., D.R. Groebe, G.W. Witherell and O.C. Uhlenbeck.** 1987. Oligoribonucleotide synthesis using T7 RNA polymerase and synthetic DNA template. Nucleic Acids Res. *15*:8783-8798.

15. **Mitchell, B.S., D. Chami and U. Schumacher.** 1992. *In situ* hybridization: a review of methodologies and applications in the biomedical sciences. Med. Lab. Sci. *49*:107-118.

16. **Oblinger, M.M. and J. Pickett.** 1992. Rapid detection of neuronal mRNAs using ([33]P) nucleotides for *in situ* hybridization. DuPont Biotech Update 172-173.

17. **Sambrook, J., E.F. Fritsch and T. Maniatis.** 1989. Molecular Cloning, 2nd ed. Cold Spring Harbor Laboratory Press, Cold Spring Harbor, NY.

18. **Shyjan, A.W., F.J. de Sauvage, N.A. Gillett, D.V. Goddel and D.G. Lowe.** 1992. Molecular cloning of a retina-specific membrane guanylyl cyclase. Neuron *9*:727-737.

19. **Simmons, D.M., J.L. Arriza and L.W. Swanson.** 1989. A complete protocol for *in situ* hybridization of messenger RNAs in brain and other tissues with radiolabeled single-stranded RNA probes. J. Histotechnol. *12*:169-181.

20. **Singer, R.H., J.B. Lawrence and C. Villnave.** 1986. Optimization of *in situ* hybridization using isotopic and non-isotopic detection methods. BioTechniques *4*:230-250.

21. **Sitzmann, J.H. and P.K. LeMotte.** 1993. Rapid and efficient generation of PCR-derived riboprobe templates for *in situ* hybridization histochemistry. J. Histochem. Cytochem. *41*:773-776.

22. **Unger, E.R., M.L. Hammer and M.L. Chenggis.** 1991. Comparison of [35]S and biotin as labels for *in situ* hybridization: Use of an HPV model system. J. Histochem. Cytochem. *39*:145-150.

23. **Viale, G. and P. Dell'Orto.** 1992. Non-radioactive nucleic acid probes: labeling and detection procedures. Liver *12*:243-251.

24. **Whitfield, H.J., Jr., L.S. Brady, M.A. Smith, E. Mamalaki, R.J. Fox and M. Herkenham.** 1990. Optimization of cRNA probe *in situ* hybridization methodology for localization of glucocorticoid receptor mRNA in rat brain: a detailed protocol. Cell. Mol. Neurobiol. *10*:145-157.

25. **Wilcox, J.N.** 1993. Fundamental principles of *in situ* hybridization. J. Histochem. Cytochem. *41*:1725-1733.

26. **Wilcox, J., K. Smith, S. Schwartz, D. Gordon.** 1986. Localization of tissue factor in the normal vessel wall and in the atherosclerotic plaque. Proc. Natl. Acad. Sci. USA *86*:2839-2843.

27. **Young, I.D., L. Ailles, K. Deugau and R. Kisilevsky.** 1991. Transcription of cRNA for *in situ* hybridization from polymerase chain reaction-amplified DNA. Lab. Invest. *64*:709-712.

28. **Zagursky, R.J., P.S. Conway and M.A. Kashdan.** 1991. Use of [33]P for Sanger DNA sequencing. BioTechniques *11*:37-38.

# CHAPTER 9

# Histologic Validation and Diagnostic Consistency of Neural Network-Assisted Cervical Screening: Comparison with the Conventional Procedure

**Mathilde E. Boon[1], L.P. Kok[2], S. Beck[1] and Myrthe R. Kok[1]**

[1]Leiden Cytology and Pathology Laboratory, Leiden; [2]Institute for Theoretical Physics, Groningen, The Netherlands

SUMMARY

*For the cytologist, neural network-assisted screening reduces the amount of redundant visual information; however, there remains the question of how this influences the diagnostic performance in large-scale screening programs. In a routine setting, 78 010 cervical smears were screened in a period of 18 months. Of these cases 42 134 were screened conventionally. The other 35 876 smears were screened using the PAPNET®-assisted method in which neural network technology was a key feature. We performed a histologic validation of the two screening processes. The PAPNET-assisted screening resulted in significantly higher positive histologic scores for carcinoma and invasive carcinoma. In addition, the accuracy rates for these lesions increased. This histologic validation indicates that, using neural network technology, the value of the cytologic detection of severe abnormalities can be improved significantly. In an additional study concerning 77 222 smears, it was found that not only did the mean positive scores of the seven screeners decrease through PAPNET, but that the coefficients of variability also (CV) became smaller as well. Due to the improvement of the performance of all cytologists involved, diagnostic consistency was enhanced by neural network technology.*

## INTRODUCTION

In many diagnostic procedures, the computer has become an indispensable piece of equipment. Recently, neural network computing has become available. Like the vertebrate brain, artificial neural networks are good at recognition, adaptation and (rough) optimization. Their architecture is highly parallel, robust and fault-tolerant. By contrast, the traditional computer is good at addition, multiplication, if (...) then (...) else (...) instructions. Its architecture is more serial (although parts are parallel) and rather fragile. One bad contact may ruin a calculation, whereas one dead neuron (for the neural network: one unit) really does not matter much. The recent book edited by Murray (7) contains working examples of neural network solutions to real problems as diverse as classification of arrythmias and flow monitoring in oil pipelines. For practicing pathologists, the literature of neural networks is not easy to comprehend, since the reader needs to be well versed in concepts such as perceptron, synaptic weight, weight space, back propagation and the like.

The PAPNET® system, developed by NSI (Suffern, NY, USA), uses neural network computing to aid screening of cervical smears. It has been described in some detail elsewhere (4,6). We wished to use the PAPNET system to:

1. Analyze conventionally prepared smears, without changing standard clinical practice;
2. Automate only the search for diagnostic images, not the diagnostic decision proper;
3. Reduce the nondiagnostic visual information; and
4. Exploit fully the ability of the cytotechnicians to recognize diagnostic information.

Several papers have been published reporting the efficacy of PAPNET for rescreening negative smears in the context of quality control (4,6), and it has been shown that the system can pick up diagnostic cells in false-negative smears (1,12). It should be noted that these were studies, and it is a well-known fact that the decision-making process of screeners working in the reality of a diagnostic setting is fundamentally different (2). Whether PAPNET should be advocated for screening depends on its applicability in a routine setting, with various persons of the staff of a diagnostic laboratory involved in the screening, without reducing sensitivity and accuracy. We analyzed the cytologic scores of a large number of routinely screened cervical smears, either conventionally processed or using the PAPNET system. Because histology is still the gold standard for cytologic screening, a histologic validation was performed. As far as we are aware, this is the first report of a large-scale use of neural network technology in diagnostic medicine.

In addition, we present the analysis of the screening performance of seven cytotechnologists involved in the screening of 21 221 smears with the conventional method and 56 001 aided by the PAPNET technology in 1993 and 1994. We will show that for the seven cytotechnologists involved, both the consistency and the positive scores increased when the PAPNET methodology was used, thus enhancing the value of our screening efforts.

## MATERIALS AND METHODS

### Histologic Validation

For the analysis of histologic validation we used 78 010 smears screened in a period of 18 months (September 1992–December 1993). All smears were made by general practitioners in a health screening program. Part of each daily load was sent to the USA to be scanned by the PAPNET scanner; the remaining smears were conventionally screened. During the first 6 months, 80 smears were sent each day to the PAPNET system; this number increased stepwise to a final 160 smears per day after 9 months. There was no selection of cases for PAPNET or for conventional screening. In total, 35 876 smears were processed through PAPNET and the remaining 42 134 were conventionally screened.

In cases diagnosed as severe dysplasia, carcinoma *in situ* or (suspicious for) invasive carcinoma, the patients were referred to the gynecologist for colposcopy and biopsy. All biopsies and operation specimens of our cases were processed at various hospital laboratories, in and outside Leiden. On request, we received the histologic sections of some of our cases (Figure 2). A total of 43 hospital laboratories were involved. Using the Dutch National Database for Pathology and information from our general practitioners, we are able to trace back almost all histologic diagnoses of our patients, allowing histologic validation of the two screening procedures.

**Diagnostic Consistency**

For our study on diagnostic consistency, we used material screened in 1993 and 1994. The performance of seven screeners, working with both PAPNET and conventional screening, was analyzed.

In 1993, five of the seven were involved in the screening; in 1994, all seven participated. In the two-year time period, 21 221 smears were screened conventionally, whereas in the same period 56 001 smears were screened via PAPNET. The smears were divided randomly between the two screening methods and between the seven screeners.

For our analysis, we calculated three Positive Scores (+Scores). The first one (Positive I) concerned slight atypia (ASCUS and AGUS according to the Bethesda classification system). The second (Positive II) concerned low-grade lesions (low-grade SIL), the third (Positive III) concerned high-grade lesions (high-grade SIL and invasive carcinoma). The Positive Scores (Positive I, II and III) for the total group of seven screeners were calculated per method for the total material of 1993 and 1994 as follows: Total number of positive smears divided by the Total number of smears screened. We refer to these as Total Conventional +Scores and Total PAPNET +Scores.

This results in six Total +Scores, for both PAPNET and Conventional screening the Total Scores are calculated for each of the three positive groups.

In addition, for each screener, the Positive Scores were calculated for each of the two screening methods as follows: number of positive smears found by the individual screener divided by the number of smears screened by that screener. We refer to these scores as Individual PAPNET +Scores and Individual Conventional +Scores. For example, for PAPNET Positive I, 12 values were calculated—five values for the screeners participating in 1993, and seven for the screeners screening in 1994. Of these scores, the mean and coefficient of variability (CV) were calculated.

In order to establish the consistency of scores for the seven screeners for each method, the differences between Individual PAPNET and Conventional +Scores were calculated. The more the Individual +Scores differ, the higher the inconsistency is. The Student's *t* test was used to establish whether the spreading of the differences was significant ($\alpha = 0.05$) for Positive I, Positive II and Positive III, respectively.

## RESULTS

### Histologic Validation

To get insight into the histologic validation of PAPNET screening, we first evaluated the cytologic scores for severe dysplasia, carcinoma *in situ* and (suspicious for) invasive carcinoma. The first positive cases diagnosed by PAPNET are shown in Figures 1–3 (see color plates in Addendum, p. A-22 and A-23). The positive cytologic scores were calculated as percentages of the screened women. For severe dysplasia, the cytologic scores were 0.29 and 0.30 for PAPNET-assisted screening versus conventional screening, respectively, and 0.21 and 0.11, for carcinoma *in situ*, respectively. The cytologic scores for (suspicious for) invasive carcinoma were 0.05 versus 0.02. The cytologic scores are listed in Table 1, in which the z values are also given. The differences for carcinoma *in situ* and for (suspicious for) invasive carcinoma proved to be significantly higher in the PAPNET series.

Of the 379 biopsied or operated patients referred to the gynecologists, we received the histology reports in 365 cases (96%) and, in many cases, the histologic sections (Figure 2). The histologic scores were calculated as percentages of the total of screened women. For severe dysplasia, they were 0.19 and 0.17, for PAPNET and conventional, respectively. For carcinoma *in situ*, the scores were 0.12 and 0.07, respectively. In total, 37 invasive carcinomas were diagnosed, 24 in the PAPNET and 13 in the conventional series, resulting in scores of 0.07 versus 0.03 (Table 2).

The accuracy rates for the three cytologic diagnoses under consideration could be calculated as percentages of cases in which the cytologic diagnosis was in complete accordance with the histological diagnosis (e.g., a cytologic diagnosis of severe dysplasia with a histologic diagnosis of severe dysplasia as well). The accuracy rates for the cytologic diagnosis of severe dysplasia for PAPNET versus conventional screening were 38% and 40%, respectively; for carcinoma *in situ*, 35% and 20%; and for (suspicious for) invasive carcinoma, 72% and 62%. Thus, based on the accuracy rates for the severe abnormalities, carcinoma *in situ* and invasive carcinoma, the results of the PAPNET series were clearly better.

Table 1. Cytologic Results: PAPNET Versus Conventional

|  | PAPNET ($n = 35\,876$) | Conventional ($n = 42\,134$) | z value[*] |
|---|---|---|---|
| Severe dysplasia | 105 (0.29%) | 127 (0.30%) | -0.22 |
| Carcinoma *in situ* | 75 (0.21%) | 46 (0.11%) | +3.53 |
| (Suspicious for) |  |  |  |
| Invasive carcinoma | 18 (0.05%) | 8 (0.02%) | +2.38 |

[*]Alpha = 0.05; z values > +1.96 indicate that the difference between PAPNET and Conventional is significant.

## Diagnostic Consistency

To study the diagnostic consistency of PAPNET screening, we first calculated the Total PAPNET +Scores of the smears screened by the seven cytotechnologists (see Table 3). Also in this material, all Positive Scores were higher for PAPNET than for Conventional. The Z test showed that for Positive I, II and III, significantly more positive smears were detected via PAPNET than when conventionally screened. The z values for Positive I, II and III were 5.05, 2.23 and 4.47 ($\alpha = 0.05$), respectively.

To get an impression of the diagnostic consistency, the CV of the Positive Scores of the seven screeners were calculated for each of the two methods (see Table 4). These calculations were based upon 12 scores per method, five for 1993 and seven for 1994. We observe that not only the mean values for PAPNET are higher than for Conventional, but also that, conversely, the CV have become smaller for PAPNET than for Conventional. For both screening methods, the CV is the highest for Positive I and the lowest for Positive III (the latter is more than 100 times smaller). The ratios of the CV of PAPNET versus Conventional are the highest for Positive III and lowest for Positive I.

The Student's *t* test, calculated over the differences between Individual +Scores, indicated that for Positive III the consistency for PAPNET was significantly higher. For Positive I and II this was not the case.

## DISCUSSION

### Histologic Validation

The time spent evaluating normal cells should be kept to a minimum such that maximum attention can be paid to the abnormal counterpart. Using PAPNET, we do not look at 300 000 or more normal cells in a slide but at 128 video tiles. We can concentrate the visual PAPNET information even more in the summary screen and accordingly can view the diagnostic information in a manner only possible when we can exploit the possibilities of the computer. We can collect "hard" information on the summary screens, such as abnor-

Table 2. Histologic Validation: PAPNET Versus Conventional

| | PAPNET ($n$ = 35 876) | Conventional ($n$ = 42 134) | z value[*] |
|---|---|---|---|
| Severe dysplasia | 68 (0.19%) | 72 (0.17%) | +0.61 |
| Carcinoma in situ | 43 (0.12%) | 29 (0.07%) | +2.34 |
| Invasive carcinoma | 24 (0.07%) | 13 (0.03%) | +2.30 |

[*]Alpha = 0.05; z values > +1.96 indicate that the difference between PAPNET and Conventional is significant.

mal-looking cells and cell groupings, or "soft" information, such as suspicious background including necrosis, old blood, clumped granulocytes, etc. (Figure 3). Based on the summary screens we can decide whether we need additional light microscopy. But in the latter cases, there is already an idea as to what the diagnosis might be (i.e., differential diagnosis small cell carcinoma versus clumps of endometrial tissue). Therefore, this kind of light microscopy is far removed from routine conventional screening, where the cytologist does not yet have a concept of the possible diagnosis and, accordingly, a smear from a cancer patient will be approached with the same state of alertness as the smear of a disease-free woman. As a result, in the PAPNET smears requiring additional light microscopy (comprising the important section of the material), the cytologist will scrutinize them with enhanced alertness. These cases contain reactive changes (2%–3%), cervical intraepithelial neoplasia (CIN) and invasive cancer (0.4%–0.6%), inflammatory disease (3%–6%), and endocervical and endometrial abnormalities (0.1%–0.5%). The remaining cases with abnormal summary screens (8%–11%) proved to be completely negative when studied light microscopically.

In the past, other methods were tested to increase the alertness of cytologists. Hindman, for instance, told his staff that he was inserting positive cases in the screening load (3). Thus the cytologists were forewarned that their performance could be checked. As a result they became overcautious because they were too much on guard about missing one of the planted positive cases. In Hindman's test, the false-positive rate (with histology as the golden standard) jumped from 4% to 22%, and even the false-negative rate was slightly raised (from 12% to 17%). We conclude that making the cytologist overanxious, as in Hindman's case, can be counterproductive. This is in sharp contrast to the situation where cytologists use the PAPNET images: here they can work without anxiety because they know that in the summary screens, they will detect the important diagnostic clues. In our earlier study we learned that, particularly in those cases that are repeatedly missed in conventional screening, the neural network caught the few cancer cells and highlighted the informative background of necrosis and old blood (1). For the cytologic diagnosis (suspicious for) invasive carcinoma, the above-described "soft" signs

Table 3. Total PAPNET and Conventional +Scores (in %) for Positive I, II and III

|  | PAPNET ($n = 56\,001$) | Conventional ($n = 21\,221$) |
|---|---|---|
| Positive I | 5.30 | 4.41 |
| Positive II | 1.72 | 1.49 |
| Positive III | 0.57 | 0.32 |

Table 4. Mean and Coefficient of Variability (CV) for Positive I, II and III (in %) Derived from Individual PAPNET and Conventional +Scores

|  | PAPNET | | Conventional | |
|---|---|---|---|---|
|  | mean | CV | mean | CV |
| Positive I | 5.07 | 3.19 | 5.01 | 4.92 |
| Positive II | 1.62 | 0.26 | 1.40 | 0.32 |
| Positive III | 0.57 | 0.02 | 0.28 | 0.04 |

are very important, and the effect of collecting these by PAPNET becomes evident when looking at Table 1: much higher scores for this category in the PAPNET series. This, however, does not mean that PAPNET leads to over-diagnosis, which is visible in the accuracy rates for this category: for the group of (suspicious for) invasive carcinoma, the accuracy rate was 72%, versus 62% in the conventional series.

In our earlier studies, we found that particularly small cancer cells hidden between leukocytes are missed by the human screener (2), and epithelial cancer fragments of the reserve cell lineage, often overlooked by conventional screening, can be concentrated on the summary screen and thus will no longer be missed. These kinds of false negatives are the dangerous ones, those in which deeply invasive cancer is found a few years later (2,4,11). It is also of interest to note that in this large material, routinely processed, the increase in PAPNET scores was most impressive for the group of invasive carcinomas—with scores of 0.07 and 0.03, respectively. Thus in those cases in which we can expect severe problems when the diagnosis is missed, PAPNET-assisted screening proved to be more accurate. This is an important point when PAPNET is used for quality control.

**Diagnostic Consistency**

To get insight into the diagnostic consistency of PAPNET, we can look at the CV among screeners. We observed that the CV decreased when PAPNET was used. This implies that the differences between the individual screeners are reduced by PAPNET. It is of interest to mention that this phenomenon was most evident with the most severe lesions (Positive III), where PAPNET proves to be significantly better for this aspect. For the clinician, this is an important finding: the outcome of the screening process, particularly for the high-grade lesions, has become less dependent on the person performing the screening of the smear. For both methods, the CV for Positive I was over 100 times greater than for Positive III; for PAPNET the figures were 3.19 versus 0.02, respectively, and for Conventional they were 4.92 versus 0.04. We conclude that these findings indicate that the best way to evaluate a new screening method is not to lump all positive diagnoses together but to use separate diagnostic categories, such that the more severe lesions can be analyzed on their own. In this way, the data are not excessively influenced by the highly variable scores of the lesser lesions.

After a learning period, which can vary, each cytotechnician performs better when the detection of abnormal cells is no longer 100% dependent on human search. Here, we stress that our analysis of routine use of the PAPNET method will provide information of the reality of screening. Studies performed in a research setting are interesting because they show the theoretical potentialities of the system (4–6,8,9), but they do not necessarily reflect the screening situation. The environment, circumstances and responsibilities of the cytotechnologists and/or diagnosticians involved in these studies differ fundamentally from the muddy realities of daily diagnostic practice.

In the PAPNET method, on one hand, the final evaluation of each smear has a light-microscopic component. On the other hand, in many positive cases, the evaluation of the summary video screen provides important visual information to the diagnostician that cannot be offered through conventional light microscopy. Consequently, the final diagnoses (thus the Positive Scores as presented in this paper) are achieved through a mixture of the old (light microscopy) and the new (looking at the video screen).

## COMMENTS

The results achieved with PAPNET-assisted routine screening surpassed even our already positive expectations based on our earlier small pilot study (10) and on rescreening false negatives from our archives (1). In our laboratory, the first case of carcinoma *in situ* and of invasive carcinoma were actually diagnosed with the aid of neural network technology: in the USA, where the system was developed, it was exclusively tested on archival material. These two "historical" cases are displayed in Figures 1–3. The application of this novel technique of visualization lead to an enhancement of the cytologic detection of severe abnormalities and of diagnostic consistency. For this reason, we decided to switch over completely to neural network-assisted screening. Since September 1994, all our smears have been sent to Amsterdam, The Netherlands, where they are scanned in the NSI scanning center opened in June 1994. In December 1995, we processed over 130 000 smears with the neural network-assisted screening method. None of our staff wants to go back to the conventional screening procedure.

## PROTOCOLS

The hardware employed by PAPNET consists of two units: the scanning station and the review station. They are fully separate instruments that can be located far apart. The scanner includes a robotic arm for loading and unloading slides, an automated microscope plus color camera, a high-speed image processor and an 80486 computer for operator interface and system control. Three objectives (50×, 200× and 400× magnification) are used in different stages of scanning. The first (low-power) scan maps slide cellularity and optimizes focusing. The second (medium-power) scan enables an algorithmic image processor using color processing and mathematical morphology to locate the set of potentially abnormal cells and clusters. One neural network then processes the individual cells; a second neural network processes the clusters. Both neural networks use a feed-forward, back-propagation architecture (5). The two neural networks used during the processing of each smear produce similarity scores for each object. These scores depend on how closely the object resembles those in the training libraries (containing a large number of positive and negative cells, and clusters, respectively). The process of training neural networks is thought to be highly analogous to the training of humans to perform a certain task. During scanning, the 64 highest scoring objects for each neural network are chosen as the most diagnostic. A final rescan collects high-resolution color images of the 64 selected objects by each of the two neural networks. These 128 cellular fields include diagnostic cells, epithelial fragments and backgrounds, which are stored on a digital tape (DAT). The object's coordinates, identifying its location on the slide, are also recorded on the tape. The 128 images contain diagnostic information that is reviewed by the cytologist.

The review station is located at the laboratory of the cytologist. It includes a PC with mouse and keyboard, a tape drive for reading the digital tapes, and a large high-resolution monitor for displaying color images. According to the protocol used in our laboratory, the cytologist should compose a summary screen of each case: it consists of the 16 most diagnostic tiles (i.e., containing the most abnormal cells) selected during examination of the 128 video images or "tiles" (Figure 1). The light microscope is calibrated such that the coordinates of the tiles match the x/y coordinates in the smear, facilitating the relocalization of abnormal cells in the smear. In line with this protocol (10), no positive diagnosis may be made merely from images on the video monitor. In our clinical work, in 20%–30% of the cases with summary screens thought to be abnormal, the cytologist had to turn to the light microscope, located immediately next to the computer, to look at the areas in the smear traced by the PAPNET system. For the final cytologic diagnosis, the visual information in the summary screen was combined with the light microscopic evaluation of the abnormal parts of the smear.

In order to exploit the features of PAPNET, the cytologist must be trained to interpret the video images and make informative summary screens. In other words, the human brain must be trained with known examples (cases with known follow-up and/or histologic validation) to appreciate the video images provided by PAPNET and to exploit the information of the summary screen.

## REFERENCES

1. **Boon, M.E. and L.P. Kok.** 1993. Neural network processing can provide means to catch errors that slip through human screening of Pap smears. Diagn. Cytopathol. *9*:411-416.
2. **Bosch, M.M.C., P.E.M. Rietveld-Scheffers and M.E. Boon.** 1992. Characteristics of false-negative smears tested in the normal screening situation. Acta Cytol. *36*:711-716.
3. **Hindman, W.M.** 1976. An effective quality control program for the cytology laboratory. Acta Cytol. *20*:233-238.
4. **Koss, L.G., E. Lin, K. Schreiber, P. Elgert and L.J. Mango.** 1994. Evaluation of the PAPNET cytologic screening system for quality control of cervical smears. Am. J. Clin. Pathol. *101*:220-229.
5. **Mango, L.J.** 1994. Computer-assisted cervical cancer screening using neural networks. Cancer Lett. *77*:155-162.
6. **Mango, L.J. and J.M. Herriman.** 1994. The PAPNET cytological screening system. Compendium of the computerized cytology and histology laboratory, p. 320-334. *In* G.L. Wied, P.H. Bartels, D.L. Rosenthal and U. Schenck (Eds.), Tutorials of Cytology, Chicago.
7. **Murray, A.J. (Ed.).** 1995. Applications of Neural Networks. Kluwer Academic Publishers, Dordrecht.
8. **Rosenthal, D.L. and L.J. Mango.** 1994. Applications of neural networks for interactive diagnosis of anatomic pathology specimens. Compendium on the computerized cytology and histology laboratory, p. 173-184. *In* G.L. Wied, P.H. Bartels, D.L. Rosenthal and U. Schenck (Eds.), Tutorials of Cytology, Chicago.
9. **Rosenthal, D.L., L.J. Mango, D.A. Acosta and R.D. Peters.** 1993. "Negative" Pap smears preceding carcinoma of the cervix: rescreening with the PAPNET system. Am. J. Clin. Pathol. *100*:331.
10. **Ouwerkerk-Noordam, E., M.E. Boon and S. Beck.** 1994. Computer-assisted primary screening of cervical smears using the PAPNET® method: comparison with conventional screening and evaluation of the role of the cytologist. Cytopathology *5*:211-218.
11. **Rylander, E.** 1977. Negative smears in women developing invasive cervical cancer. Acta Obstet. Gynecol. Scand. *56*:115-118.
12. **Sherman, M.E., L.J. Mango, D. Kelly, G. Paull, V. Ludin, C. Copeland, D. Solomon and M.H. Schiffman.** 1994. PAPNET analysis of reportedly negative smears preceding the diagnosis of a high-grade squamous intraepithelial lesion or carcinoma. Mod. Pathol. *7*:578-579.

# CHAPTER 10

# Neuronal Tract Tracing Methods: Recent Developments for Light and Electron Microscopy

## Gloria E. Meredith

Department of Anatomy, Royal College of Surgeons in Ireland, Dublin, Ireland

## SUMMARY

*Anatomical characterization of brain circuitry provides the framework for the interpretation of physiological and behavioral data, and this chapter summarizes the latest techniques that can be used to dissect such circuits. Of the many experimental techniques available, the most versatile are able to reveal simultaneously at least two of the following: neuronal morphology, receptor distribution, neurotransmitter content of neurons and synaptic connections. The past twelve or so years have seen major advancements in these techniques, such as the development of* Phaseolus vulgaris-leucoagglutinin, biocytin and compounds conjugated to biotin, which are highly versatile, as well as sensitive anterograde tracers of neuronal pathways and their collaterals. Retrograde tracers such as Fluoro-gold and the B subunit of cholera toxin not only label populations more completely than horseradish peroxidase but also label dendritic trees in a "Golgi"-like manner. Viruses transported transneuronally label second- and even third-order neurons. Retrograde tracing may be used in conjunction with intracellular filling in fixed tissue and, when combined with anterograde labeling of suspected synaptic inputs, gives a combination that truly represents the state of the art for neuroanatomy, providing, as it does, insight into complete brain microcircuits.*

## INTRODUCTION

The roots of modern tract tracing in the central nervous system are found in the second half of the nineteenth century when investigators followed pathways by dissection. Since that time new methods have appeared as our understanding of neurons and their physiological functions have increased. For example, myelin stains were first developed in 1885 by Marchi and Algeri following Augustus Waller's 1850 demonstration of "Wallerian" degeneration. Silver methods, which were then developed to impregnate degenerating axons (6,49), improved further in the 1950s (46) and the 1960s (15).

More recently, detailed knowledge of the molecular and structural basis of axonal transport has led investigators to compounds that are actively transported *in vivo* and then revealed, postmortem, with fluorescent microscopy or immunocytochemistry. Examples of molecules that are endocytosed and actively transported include horseradish peroxidase (HRP)-based compounds (40,45), tritiated amino acids (14,38), and some fluorochromes, lectins and toxins (37,39,53). Endocytosis occurs in one of three ways (72): fluid-phase, e.g., HRP; receptor-mediated, e.g., antibodies to transmitter-related enzymes; or adsorptive, e.g., wheatgerm agglutinin (WGA) and cholera toxin B subunit (CTB; 3,19,45,70).

A "good" tracing molecule: *(i)* should not be taken up by axons of passage; *(ii)* should fill the axons along their complete course; *(iii)* should not diffuse rapidly or widely at the injection site; and finally, *(iv)* should be broken down slowly once inside the neuron and persist in perikarya and dendrites and/or axons and their terminal *in vivo* and, of course, postmortem, after fixation.

### Tracing Molecules, Immunocytochemistry and Dual Labels

The methods described in the following sections are by no means the only ones available but have been selected on the basis of their ability to meet the criteria above, their reliability and their relative ease of application. These tracers are administered either with pressure or microiontophoresis, and each tracer is delivered perferentially by one method or the other. Pressure injections can be made manually or with a device that regulates precisely the

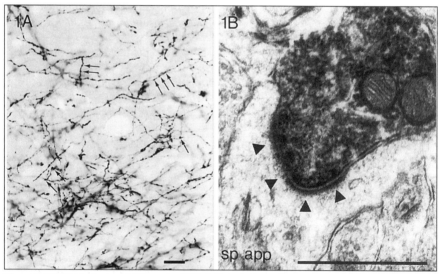

**Figure 1.** (A) Light micrograph illustrating PHA-L fibers in the nucleus accumbens following stereotactic placement of the tracer in the ventral subiculum of the rat. Arrows point out numerous varicosities along fibers. (B) Electron micrograph of a subicular terminal in nucleus accumbens labeled with PHA-L. Note the asymmetrical contact (arrowheads) onto an unlabeled spine. Micrograph (B) has been reproduced from Meredith and Wouterlood (41) with permission of Wiley and Sons. sp app, spine apparatus. Scale bar in A = 10 μm and in B = 0.5 μm.

amount of tracer delivered (see Protocol 1). Iontophoretic injections are an ideal way to deliver many tracers of small molecular weight (see Protocols 1–3). Recent advances in microiontophoresis include the delivery of tracers from micropipets that have their lumens coated with silicone to reduce surface tension (16).

## Anterograde and Retrograde Tracers

The development of *Phaseolus vulgaris*-leucoagglutinin (PHA-L) from red kidney beans (23,24) and, later, low-molecular-weight, highly soluble compounds conjugated to biotin, such as the dextran amines (3,17,25,35, 53,69) and lysine (biocytin; 36), have revolutionized anterograde tracing of neural pathways. All these tracers are highly specific and sensitive; they fully label terminal fields for light microscopic (LM) study and terminal and *en passant* endings for electron microscopic (EM) analysis (Figures 1–3). Moreover, PHA-L is picked up and transported by injured axons of passage only after injections at high current (cf. 18,28). However, the biotinylated dextran amines are more readily transported by axons of passage (18). Both PHA-L (Figure 1, A and B) and the compounds conjugated to biotin retain well their antigenicity in fixed tissue, and can be easily combined with neurotransmitter immunostaining (Figures 2, C-D and 3, see color plates in Addendum, p. A-24 and A-25).

The lectin PHA-L primarily contains leucoagglutinin or L subunits but also has a significant proportion of erythroagglutinin or E subunits (59). The L subunit, however, is the portion that is actively transported (23). According to Groenewegen and Wouterlood (28), the pH of the solution should be as close as possible to PHA-L's isoelectric point of pH 5.0–5.1, so that the lectin maintains its positive charge in the electrical field during iontophoresis.

Despite the fact that pressure injections are easy to make, PHA-L is best administered iontophoretically (23,28,59) from glass pipets with broken tips (Protocol 1). The appearance of the injection site is largely dependent upon the pipet tip size and the strength of the current. An injection into a nonlayered structure produces a round to oval site with a dense central core of cells filled with a granular reaction product. The spread of tracer seems to be more restricted in layered structures (Figure 2A); nevertheless, cells fill completely at those sites (Figure 2B). The mechanism whereby PHA-L is taken up is not fully understood, but the compound seems capable of labeling different neuronal populations throughout the central nervous system in a variety of vertebrates. In addition, it is transported over long distances by increasing the survival time (23,28), and, because it is not taken into the lysosomal system, it is stable for extended periods (28). This compound is transported slowly at a rate of 5–6 mm per day (23), primarily in an anterograde manner, although there are neural systems where retrograde transport also occurs (55).

Tissue taken from animals injected with PHA-L *in vivo* should be well fixed, preferably by perfusion (Protocol 1), either with a solution of paraformaldehyde alone or a mixture of paraformaldehyde and glutaraldehyde. The tracer PHA-L remains stable in the tissue after many different fixations, and the choice of fixative ultimately depends on the goal of the experiment. PHA-L is best revealed with an immunocytochemical reaction (Figures 1A and 3A) on free-floating tissue sections (Protocol 1). The addition of a relatively high percentage of buffered Triton® X-100 to the incubations is necessary to enhance antibody penetration and reveal sufficient detail for LM studies (Figure 1A). However, in EM investigations of PHA-L-labeled terminals (Figure 1B), the use of such detergents is precluded as they disrupt cellular membranes. Other "tricks", such as a reduction in the extent or time for fixation, the employment of a freeze-thaw procedure or a pretreatment in detergent, can all aid antibody penetration for ultrastructural studies (see Protocol 5). The most evident drawbacks of PHA-L include its capricious nature and its lability when tagged for immunofluorescence. However, immunofluorescent labels are possible, and Gerfen and colleagues (24) recommend a protocol that uses a rabbit anti-goat IgG conjugated to fluorescein isothiocyanate (FITC).

Compounds that are conjugated to biotin, such as biocytin and the biotinylated dextran amine (BDA), are also sensitive anterograde tracers (3,17,18, 36,59,71). They are particularly easy to use since they can be directly and rapidly detected with an ABC reaction, which also makes them well suited for ultrastructural studies. On the other hand, fluorescently labeled dextrans, such as Fluoro-ruby (tetramethyl rodamine-dextran-amine; 10,18,51,53) can be studied directly in formalin-fixed material using a fluorescent microscope and can be combined with other fluorescent tracers that are visible at different wavelengths (53). Fluoro-ruby can be purchased from the same supplier as BDA (see Protocol 2) and is made up as a 10% solution in 0.1 M phosphate buffer (18). It is injected iontophoretically with a pulsed-positive DC current (4 μA; 18,53,69) and immunoreacted with a rabbit anti-TRITC antiserum (Nuclilabs, Ede, The Netherlands). However, care is required as the reaction may label blood vessels. In addition, Fluoro-ruby is capable of labeling motoneurons from a cut peripheral nerve but not when given as an intramuscular injection (51).

Biocytin is delivered best by pressure (see Protocol 2) but is also taken up effectively with iontophoresis (29) with pipet tip diameter of 10 μm and with DC-positive current, 2–3.5 μA, pulsed for 7–10 min. It is dissolved in Tris-HCl buffer (36,59) or in ethanol (5% solution; 50) and iontophoretic injections of a 5% solution in saline can also be made. The uptake mechanisms for biocytin and the dextran amines are not clear. However, the transport of biocytin is fast, as some connections are labeled in 4–8 h (33). The size of the biocytin injection site appears to be dependent upon the cellular organization (Figure 2C) as with PHA-L, but it is not directly correlated to the volume of the tracer released (59). Biocytin is primarily transported anterogradely, but some retrograde transport occurs in some systems. Figure 2D illustrates neurons labeled retrogradely after a biocytin injection in the piriform cortex (site in Figure 2C). Neurons labeled close to the injection site often have well-filled processes (arrows in Figure 2D).

The dextran amines are transported both anterogradely and retrogradely, with the anterograde direction occurring most commonly (35,69). They need at least 6 days to label targets over short distances. However, the label is not diminished even after 30 days (53). The injection sites resemble those of PHA-L, but, according to Dolleman-van der Weel et al. (18), those of Fluoro-ruby are more restricted than those of PHA-L or BDA.

Molecules such as HRP, alone or conjugated to WGA, Fluoro-gold and CTB, are versatile and sensitive retrograde tracers even though more neurons are labeled more completely with Fluoro-gold and CTB than with HRP (11,40,51,70). Biotin amplification of HRP signals can be carried out with a protocol by Adams (2). Nevertheless, the tracer HRP and many of its applications have been reviewed extensively (cf. 40), beginning with the excellent guide published by Mesulam (45).

Fluoro-gold, a cationic substituted (*trans*)-stilbene compound, was introduced for use in retrograde transport studies (52; Protocol 3). It labels cells and their processes very completely in a Golgi-like manner (11) and has recently been shown to be effective for verifying, quantitatively, the extent of neuronal loss following cytotoxic lesions (13); it is also a good tracer for EM studies when visualized with an immunocytochemical reaction (66; see Protocol 3). It is dissolved from powder and can be pressure-injected (11; total volume 0.1–0.2 µL) or iontophoretically delivered (26; Protocol 3). The size of the injection site is dependent upon the quantity injected. However, large quantities injected with pressure, i.e., greater than about 0.5 µL, can cause necrosis and should therefore be avoided. Fluoro-gold is only taken up by cut or damaged axons (52) and large diameter pipet tips (25–40 µm) lesion fibers effectively at the injection site.

Fluoro-gold is retrogradely transported to the cell body as small granules presumably in lysosomes (66). These gradually accumulate throughout the cytoplasm but not in the nucleus over a period of 2 days or more (66). The first accumulations are sparse, but longer survival times increase the number of granules in the perikaryon and dendrites (52). Survival times up to 2 months have been successfully employed, but 4–15 days are recommended (unpublished observations). Animals injected with Fluoro-gold may be perfused with a variety of fixatives, except those containing heavy metals (52); 4% formaldehyde is recommended and is compatible with immunocyto-chemical reactions. Up to 1% glutaraldehyde may also be added to the perfusion fluids if ultrastructural preservation is important (11; see Protocol 3).

CTB, either on its own (3,5,20) or conjugated to HRP (CTB-HRP; 63,70), to gold (CTB-gold; 39), to FITC or to biotin is an excellent tracer. It can be purchased alone or conjugated to other molecules from Sigma Chemical (St. Louis, MO, USA or Poole, Dorset, UK); Llewellyn-Smith and colleagues

(39) give full details of how to conjugate CTB to gold. This compound, which labels numerous neurons retrogradely even from very restricted injection sites, labels neuronal processes fully and, indeed, more completely than with Fluoro-gold (5). Complete injection parameters are found in Llewellyn-Smith et al. (39) and Trojanowski et al. (63). CTB is transported in both directions but has the drawback of being readily picked up by axons of passage (5,20).

Fluoro-gold and CTB are highly versatile. They label most central nervous system pathways in most vertebrates as our experience with amphibians shows (26). Both tracers can be used in combination with other fluorescent tracers, HRP, PHA-L, $^3$H-adenosine, biotinylated amines, and in the immunodetection of neurotransmitters (3,5,9,39,52). Antibodies to Fluoro-gold and CTB produce permanent records of the labeled cells for both LM and EM studies (3,5,11,39,66). However, CTB-gold requires silver intensification to be visible with LM (39).

## Transneuronal Tracers

Transneuronal transport of certain materials was demonstrated in the central nervous system in the early 1970s (27). Since that time, investigators have found that certain lectins such as WGA can cross synapses from axon terminals (64,70), except in certain systems (48). Furthermore, the toxins of cholera and tetanus (C fragment), which can be conjugated to HRP, and both pseudorabies and alpha herpe viruses have been introduced as transsynaptic tracers (21,62,65). The process of transneuronal transport of lectins and toxins is limited, capricious and highly dependent upon the system; however, when they cross synapses, they do so quickly, in 24–48 h (65), depending perhaps upon the type of synapse involved (48). Their mode of transport seems to involve exocytosis followed by adsorptive endocytosis, whereas that of viruses does not involve lytic processes but relies instead on the passage of virus across the synapse (62). Molecules such as HRP that are taken up with fluid-phase endocytosis seem incapable of transneuronal transport (64).

Use of the lectin WGA (Sigma Chemical), for transneuronal transport requires either pressure (64) or iontophoretic (40) injection of a 0.5%–5% solution of WGA in distilled water. Following transneuronal transport, the WGA signal is inevitably reduced, whereas viruses replicate in the cell body and therefore increase in quantity in the second-order neuron (65). The exact mode of transfer of viruses varies, and there is evidence of glial responses after viral infection. It is important to select a viral strain on the basis of its circuit-specific transport capabilities and to observe survival times strictly (62,65).

Using viruses for tracing multiple synaptic pathways requires a special level of laboratory safety. Animals injected with the virus must be kept isolated, and all work should be carried out in fume hood space dedicated to that purpose alone. The viruses are normally stored at -70°C in a freezing compartment dedicated to storage of the virus. Serum titers of research workers should be monitored and appropriate dress for handling hazardous materials should be worn (see Reference 62 for details of precautions that should be taken).

Viral administration is generally direct into the esophagus or stomach or into a peripheral sense organ such as the eye. The quantity of virus is measured in terms of plaque-forming units (pfu) per mL; Ugolini et al. (65) and Strick and Card (62) recommend a titer higher than $10^8$ pfu/mL to ensure infection of a synaptically linked circuit. The injection volume seems to be less critical than with other tracers, and these viruses can be detected immunohistochemically with polyclonal antisera directed against the various viral strains. Both avidin-biotin-peroxidase complex (ABC) and peroxidase antiperoxidase (PAP) procedures can be used to reveal the antigen. Detailed infection and immunohistochemical procedures can be found in the excellent review of Strick and Card (62).

### Dual Labels for Light and Electron Microscopy

Anterogradely transported molecules such as PHA-L and biocytin can be combined with retrograde tracers to study interactions between cells and their afferent inputs (e.g. 1,18,28,57). More commonly, anterograde and retrograde tracers are combined with the immunodetection of neuroactive substances (Figure 3, A and B; References 3,41,58,66; Protocol 2). Immunoreactions to reveal two different neurotransmitters are also employed at LM and EM levels to identify synaptic interactions between populations of neurochemically distinct cells (Figures 4 and 5 and References 44,54,67). Combined methodologies must adhere to three important principles if they are to be successful: *(i)* tracers, antisera and chromogens applied to the tissue must all be compatible; *(ii)* fixatives should be compatible with the antigens in question and *(iii)* the final markers, i.e., chromogens, gold particles, etc., must be distinguishable at LM and/or EM levels.

If Fluoro-gold and PHA-L are combined with other agents, it is important to use a high percentage of formaldehyde (> 2%) and less than 1.0% glutaraldehyde if they are to retain their antigenicity. In addition, Fluoro-gold can be diluted when combined with other fluorochromes that could be masked by its intensity (52). The best combination of chromogens for LM work probably are diaminobenzidine (DAB) with a DAB-metal conjugate, because the precipitates differ distinctly in color (3,32). For example, DAB, when used with cobalt-labeled DAB yields brown and black precipitates (3), respectively, and DAB combined with nickel-enhanced DAB produces brown and dark blue precipitates (32,41), respectively. Figure 3, A and B, show PHA-L-labeled fibers colored by DAB-nickel deposits and choline acetyltransferase-immunoreactive neurons filled with the DAB precipitate. The preservation of structural detail in this dual-label tissue is excellent when compared to tissue immunoreacted with a single label (compare Figures 1A and 3B). As a general rule, the nickel-enhanced DAB must be used first because the metal-enhanced chromogen can mask the DAB reaction product if applied secondarily (see Comments).

Combinations of PHA-L with other tracers such as HRP or BDA require that care is taken not to label inadvertently the two compounds in a single reaction. Sequential staining is best when combining PHA-L, biocytin, BDA and Fluoro-ruby (18), but simultaneous incubations can be used if the primary antisera directed against PHA-L and the ABC mixture to reveal the other tracer, are combined in the first incubation, followed by a chromogen reaction, i.e., DAB-nickel for the tracer conjugated to biotin. The remaining antibody incubations for PHA-L should follow (see e.g. 57). However, when using HRP or WGA-HRP in combination with other tracers, the HRP must be reacted before the second antigen is incubated to avoid cross-reactions. Dual-label work requires knowledge of which molecule should be reacted first, e.g., PHA-L and Fluoro-gold reveal more detail if they are reacted before the second antigen.

Ultrastructural studies demand chromogens that are electron-dense and readily distinguishable from one another (Protocols 2,4). The combinations most commonly employed are *(i)* benzidine hydrochloride (BDHC) and DAB (41,57), *(ii)* gold particles of different sizes (68) or silver-enhanced gold (SIG) combined with DAB (Figure 4, see color plate in Addendum, p. A-26 and Figure 5; Protocol 2), and *(iii)* DAB and silver-enhanced DAB (Figure 7, C and D; Protocol 4). Combinations of gold with or without DAB are particularly appealing since different size gold particles can be used together in the same tissue, and DAB can be combined with SIG for correlative LM-EM studies (44,68).

### Tract Tracing Combined with Intracellular Filling of Neurons in Fixed Slices

Studies of intracellularly filled neurons and their labeled afferent inputs truly represent the state of the art for neuroanatomical investigations (see Figure 6 and Protocol 4). Neurons that have been labeled *in vivo* with a retrograde fluorescent tracer can be intracellularly injected in lightly fixed slices. When that work is combined with anterograde tracing, also *in vivo*, of a prospective input to the "identified" projection cell, the end result is visualization of key synaptic features of a brain microcircuit (4,35,43,71).

The ideal substance for intracellular injection must be visible before and after histochemical processing, must not diffuse out of the cell, and must withstand (immuno) histochemical processing. The development of Lucifer yellow (LY; References 60,61), a 4-aminonaphthalimide dye, called sulphoflavine in the wool industry, revolutionized intracellular filling. Neurons injected with LY (Figure 6A, see color plate in Addendum, p. A-27) undoubtedly fluoresce brighter and appear more completely filled than when most other dyes are used, with the possible exception of the recently developed mini-ruby substance (see Reference 47). Fluorescent dyes for intracellular studies have the drawback of "fluorescence fading", a phenomenon presumably due to the dye in "its excited singlet state" interacting in a deleterious manner with proteins of the cell (60). Much of the detail in these fluorescent preparations is lost literally "before your eyes" (34). This is a particularly acute problem for investigators who want to study neuronal morphology and connections quantitatively. Therefore, LY-filled neurons should

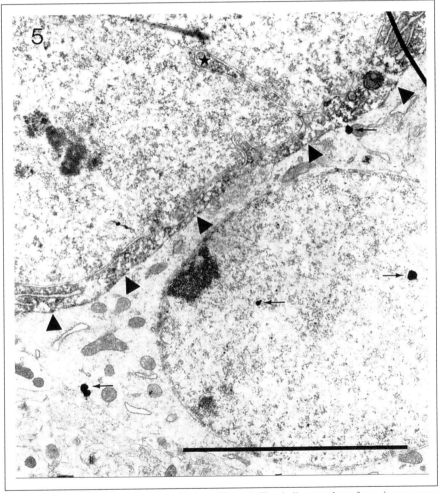

**Figure 5. Electron micrograph of two cells from Figure 4.** The choline acetyltransferase-immunoreactive neuron that contains the amorphous DAB reaction product is uppermost in the illustration. Note that the DAB deposit is only found in the thin rim of cytoplasm and does not cover the nucleus. A star marks a deep indentation in the nuclear envelope. The cell pictured in the lower half of the micrograph contains the SIG granules (small arrows) both in the nucleus and in the cytoplasm. Large filled arrowheads mark the boundary between these two cells and show how they appose one another without intervening glial processes. Scale bar = 5 μm.

be converted or immunoreacted (Figure 6B) to produce a permanent record for study (8,9,43,71) or analyzed using software adapted for use with confocal scanning microscopy (Reference 12 and see Protocol 4). Photoconversion is a useful procedure, but LY immunocytochemistry is an improvement since many cells can be processed at the same time (42). In addition, the injection of biotinylated LY shortens the time for immunocytochemistry since only the ABC reaction is necessary. There are two critical parameters for intracellular injection (Protocol 4): *(i)* the condition of the tissue and *(ii)* the quality of the electrode. The ideal fixed slice is translucent, unless myelinated fiber bundles are present, and is one in which the electrode is easily advanced to any depth (for details, see 42). Properly fixed tissue retains its ability to be injected intracellularly for up to a month, but the optimum life span is 1–10 days.

A fluorescent tracer compatible with the wavelength of LY, such as Fast Blue (Figure 5A and References 35,42,71), Fluoro-gold (9) or DiI (9,30), can be injected, *in vivo*, into a target. It is important to assess the uptake potential, pathway selectivity, toxicity and excitation wavelength before selecting a compound for retrograde transport (9,42). Postmortem, the retrogradely labeled cells can be impaled in fixed slices and filled with LY, as seen in Figure 6A, and thereafter immunoreacted with antibodies against LY (Figure 6B) to produce a permanent record (Protocol 4), also for EM analysis (Figure 7, A and D). My colleagues and I have found Fast Blue to be ideally suited as a retrograde label in this work (35,42,43,71).

Intracellular filling may be further combined with anterograde tracing using PHA-L (Figure 6C), BDA (35,71) or DiI (9,30) or with immunohistochemical labels for neuroactive substances, i.e., neurotransmitters, enzymes, receptors or synaptic vesicle proteins (4,43) for LM (Figure 6C), EM (Figure 7, C and D) or confocal laser scanning microscopic (4) analyses. The tracer, PHA-L, combines easily with LY intracellular filling for LM studies (Figure 5C) but is not compatible for EM work (71); BDA, on the other hand, is easily combined with LY immunohistochemistry at the ultrastructural level (35). Dual-label immunohistochemical reactions are successful when the LY-filled cell is immunoreacted first and the DAB precipitate is silver-intensified (Figure 7C); thereafter, the second antigen is localized with a DAB reaction (Figure 7D and Protocol 4).

## FUTURE DEVELOPMENTS

Strategies for the future lie in the discovery of more procedural options to establish connections in brain circuits. Identification of synaptic interactions requires the application of markers that can be readily distinguished from one another in confocal laser scanning and EM studies. Work with fluorescent tracers requires that labels are visualized at different wavelengths (4,56) or immunohistochemical reactions employing chromogens of different colors

(3,18,41,57). Indeed, Vector Laboratories (Burlingame, CA, USA or Peterborough, UK) now produces substrates in four different colors for LM immunohistochemistry. However, most of these cannot be distinguished ultrastructurally. Few EM studies employ more than a single marker because the procedures involved are difficult, capricious and insufficiently sensitive due

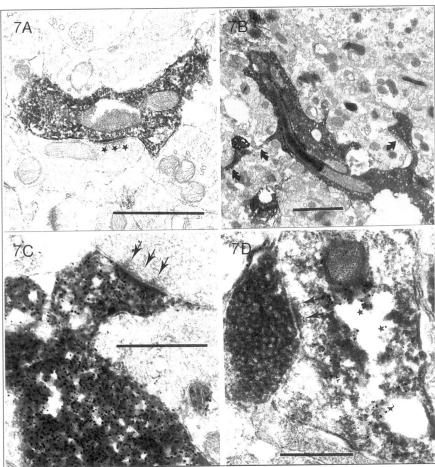

**Figure 7. Electron micrographs of cellular processes that have been intracellularly filled with LY and immunoreacted.** (A and B) illustrate single-label immunocytochemistry using DAB as the chromogen. In (A), stars mark the unlabeled bouton that is making an asymmetrical contact with the LY-filled dendrite. In (B), curved arrows point to the well-filled spines along a dendritic segment. Micrographs (C) and (D) illustrate how dual label immunocytochemistry can be used to label different elements and show their synaptic interactions. The cell was first identified in the slice preparation as a projection neuron by its Fast Blue label, then intracellularly injected with LY and immunoreacted. The DAB chromogen was further enhanced with silver. Gold toning produces the granular deposit visible in (C) and in (D), where the granules are marked with stars. The second immunoreaction was carried out for tyrosine hydroxylase and in (D), a terminal filled with the amorphous DAB deposit is making an asymmetrical contact (arrowheads) with a LY-immunoreacted dendrite. The arrow in (C) points to an unlabeled contact onto a LY-immunoreacted spine. Scale bars in A and B = 1 μm and in C and D = 0.5 μm.

to the poor penetration of antibodies in material prepared without detergents (but see Protocol 5). Triple labels are particularly tedious. It is possible, however, to combine SIG, DAB and axonal degeneration at the EM level (unpublished observations). Vector Laboratories has recently produced a peroxidase substrate, SG, which seems to be distinguishable from DAB and BDHC at the EM level (31) and Smith and Bolam (57) describe a protocol for two chromogens, DAB and BDHC, that can be combined with post-embedding gold, thereby giving the desired triple label. The development of double and triple labels for ultrastructural studies is still in its infancy but the potential to combine tract tracing with other techniques is great. Moreover, the rapid expansion in the use of confocal scanning laser microscopy (4,12) is important for analyses of the brain's microcircuits without working at the ultrastructural level. Further experimental methods that remain largely unexplored are those that combine tracing procedures with methods used by molecular biologists.

## SELECTED PROTOCOLS

**Anesthesia:** White Wistar rats are used in all experiments and recover well from surgery if anesthetized with a mixture of 4 parts ketamine and 3 parts xylazine (1 mL/kg body weight). Preceding perfusions, animals are anesthetized with an overdose of sodium pentobarbital (60 mg/kg body weight). For anesthetic choices other than these, the reader is referred to the recent, excellent review of Flecknell (22).

## 1. Tracing with PHA-L

*Equipment and supplies*

**Iontophoretic injection device.** A high voltage type that can overcome the high impedance of small tips. Good choice: Midguard® Constant Current Source (Stoelting, Wood Dale, IL, USA).

**Silver wire** to insert into the pipets for iontophoresis.

**Pipet puller.** Vertical or horizontal puller will do since electrode tips will be broken.

**Borosilicate glass tubing** with a glass fiber filament fused into the lumen, o.d. 1.5 mm. Tips should be broken (see Comments) to approximately 20–40 μm for injection sizes of about 300–500 μm in diameter.

**Vibraslice®, Vibratome® or freezing microtome.**

**Shaker.** Preferably horizontal with speed control. Orbital shakers are less appropriate but can be used.

**PHA-L.** This tracer comes as a powder from Vector Laboratories and should be used as a 2.5% solution in 0.5 M Tris-buffer, pH 7.4, with 0.9% saline (TBS, 28) or in 0.1 M sodium phosphate buffer (PB; Reference 23), pH 7.4 or in 0.1 M PB, pH 8.0, with 0.9% saline (PBS; Reference 59). The pipet may be filled by attaching plastic tubing to its open end, lowering the tip into the PHA-L solution and filling it by suction using a syringe attached to the tubing (59). The pipet may also be filled with a 5- to 10-μL Hamilton syringe, using care to ensure that the solution follows the fiber all the way into the tip. Each pipet should contain 3–5 μL of PHA-L. It is stored in small aliquots, e.g., 10 μL, sealed with Parafilm® in the freezer for extended periods or kept at 4°C for 2–3 months. It can be defrosted and refrozen often.

*Injection, survival and perfusion*

**Injection parameters:** PHA-L is injected for 15–20 min but no longer than 30 min, with alternating 5 μA (3–10 μA) DC current, 7 s on, 7 s off (Midguard settings). The current is delivered at frequencies of 0.14–1.0 Hz.

**Survival time:** 7–10 days depending on the distance the tracer is to travel.

**Perfusion:**

(1) Ringer wash (descending aorta clamped to increase the quantity of fluid passing through the brain).

(2) Fixatives (see Comments)

    (A) 500 mL (for a 200–400 g rat) of a fixative mixture of 3%–4% paraformaldehyde and 0.5%–1% glutaraldehyde in 0.1 M PB, pH 7.4.

    (B) The same mixture (300–400 mL) without glutaraldehyde should follow the first fixative.

    (C) A postfixation of the brain is only needed if it appears soft upon removal.

(3) Cutting sections

The brain should be transferred to 0.1 M PB in preparation for cutting on a vibrating microtome. It should be allowed to sink in graded solutions of 10%–20%–30% sucrose in 0.1 M PB, if it is to be cut on a freezing microtome. Sections are cut at 25–40 μm in thickness.

*Single- and dual-label immunohistochemistry for light microscopy*

**Single label** (see Comments):

(1) Free-floating sections should be rinsed thoroughly (3 × 10 min), with vigorous agitation, in 0.05 M TBS with 0.5% Triton X-100 added (TBS-T), pH 8.0.

(2) Incubate, with gentle agitation, in goat anti-PHA-L (Sigma Chemical), 1:8000 in TBS-T, 16 h at room temperature or 72 h at 4°C, followed by 3 × 10 min rinses in TBS or TBS-T.

(3) Incubate in donkey anti-goat IgG (Nordic Immunology, Tilburg, The

Netherlands), 1:200 in TBS-T for 90 min at room temperature followed by $3 \times 10$ min rinses in TBS or TBS-T.

(4) Incubate in goat peroxidase anti-peroxidase complex (goat-PAP), 1:2000 in TBS-T for 90 min at room temperature. Rinse $3 \times 10$ min rinses in 0.05 M Tris-HCl, pH 7.6.

(5) React the sections 7–20 min in 0.0125 g 3,3'-diaminobenzidine (DAB) in 25 mL Tris-HCl, pH 7.6 (filter the solution) with 8.5 µL $H_2O_2$, which produces a brown reaction product. To produce a blue-black deposit, react the sections for 7–25 min in 0.225 g ammonium nickel sulphate (careful! toxic!), 0.0075 g DAB in 50 mL Tris-HCl, pH 7.6 (do not filter), with 10 µL $H_2O_2$.

(6) Rinse sections in Tris-HCl, mount on slides from a 0.2% gelatin solution (0.1 g gelatin/50 mL 0.05 M Tris-HCl, pH 8.0; heat and stir until in solution), dehydrate and coverslip.

**Dual label, PHA-L with antibodies directed against choline acetyltransferase (ChAT)** (see Comments):

(1) Free floating sections are rinsed, with vigorous agitation, $3 \times 10$ min in 0.1 M PB with 0.9% saline added with 0.3% Triton X-100 (PBS-T), pH 8.0.

(2) Incubate, with gentle agitation, a mixture of goat anti-PHA-L, 1:8000, and monoclonal rat anti-ChAT (Boehringer Mannheim, Lewes, Sussex, UK, diluted 1:250, or Incstar, Stillwater, MN, USA, diluted 1:100), in 0.01 M PB with 0.9% saline and 0.3% Triton X-100 for 72 h at 4°C followed by $3 \times 10$ min rinses in 0.1 M PBS.

(3) Incubate sections in donkey anti-goat IgG, 1:100, and rabbit anti-rat, 1:50, in 0.01 M PBS-T for 90 min at room temperature followed by $3 \times 10$ min rinses in 0.1 M PBS.

(4) Incubate in goat-PAP, 1:800, in 0.01 M PBS-T, 90 min at room temperature. Rinse $3 \times 10$ min in 0.05 M Tris-HCl, pH 8.0.

(5) React the sections for 7–20 min in 0.225 g ammonium nickel sulphate (careful! toxic!), 0.0075 g DAB in 50 mL 0.05 M Tris-HCl, pH 8.0 (do not filter), and 10 µL $H_2O_2$. Follow with one rinse in Tris-HCl and $3 \times 10$ min rinses in 0.1 M PBS, pH 8.0.

(6) Incubate the sections in monoclonal rat PAP (Sternberger-Meyer, MD, USA), 1:200, in PBS-T for 2 h at room temperature, followed by $3 \times 10$ min rinses in 0.05 M Tris-HCl, pH 7.6.

(7) React sections for 5–15 min in 0.0125 g DAB in 25 mL Tris-HCl, pH 7.6 (filter the solution) and add 8.5 µL $H_2O_2$.

(8) Rinse sections in Tris-HCl; mount on slides from a 0.2% gelatin solution, dehydrate and coverslip.

## 2. Tracing with Biocytin or BDA

*Equipment and supplies*

**Pressure injection equipment** for biocytin. The simplest method is to attach plastic tubing to the open end of a pipet and attach a syringe to the opposite end of the tubing; the solution can then be expelled with pressure on the syringe. However, the quantities ejected are difficult to control. A control air pressure system is recommended. Good choice: Nanoliter Injector (World Precision Instruments, Sarasota, FL, USA and Aston, Stevenage, Hertfordshire, UK).

**Iontophoretic injection device** for BDA. A high-voltage type that can overcome the high impedance of small tips (see Protocol 1).

**Silver wire.**

**Pipet puller.**

**Borosilicate glass tubing** with a glass fiber filament fused into the lumen, o.d. 1.0–1.5 mm. Tips should be broken (see Comments) to approximately 25–30 µm to inject biocytin with pressure and 15–20 µm to inject BDA with iontophoresis. Biocytin injection sites are approximately 100–200 µm in diameter, whereas BDA sites are generally larger (200–500 µM).

**Vibraslice, Vibratome or freezing microtome.**

**Shaker,** preferably horizontal with speed control. Orbital shakers are less appropriate but can also be used.

**Biocytin** comes as a powder from Sigma Chemical. It is made up as a 5% solution with 0.05 M Tris-HCl, pH 7.6 (0.025 g biocytin in 500 µL buffer). The solution is cloudy and remains so even after thorough mixing. Store in small aliquots, e.g., 15–20 µL, sealed with Parafilm in the freezer. If the powder appears to come out of solution upon defrosting, draw the liquid up into a Hamilton syringe and mix again with the powder. Once defrosted, it is best not to refreeze the biocytin mixture.

**BDA** comes as a powder from Molecular Probes (Eugene, OR, USA). It is made up as 10% solution in 0.01 M PB, pH 7.4. It mixes readily and can be stored in small aliquots, e.g., 15–20 µL, sealed in the freezer.

*Injection, survival and perfusion*

**Injection parameters:**

(1) Biocytin is pressure-injected at a rate of 0.1 µL/60–90 s with a total quantity of 0.25–0.4 µL depending on the desired size of the injection site. Following the injection, leave the pipet in place 10–15 min to prevent leakage of the mixture along the electrode tract.

(2) BDA is injected iontophoretically (9) using a pulsed-positive DC current (5 µA) for 5 min (7 s on, 7 s off at Midguard settings).

**Survival time:** Optimum is 24–48 h for biocytin, which deteriorates rapidly once transported, and 7–10 days for BDA, which does not degrade as fast and can be left up to 30 days.

**Perfusion:**

(1) Ringer wash (descending aorta clamped)

(2) Fixatives

    (A) 500 mL (for a 200–400 g rat) of a fixative mixture of 3%–4% paraformaldehyde and 0.05% (BDA, Reference 18) or 0.1%–0.2% glutaraldehyde (biocytin, Reference 59) (see Comments) in 0.1 M PB, pH 7.4.

    (B) Brain should be removed and postfixed for 1 h in the same fixative at room temperature.

(3) Cutting sections: Place the brain in 0.1 M PB in preparation for cutting on a vibrating microtome or allow it to sink in graded solutions of 10%–20%–30% sucrose in 0.1 M PB, if the tissue is to be cut on a freezing microtome for LM.

*Single and dual labels for light and electron microscopy (see Comments).*

**Single label:**

(1) Free-floating sections should be rinsed, with vigorous agitation, $3 \times 10$ min, in 0.01 M PBS, pH 8.0.

(2) Incubate with gentle agitation for 1.5 h at room temperature (or 6 h at 4°C—a handling recommended if tissue is to be used for EM) in the standard ABC kit (Vector Laboratories): mix thoroughly, following manufacturer's instructions and use 0.01 M PBS; mix 30 min before incubation is to begin.

(3) Rinse once in 0.01 M PBS, pH 8.0, followed by $3 \times 10$ min rinses in 0.05 M Tris-HCl, pH 7.6.

(4) React the sections 7–20 min in 0.0125 g DAB in 25 mL Tris-HCl, pH 7.6 (filtered) with 8.5 µL $H_2O_2$, which produces a brown reaction product (ideal for EM). To produce a blue-black deposit, react the sections for 7–25 min in 0.225 g ammonium nickel sulphate (careful! toxic!), 0.0075 g DAB in 50 mL Tris-HCl, pH 7.6 (do not filter) with 10 µL $H_2O_2$. Ding and Elberger (17) recommend reacting the sections in tetramethylbenzidine-sodium tungstate as chromogen if fetal/neonatal tissue is used.

(5) Rinse sections in Tris-HCl, pH 8.0, mount on slides from a 0.2% gelatin solution, dehydrate and coverslip or follow standard embedding procedures for EM.

**Dual label:**

Dual labels are carried out with ease using these tracers since sections need treatment with only the ABC kit to reveal the biocytin or BDA, and this treatment can occur before, in conjunction with, or following incubations for the antibody series to reveal the second antigen of choice.

For EM, two electron-dense labels that can be readily identified under the electron beam, must be used (44,71). The first is generally DAB and the second can either be a chromogen such as benzidine dihydrochloride (BDHC; careful! carcinogenic!), silver-enhanced DAB (see Protocol 4) or gold particles (SIG):

(1) Rinse sections $3 \times 10$ min in PBS and incubate in the primary antibody for the neurochemical of interest, e.g., monoclonal mouse anti-leucine[5]enkephalin (ENK; Sera-Lab, Ltd., Crawley Down, Sussex, UK), diluted 1:100 in 0.01 M PBS, 72 h at 4°C.

(2) Rinse sections $3 \times 10$ min in PBS and incubate in colloidal gold-labeled goat anti-mouse IgG (5-nm gold particles; Janssen Life Sciences, Olen, Amersham, Belgium), diluted 1:50 in PBS, for 6–8 h at 4°C.

(3) Rinse sections $3 \times 10$ min in PBS and place in a 2% glutaraldehyde solution in 0.1 M PB for 30 min in order to secure the binding of the secondary antibodies.

(4) Rinse sections briefly in 1% sodium acetate and silver intensify the colloidal gold by incubating sections using the IntenSE-M kit (Janssen Life Sciences/Amersham).

(5) Rinse sections briefly in 1% sodium acetate and then $3 \times 10$ min in PBS and incubate, with gentle agitation, 6 h at 4°C in the standard ABC kit (Vector Laboratories): mix thoroughly, following instructions and use 0.01 M PBS; mix 30 min before incubation is to begin.

(6) Rinse once in 0.01 M PBS, pH 8.0, followed by $3 \times 10$ min rinses in 0.05 M Tris-HCl, pH 7.6.

(7) React the sections 7–20 min in 0.0125 g DAB in 25 mL Tris-HCl, pH 7.6 (filtered) with 8.5 µL $H_2O_2$.

(8) Embed sections for EM according to standard procedures (7).

## 3. Tracing with Fluoro-gold

*Equipment and supplies*

**Iontophoretic injection device** (see Protocol 1) and silver wire to insert into pipet.

**Pipet puller.** Vertical or horizontal puller is acceptable as electrode tips will be broken.

**Borosilicate glass tubing** with a glass fiber filament fused into the lumen, o.d. 1.5 mm. Tips should be broken (see Comments) to approximately 25–40 µm for injection sizes of approximately 400–500 µm in diameter.

**Vibraslice, Vibratome or freezing microtome.**

**Shaker,** preferably horizontal with speed control. Orbital shakers are less appropriate but can be used.

**Fluoro-gold** is a powder available from Fluorochrome (Englewood, CO, USA). It is dissolved to make a 2% solution in 0.1 M sodium cacodylate-trihydrate (cacodylate acid—contains arsenic! toxic!).

*Injection, survival and perfusion*

**Injection parameters:** Fluoro-gold is injected for 10–15 min with alternating 5–10 µA DC current, 7 s on, 7 s off (Midguard settings); lower currents yield reduced injection site size.

**Survival time:** 5–15 days depending on the distance the tracer is to travel; this compound is stable after injection for time periods up to 30 days.

**Perfusion:**

(1) Ringer wash (descending aorta clamped)

(2) Fixatives (see Comments)

    (A) 500 mL (for a 200–400 g rat) of a fixative mixture of 4% paraformaldehyde, 0.5%–2.0% glutaraldehyde and 15% saturated picric acid in 0.1 M PB, pH 7.4.

    (B) Postfix the brain in the same fixative solution if it is soft upon removal.

(3) Cutting sections: The brain should be transferred to 0.1 M PB in preparation for cutting on a vibrating microtome or allow it to sink in graded solutions of 10%–20%–30% sucrose in 0.1 M PB, if it is to be cut on a freezing microtome. Sections are cut at 25–40 µm in thickness.

*Epifluorescent illumination and Fluoro-gold*

Fluoro-gold can be seen immediately after sections have been cut. The compound fades slowly and combines easily with other tracers, providing they are labeled with a compound seen with a different wavelength (9,10), e.g., TRITC filter (Zeiss filter combination 15; Carl Zeiss, Thornwood, NY, USA). Labeling follows incubation with secondary antibodies such as goat anti-rabbit IgG (G + L) Rhodamine (TAGO, Burlingame, CA, USA).

Tissue injected with fluorescent tracers and/or tracers immunolabeled with fluorescent tags requires the use of a non-fluorescent mounting medium. Fluoromount G (Fisher Biotech, Orangeburg, NY, USA) is soluble in most aqueous buffers and can be used effectively. Sections mounted with this medium can be stored on glass slides in the freezer for months with little or no fading.

*Single label for light and electron microscopy*

(1) Rinse sections, with vigorous agitation, 3 × 10 min, in 0.01 M PBS and

incubate, with gentle agitation, in 0.3% Triton X-100 in 0.01 M PBS for 30 min at room temperature.

(2) Rinse sections, with vigorous agitation, $3 \times 10$ min, in 0.01 M PBS and incubate, with gentle agitation, in rabbit anti-Fluoro-gold, 1:2000 (however, a series of dilutions using the commercial antibodies from Fluorochrome, Inc. should be tested; see also Reference 66) in 0.01 M PBS for 72 h at 4°C.

(3) Rinse $3 \times 10$ min in PBS. Incubate sections in a biotinylated donkey anti-rabbit, 1:200, in PBS for 8 h at 4°C.

(4) Rinse $3 \times 10$ min in PBS and incubate sections in standard ABC kit (mix thoroughly, following manufacturer's instructions and use 0.05 M TBS/TBS-T; mix 30 min before incubation is to begin) for 6–8 h at 4°C.

(5) React the sections 7–20 min in 0.0125 g DAB in 25 mL Tris-HCl, pH 7.6 (filtered) with 8.5 µL $H_2O_2$, which produces a brown reaction product that is electron-dense (ideal for EM study; see Reference 66).

(6) Follow standard embedding procedures for EM (7).

## 4. Intracellular Filling of Neurons in Fixed Slices

*Equipment and supplies*

**Fixed stage microscope with epifluorescent illumination.** Good choices: Nikon Optophot® 2UD (Melville, NY, USA); Zeiss Axioskop® FS; or the less-expensive Micro Instruments M2A® (MicroInstruments Ltd., Oxford, UK).

**Filters for epifluorescent illumination** to visualize LY and Fast Blue in the same section: Nikon BV-2A; Zeiss wide-bandpass filter combination 18 (395- to 425-nm excitation, 425-nm emission, 450-nm longpass filter); M2A microscopes use Nikon BV-2A filter.

**Iontophoresis unit capable of delivering 1-3 nA of current** while providing good feedback to control electrode resistance. Good choice: Model 260, Dual Microiontophoresis Current Generator from World Precision Instruments.

**Micromanipulator** preferably with joy stick and motor-driven. Good choice: DC3001 from World Precision Instruments.

**Vibration isolation system.** Good choice: a table from Technical Manufacturing Corp., Peabody, MA, USA).

**Pressure or iontophoretic injection devices** if tracing work is to accompany intracellular injection (see Protocols 1 and 2 for suggested devices).

**Vertical or horizontal pipet puller**

**Borosilicate glass tubing** with a glass-fiber filament fused into the lumen, o.d. 1.0–1.5 mm. Pipets should have long necks and small tip diameters

($< 0.5$ µm ) with an impedance of 90–300 MΩ, when filled with LY.

**Pipets** for tracer injection with broken tips (Protocols 1 and 2).

**Silver, platinum or silver chloride-coated wire.**

**Plexiglass dish** made to fit the slide holder on microscope stage. The dish is partially cut away and may or may not have a Sylgard® base. Without the base, a glass slide is used for the slice (42). The circuit with the electrode is completed with a silver wire embedded in the base of the dish.

**Vibraslice, Vibratome or freezing microtome.**

**Shaker,** preferably horizontal with speed control. Orbital shakers are less appropriate but can also be used.

**LY** is available as a powder from Sigma Chemical. It is made into a 3%–8% solution but a 4% aqueous solution is preferred. It is stored in the dark at 4°C for 90–120 days. If it thickens or changes color before 90 days, it should be made afresh.

**Biotinylated LY** is available from Molecular Probes and is also made up as a 4% solution. Storage is difficult and thus, this solution should be made fresh, as often as possible.

**Fast blue** for retrograde tracing is available as a powder from Sigma Chemical and is made up in a 2% solution with 0.1 M PB. It is stored in the dark at 4°C for several months.

**Tracers** for anterograde tracing (PHA-L, BDA).

**4,6-diamino-2-phenylindole (DAPI)** to be used as a counterstain for nuclei in the slice is available from Sigma Chemical. It is a powder and is mixed with 0.1 M sodium phosphate buffer in a 1% stock solution. The solution is diluted further to a concentration of $10^{-7}$ when used in a bath to stain neurons in a slice.

*Injection, survival and perfusion*

**Injection of retrograde and anterograde tracers** (optional): One microliter of Fast Blue is pressure-injected from a Hamilton syringe into the target zone of neurons that are to be labeled intracellularly, thus making the cells visible in slices under epifluorescence. During the same surgery, PHA-L or BDA (see Protocols 1 and 2 and Reference 35) can be iontophoretically injected into an area thought to provide an innervation of the neurons that are to be injected intracellularly.

**Survival time:** 7–14 days depending on the distance to be traveled by the tracer(s); 7 days is adequate for the transport of Fast Blue. Slices can also be prepared from brains of animals that had not been previously operated.

**Perfusion:**

(1) Ringer wash (descending aorta clamped)

(2) Fixatives

    (A) 500 mL (for a 200–400 g rat) of a fixative mixture of 3%–4% paraformaldehyde, 0.5%–1% glutaraldehyde and 15% saturated picric acid in 0.1 M PB, pH 7.4.

    (B) Postfixation is only needed if the brain is soft upon removal (see Comments).

**Injection parameters for intracellular injection:**

(1) A slice, 100–300 μm thick (cut on a vibrating microtome), is placed on a glass slide in the plexiglass dish or directly on the Sylgard® base.

(2) If Fast Blue has not been injected, neurons may be made visible by incubating the slice for 10 min or less, in $10^{-7}$ M solution DAPI, prepared fresh from a stock solution.

(3) Cover the slice with a Millipore filter paper and cut a hole to reveal the region where the cells are labeled with Fast Blue or DAPI. Small weights hold the paper and slice in place and immerse the slice in 0.1 M PB, pH 7.4–7.6.

(4) Pipet that is to serve as the electrode is filled, i.e., with 2 μL of a 2%–8% solution of LY or a 4%–5% solution of biotinylated LY.

(5) Electrode connection to the recording circuit is usually accomplished with a silver, platinum or a silver chloride-coated wire inserted in the pipet.

(6) Electrode should approach the stage at an angle, between 30° and 65° (45° appears optimum) and a holding (positive) current should be applied during electrode approach to prevent dye leakage into the extracellular space. A Fast Blue- or DAPI- labeled cell is selected, impaled and injected by applying a negative DC current of 1–3 nA for 1–10 min (see Comments). Optimum electrode resistance for dye ejection is between 100 and 250 MΩ.

*Single and dual labels for light and electron microscopy*

**Single label:**

(1) Following the injection of several cells, the slice is removed. If cells were injected with biotinylated LY in a slice of 100-μm thickness:

    (A) The slice is rinsed in 0.01 M PBS 3 × 10 min and incubated, with gentle agitation, for 1.5 h at room temperature or 6 h at 4°C using a standard ABC kit (Vector Laboratories): mix thoroughly following manufacturer's instructions and use 0.01 M PBS; mix 30 min before incubation is to begin.

(2) If cells were injected with LY in a slice thicker than 100 μm, the slice must be resectioned:

(A) Immerse the slice in gradations of 10–20–30% sucrose buffer (0.1 M PB) until it sinks or follow Protocol 5B.

(B) Place the slice flat on a plateau of frozen 30% sucrose buffer on the stage of a freezing microtome. Cut sections of 25–40 μm and place them, serially, in a dish with small wells for free-floating immunocytochemistry.

(3) Immunocytochemistry:

(A) Rinse sections vigorously in 0.05 M TBS (EM) or for LM, TBS-T (with 0.5% Triton X-100), 3 × 10 min and incubate, with gentle agitation, in a 20% solution of a normal serum of the species in which the secondary antibodies were raised, e.g., swine normal serum, for 45 min at room temperature.

(B) Rinse in TBS (EM) or TBS-T (LM), 3 × 10 min and incubate in rabbit anti-LY (Molecular Probes; serial dilutions should be tested at 1:100, 1:500 and 1:1000) with 1% normal swine serum added, in TBS (EM) or TBS-T (LM), for 72 h at 4°C.

(C) Rinse sections in TBS (EM) or TBS-T (LM), 3 × 10 min and incubate in swine anti-rabbit IgG, diluted 1:100 with 1% normal swine serum added, in TBS (EM) or TBS-T (LM), for 8–10 h at 4°C.

(D) Rinse in TBS (EM) or TBS-T (LM), 3 × 10 min and incubate sections in rabbit PAP, diluted 1:800 in TBS for 8 h at 4°C.

(E) Rinse sections in Tris-HCl, pH 7.6, 3 × 10 min and then react them for 7–20 min in 0.0125 g DAB in 25 mL Tris-HCl, pH 7.6 (filtered) with 8.5 μL $H_2O_2$ added.

**Dual label for LY and BDA:**

(1) LM (see also Reference 35): the incubations to reveal LY should be completed with a nickel-enhanced DAB reaction: react the sections for 7–25 min in 0.225 g ammonium nickel sulphate (careful! toxic!), 0.0075 g DAB in 50 mL Tris-HCl, pH 7.6 (do not filter) with 10 μL $H_2O_2$. The ABC incubation for BDA (see Protocol 2) should be completed with a DAB reaction (see E above).

(2) EM (see also Reference 35):

(A) The DAB reaction product in the LY-injected neurons (see E above) must first be silver-intensified (71). Rinse the sections 5 × 5 min in 0.02 M PB, pH 7.2.

(B) Transfer sections to a developer made with distilled water and the following compounds added in the order given (only use spotlessly clean glassware): 0.2% paraformaldehyde, 1.0%

tungstosilicic acid, 0.2% $AgNO_3$, 0.2% $NH_4NO_3$. With continuous agitation, add slowly an equal volume of 5% $Na_2CO_3$ in distilled water.

(C) React the sections in the developer in the dark for 5–15 min or until desired degree of intensification (color change from brown to black).

(D) Rinse sections $2 \times 1$ min in 0.02 M PB and place in 2.5% $Na_2S_2O_3$ in distilled water for 1–3 min.

(E) Place the sections in 0.05% $HAuCl_4$ in distilled water (gold chloride) for 45 min at 4°C.

(F) Rinse sections $3 \times 5$ min in 0.02 M PB and $2 \times 5$ min in PBS and incubate, with gentle agitation, for 6 h at 4°C in the standard ABC kit: mix thoroughly following manufacturer's instructions and use 0.01 M PBS, mix 30 min before incubation is to begin.

(G) Rinse once in 0.01 M PBS, pH 8.0, followed by $3 \times 10$ min in 0.05 M Tris-HCl, pH 7.6.

(I) React the sections 7–20 min in 0.0125 g DAB in 25 mL Tris-HCl, pH 7.6 (filtered) and add 8.5 μL $H_2O_2$.

(J) Embed sections for EM according to standard procedures (7).

*Analysis*

Cells reacted with a chromogen such as DAB or nickel-enhanced DAB can be reconstructed using hardware (computer and digitizer) with any number of different softwares. These programs provide valuable quantitative data on the cell bodies and dendritic trees of intracellularly filled neurons. Good choice: Neurolucida™ (MicroBrightfield, Colchester, VT, USA).

Cells filled in slices can also be quantitatively analyzed using 3-D confocal fluorescence microscopy (12). New lowcost software for 3-D data collection and imaging is available from Autoquant*® Imaging (Troy, NY, USA).

## 5. Methods to Enhance the Penetration of Antibodies for Light and Electron Microscopy.

**A.** *Detergent pretreatment (71)*

(1) Rinse sections with vigorous agitation in a buffered saline (PBS or TBS), $3 \times 10$ min and incubate sections in buffer with 0.05% Triton X-100 added for 30 min with gentle agitation at 4°C.

(2) Rinse sections thoroughly in buffer (PBS or TBS) before starting incubations with the series of antibodies.

**B.** *Freeze-thaw treatment (71)*

(1) Place sections or the slice in an ascending series of dimethyl sulphoxide solutions (5%, 10%, 20% in 0.1 M PB, pH 7.4–7.6) for 10 min each.

(2) Refresh the 20% solution once and then place the tissue in a stainless steel beaker containing isopentane (2-methylbutane) already chilled in a container of liquid nitrogen.

(3) Remove tissue with a wooden spoon and allow it to defrost slowly on a glass slide at 4°C. Rinse the tissue well and carry on with incubations for the series of antibodies.

## Comments

*Protocol 1*

Pipet tips should be broken; those with tips smaller than 10 µm quickly become clogged and produce little labeling, while those with tips larger than 35 µm produce diffuse deposits that are transported poorly, if at all. If a tip becomes clogged during iontophoresis, repeated alteration of the polarity of the circuit should "unclog" it; if that fails, a new pipet must be used.

For PHA-L immunohistochemistry, it is important to do two perfusions of the brain, the first with and the second without glutaraldehyde, as PHA-L reacts poorly to double bonds and reactions are less effective if glutaraldehyde is not washed out. Other fixatives, such as saturated picric acid, can also be added to the aldehydes in concentrations up to 15%. Reactions for PHA-L can be carried out with antibodies raised either in goat or rabbit. My colleagues and I have had the best results using the goat primary antiserum available from Vector Laboratories. A peroxidase antiperoxidase (PAP) complex can follow the bridging IgG antiserum or an avidin-biotin (ABC) reaction can be used if a biotinylated secondary antibody is employed. We have had good results with both. The PAP procedure is less expensive than the ABC reaction but may produce higher background staining.

For the dual label work that combines ChAT antibodies with PHA-L, workers should be aware that the monoclonal rat anti-ChAT from Incstar is my choice for work with lower vertebrates, and monoclonal rat anti-ChAT from Boehringer-Mannheim (Indianapolis, IN, USA or Lewes, Sussex, UK) is my preference for mammalian tissue.

*Protocol 2*

Pressure injections of biocytin with the Nanoliter Injection System of World Precision Instruments are easy if tips are broken to 25–40 µm. Tips that are too small (< 25 µm) tend to clog, and those that are too large can introduce air into the system. Injections require release of the material at spaced intervals of at least one minute. Failure to do this can cause necrosis at the injection site. The optimal survival time for biocytin is 24–48 h (59). Longer times are counterproductive because biocytin degrades rapidly. Moreover, some glutaraldehyde must be present in the fixative to ensure success in revealing biocytin-labeled structures. Biocytin and BDA cannot be combined effectively since both tracers are labeled with an ABC reaction. However, both can be combined with PHA-L. Dual labels that combine biocytin or BDA with antibodies directed against a specific neurotransmitter or peptide require that the first reaction is carried out with nickel-enhanced DAB and the second reaction with DAB. The DAB-nickel compound must precede the DAB reaction if the latter is not to be masked by the metal. Also, the DAB-nickel chromogen cannot be distinguished from DAB with the EM. Therefore, different combinations of heavy metals or silver-enhanced reaction products should be employed.

*Protocol 3*

Pipet tips should be broken to at least 25 µm, since Fluoro-gold is only taken up by damaged structures. High glutaraldehyde should be avoided in perfusions unless the tissue is being prepared for ultrastructural studies, as this fixative emits a fluorescent halo under epifluorescent illumination.

*Protocol 4*

Fixation is critical when preparing slices. If the material will be processed for EM, the post-fixation period should be kept to a minimum, and work with the slice should take place on the same day as the perfusion of the animal. For LM, slices may be maintained at 4°C much longer. It is important to remember that underfixed tissue results in clogged electrode tips but can be improved by a 15–45 min immersion fixation in 4% paraformaldehyde at 4°C. In overfixed tissue, cells are difficult to penetrate and often leak dye profusely. This problem may be solved by rinsing or leaving the tissue in buffer at 4°C for 2–3 days. The best solution, however, is to start over again (31).

In working with slices, pipets may "leak" dye readily from the tip, even with a small positive holding current, which should always be used when approaching a cell for injection. When this occurs, it usually means that the tip has broken and the pipet needs replacing (31). Cells to be injected should be selected from the depths of a slice and not near the surface where dendrites may be truncated during slice preparation and will leak dye when being filled.

**ACKNOWLEDGMENTS**

The work was funded by the Department of Anatomy and Embryology at the Vrije Universiteit in Amsterdam, The Netherlands, and by a Research

Grant from the Royal College of Surgeons in Ireland. I am grateful to Ms. A. Pattiselanno for her technical assistance, Dr. H.T. Chang for preparing the material seen in Figures 4 and 5, Dr. C.A. Ingham for the preparation seen in Figure 2, C and D, and the audio-visual department at the Royal College of Surgeons in Ireland for their photographic assistance. Finally, my thanks to Dr. B.L. Roberts for critical comments on the manuscript.

## REFERENCES

1.**Aantal, D.G., T.F. Freund, P. Somogyi and R.A.J. McIlhenney.** 1990. Simultaneous anterograde labelling of two afferent pathways to the same target area with *Phaseolus vulgaris* leucoagglutinin and *Phaseolus vulgaris* leucoagglutinin conjugated to biotin or dinitrophenol. J. Chem. Neuroanat. *3*:1-9.

2.**Adams, J.C.** 1992. Biotin amplification of biotin and horseradish peroxidase signals in histochemical stains. J. Histochem. Cytochem. *40*:1457-1463.

3.**Alisky, J.M. and D.L. Tolbert.** 1994. Differential labeling of converging afferent pathways using biotinylated dextran amine and cholera toxin subunit B. J. Neurosci. Methods *52*:143-148.

4.**Belichenko, P.V. and A. Dahlstrom.** 1994. Dual channel confocal laser scanning microscopy of Lucifer Yellow-microinjected human brain cells combined with Texas red immunofluorescence. J. Neurosci. Methods *52*:111-118.

5.**Berendse, H.W., H.J. Groenewegen and A.H.M. Lohman.** 1992. Compartmental distribution of ventral striatal neurons projecting to the mesencephalon in the rat. J. Neurosci. *12*:2079-2103.

6.**Bielschowsky, M.** 1904. Die silberimprägnation der neurofibrillen. J. Psychol. Neurol. (Lpz.) *3*:169-188.

7.**Bolam, J.P.** 1992. Preparation of central nervous system tissue for light and electron microscopy, p. 1-29. *In* P. Bolam (Ed.), Experimental Neuroanatomy. A Practical Approach. Oxford University Press, Oxford.

8.**Buhl, E.H. and J. Lübke.** 1989. Intracellular Lucifer yellow injection in fixed brain slices combined with retrograde tracing, light and electron microscopy. Neuroscience *28*:3-16.

9.**Buhl, E.H.** 1993. Intracellular injection in fixed brain slices: a highly versatile tool to examine neuronal geometry in combination with other neuroanatomical techniques, p. 27-46. *In* G.E. Meredith and G.W. Arbuthnott (Eds.), Morphological Investigations of Single Neurons In Vitro. IBRO Handbook Series: Methods in the Neurosciences, Vol. 16. Wiley and Sons, Chichester.

10.**Chang, H.T.** 1993. Immunoperoxidase labeling of the anterograde tracer Fluoro-Ruby (tetramethyl-rhodamine-dextran amine conjugate). Brain Res. Bull. *30*:115-118.

11.**Chang, H.T., H. Kuo, J.A. Whittaker and N.G.F. Cooper.** 1990. Light and electron microscopic analysis of projection neurons retrogradely labeled with Fluoro-gold: notes on the application of antibodies to Fluoro-gold. J. Neurosci. Methods *35*:31-37.

12.**Cohen, A.R., B. Roysam and J.N. Turner.** 1994. Automated tracing and volume measurements of neurons from 3-D confocal fluorescence microscopy data. J. Microsc. *173*:103-114.

13.**Corodimas, K.P., J.S. Rosenblatt, T.M. Matthews and J.I. Morrell.** 1995. Neuroanatomical tract tracing provides histological verification of neuron loss following cytotoxic lesions. J. Neurosci. Methods *56*:71-75.

14.**Cowan, W.M., D.I. Gottlieb, A.E. Hendrickson, J.L. Price and T.A. Woolsey.** 1972. The autoradiographic demonstration of axonal connections in the central nervous system. Brain Res. *37*:21-51.

15.**de Olmos, J.S.** 1969. A cupric-silver method for impregnation of terminal axon degeneration and its further use in staining granular argyrophilic neurons. Brain Behav. Evol. 2:213-237.

16.**Dianyi, Y. and F.J. Gordon.** 1994. A simple method to improve the reliability of iontophoretic administration of tracer stubstances. J. Neurosci. Methods *52*:161-164.

17.**Ding, S-L. and A.J. Elberger.** 1995. A modification of biotinylated dextran amine histochemistry for labeling the developing mammalian brain. J. Neurosci. Methods *57*:67-75.

18.**Dolleman-Van der Weel, M.J., F.G. Wouterlood and M.P. Witter.** 1994. Multiple anterograde tracing, combining *Phaseolus vulgaris* leucoagglutinin with rhodamine- and biotin-conjugated dextran amine. J. Neurosci. Methods *51*:9-21.

19.**Dumas, M., M.E. Schwab and H. Thoenen.** 1979. Retrograde axonal transport of specific macromolecules as a tool for characterizing nerve terminal membranes. J. Neurobiol. *10*:179-197.

20.**Ericson, H. and A. Blomqvist.** 1988. Tracing of neuronal connections with cholera toxin subunit B: light and electron microscopic immunohistochemistry using monoclonal antibodies. J. Neurosci. Methods *24*:225-235.

21. **Fishman, P.S. and D.R. Carrigan.** 1987. Retrograde transneuronal transfer of the C-fragment of tetanus toxin. Brain Res. *406*:275-279.

22. **Flecknell, P.** 1995. Anaesthesia of animals in neuroscience. IBRO News *23*:5-8.

23. **Gerfen, C.R. and P.E. Sawchenko.** 1984. An anterograde neuroanatomical tracing method that shows the detailed morphology of neurons, their axons and terminals: immunohistochemical location of an axonally transported plant lectin, *Phaseolus vulgaris* leucoagglutinin (PHA-L). Brain Res. *290*:219-238.

24. **Gerfen, C.R., P.E. Sawchenko and J. Carlsen.** 1989. The PHA-L anterograde axonal tracing method, p. 19-47. *In* L. Heimer and L. Záborszky (Eds.), Neuroanatomical Tract-Tracing Methods 2. Recent Progress. Plenum Press, New York.

25. **Glover, J.C., G. Petursdottir and K.S. Jansen.** 1986. Fluorescent dextran-amines used as axonal tracers in the nervous system of the chicken embryo. J. Neurosci. Methods *18*:243-254.

26. **González, A., G.E. Meredith and B.L. Roberts.** 1993. The organization of efferent neurons innervating octavolateral sense organs in three amphibians, determined with retrograde labelling and choline acetyltransferase immunohistochemistry. J. Comp. Neurol. *332*:258-268.

27. **Grafstein, B.** 1971. Transneuronal transfer of radioactivity in the central nervous system. Science *172*:177-179.

28. **Groenewegen, H.J. and F.G. Wouterlood.** 1990. Light and electron microscopic tracing of neuronal connections with Phaseolus vulgaris-leucoagglutinin (PHA-L), and combinations with other neuroanatomical techniques, p. 47-124. *In* A. Björklund, T. Hökfelt, F.G. Wouterlood and A.N. van den Pol (Eds.), Analysis of Neuronal Microcircuits and Synaptic Interactions. Handbook of Chemical Neuroanatomy, Vol. 8. Elsevier Science, Amsterdam.

29. **Hall, W.C. and P. Lee.** 1993. Interlaminar connections of the superior colliculus in the tree shrew. I. The superficial gray layer. J. Comp. Neurol. *332*:213-223.

30. **Honig, M.G. and R.I. Hume.** 1989. DiI and DiO: versatile fluorescent dyes for neuronal labelling and pathway tracing. TINS *12*:333-343.

31. **Hussain, Z.** 1995. The Limbic Connections and Neuronal Populations of the Nucleus Accumbens: A Subterritorial Study with Specific Attention to the Interneuronal Population. Oxford University, UK. Thesis.

32. **Hsu, S-M. and E. Soban.** 1982. Color modification of diaminobenzidine (DAB) precipitation by metallic ions and its application for double immunohistochemistry. J. Histochem. Cytochem. *30*:1079-1082.

33. **Izzo, P.N.** 1991. A note on the use of biocytin in anterograde tracing studies in the central nervous system: application at light and electron microscopic level. J. Neurosci. Methods *36*:155-166.

34. **Johnson, G.D., R.S. Davidson, K.C. McNamee, G. Russell, D. Goodwin and E.J. Holborow.** 1982. Fading of immunofluorescence during microscopy: a study of the phenomenon and its remedy. J. Immunol. Methods *55*:231-242.

35. **Jorritsma-Byham, M.P. Witter and F.G. Wouterlood.** 1994. Combined anterograde tracing with biotinylated dextran-amine, retrograde tracing with Fast Blue and intracellular filling of neurones with Lucifer Yellow: an electron microscopic method. J. Neurosci. Methods *52*:153-160.

36. **King, M.A., P.M. Louis, B.E. Hunter and D.W. Walker.** 1989. Biocytin: a versatile anterograde neuroanatomical tract-tracing alternative. Brain Res. *497*:361-367.

37. **Kuypers, H.G.J.M., M. Bentivoglio, D. Van Der Kooy and C.E. Catsman-Berrevoets.** 1979. Retrograde axonal transport of fluorescent substances in the rat's forebrain. Neurosci. Lett. *6*:127-135.

38. **LeVay, S. and H. Sherk.** 1983. Retrograde transport of [$^3$H]proline: a widespread phenomenon in the central nervous system. Brain Res. *271*:131-134.

39. **Llewellyn-Smith, I.J., J.B. Minson, A.P. Wright and A.J. Hodgson.** 1990. Cholera toxin B-gold, a retrograde tracer that can be used in light and electron microscopic immunocytochemical studies. J. Comp. Neurol. *294*:179-191.

40. **Llewellyn-Smith, I.J., P. Pilowsky and J.B. Minson.** 1992. Retrograde tracers for light and electron microscopy, p. 31-59. *In* P. Bolam (Ed.), Experimental Neuroanatomy. A Practical Approach. Oxford University Press, Oxford.

41. **Meredith, G.E. and F.G. Wouterlood.** 1990. Hippocampal and midline thalamic fibers and terminals in relation to the choline acetyltransferase-immunoreactive neurons in nucleus accumbens of the rat: a light and electron microscopic study. J. Comp. Neurol. *296*:204-221.

42. **Meredith, G.E. and G.W. Arbuthnott.** 1993. The challenge of *in vitro* preparations for morphological investigations, p. 1-25. *In* G.E. Meredith and G.W. Arbuthnott (Eds.), Morphological Investigations of Single Neurons In Vitro. IBRO Handbook Series: Methods in the Neurosciences, Vol. 16. Wiley and Sons, Chichester.

43. **Meredith, G.E. and F.G. Wouterlood.** 1993. Identification of synaptic interactions of intracellularly injected neurons in fixed brain slices by means of dual-label electron microscopy. Microsc. Res. Tech. *24*:31-42.

44. **Meredith, G.E. and H.T. Chang.** 1994. Synaptic relationships of enkephalinergic and cholinergic neurons in the nucleus accumbens of the rat. Brain Res. *667*:67-76.
45. **Mesulam, M-M.** 1982. Principles of horseradish peroxidase neurohistochemistry and their applications for tracing neural pathways—axonal transport, enzyme histochemistry and light microscopic analysis, p. 1-151. *In* M-M. Mesulam (Ed.), Tracing Neural Connections with Horseradish Peroxidase. Wiley and Sons, Chichester.
46. **Nauta, W.J.H. and P.A. Gygax.** 1951. Silver impregnation of degenerating axon terminals in the central nervous system (1) technic (2) chemical notes. Stain Technol. *26*:5-11.
47. **Ohm, T.G. and S. Dickmann.** 1994. The use of Lucifer Yellow and Mini-Ruby for intracellular staining in fixed brain tissue: methodological considerations evaluated in rat and human autopsy brains. J. Neurosci. Methods *55*:105-110.
48. **Porter, J.D., B.L. Guthrie and D.L. Sparks.** 1985. Selective retrograde transneuronal transport of wheat germ agglutinin-conjugated horseradish peroxidase in the oculomotor system. Exp. Brain Res. *57*:411-416.
49. **Ramón y Cajal, S.** 1904. Quelques méthodes de coloration des cylindres, axes, des neurofibrilles et des nids nerveux. Trav. Lab. Rech. Biol. *3*:1-7.
50. **Raos, V. and M. Bentivoglio.** 1993. Cross talk between the two sides of the thalamus through the reticular nucleus: a retrograde and anterograde tracing study in the rat. J. Comp. Neurol. *332*:145-154.
51. **Richmond, F.J.R., R. Gladdy, J.L. Creasy, S. Kitamura, E. Smits and D.B. Thomson.** 1994. Efficacy of seven retrograde tracers, compared in multiple-labelling studies of feline motoneurones. J. Neurosci. Methods *53*:35-46.
52. **Schmued, L.C. and J.H. Fallon.** 1986. Fluoro-gold: a new fluorescent retrograde axonal tracer with numerous unique properties. Brain Res. *377*:147-154.
53. **Schmued, L., K. Kyriakidis and L. Heimer.** 1990. In vivo anterograde and retrograde axonal transport of the fluorescent rhodamine-dextran-amine, Fluoro-ruby, within the CNS. Brain Res. *526*:127-134.
54. **Sesack, S.R. and V.M. Pickel.** 1992. Dual ultrastructural localization of enkephalin and tyrosine hydroxylase immunoreactivity in the rat ventral tegmental area: multiple substrates for opiate-dopamine interactions. J. Neurosci. *12*:1335-1350.
55. **Shu, S.Y. and G.M. Peterson.** 1988. Anterograde and retrograde axonal transport of *Phaseolus vulgaris*-leucoagglutinin (PHA-L) from the globus pallidus to the striatum of the rat. J. Neurosci. Methods *25*:175-180.
56. **Skirboll, L.R., K. Thor, C. Helke, T. Hökfelt, B. Robertson and R. Long.** 1989. Use of retrograde fluorescent tracers in combination with immunohistochemical methods, p. 5-18. *In* L. Heimer and L. Záborszky (Eds.), Neuroanatomical Tract-Tracing Methods 2. Recent Progress. Plenum Press, New York.
57. **Smith, Y. and J.P. Bolam.** 1991. Convergence of synaptic inputs from the striatum and the globus pallidus onto identified nigrocollicular cells in the rat: a double anterograde labeling study. Neuroscience *44*:45-73.
58. **Smith, Y., B.D. Bennett, J.P. Bolam, A. Parent and A.F. Sadikot.** 1994. Synaptic relationships between dopaminergic afferents and cortical and thalamic input in the sensorimotor territory of the striatum in monkey. J. Comp. Neurol. *344*:1-19.
59. **Smith, Y.** 1992. Anterograde tracing with PHA-L and biocytin at the electron microscopic level, p. 61-79. *In* P. Bolam (Ed.), Experimental Neuroanatomy. A Practical Approach. Oxford University Press, Oxford.
60. **Stewart, W.W.** 1978. Functional connections between cells as revealed by dye-coupling with a highly fluorescent naphthalimide tracer. Cell *14*:741-759.
61. **Stewart, W.W.** 1981. Lucifer dyes—highly fluorescent dyes for biological tracing. Nature *292*:17-21.
62. **Strick, P.L. and J.P. Card.** 1992. Transneuronal mapping of neural circuits with alpha herpesviruses, p. 81-101. *In* P. Bolam (Ed.), Experimental Neuroanatomy. A Practical Approach. Oxford University Press, Oxford.
63. **Trojanowski, J.Q., J.O. Gonatas and N.K. Gonatas.** 1981. Conjugates of horseradish peroxidase (HRP) with cholera toxin and wheat germ agglutinin are superior to free HRP as orthogradely transported markers. Brain Res. *223*:381-385.
64. **Trojanowski, J.Q. and M.L. Schmidt.** 1984. Interneuronal transfer of axonally transported proteins: studies with HRP and HRP conjugates of wheat germ agglutinin, cholera toxin and the B subunit of cholera toxin. Brain Res. *311*:366-369.
65. **Ugolini, G., H.G.J.M. Kuypers and A. Simmons.** 1987. Retrograde transneuronal transfer of herpes simplex virus type 1 (HSV 1) from motoneurons. Brain Res. *422*:242-256.
66. **Van Bockstaele, E.J., A.M. Wright, D.M. Cestari and V.M. Pickel.** 1994. Immunolabeling of retrogradely transported Fluoro-Gold: sensitivity and application to ultrastructural analysis of transmitter-

specific mesolimbic circuitry. J. Neurosci. Methods 55:65-78.

67. **Van Bockstaele, E.J., K.N. Gracy and V.M. Pickel.** 1995. Dynorphin-immunoreactive neurons in the rat nucleus accumbens: ultrastructure and synaptic input from terminals containing substance P and/or dynorphin. J. Comp. Neurol. *351*:117-133.

68. **Van den Pol, A.N. and C. Decavel.** 1990. Synaptic interactions between chemically defined neurons: dual ultrastructural immunocytochemical approaches, p. 199-271. *In* A. Björklund, T. Hökfelt, F.G. Wouterlood and A.N. van den Pol (Eds.) Analysis of Neuronal Microcircuits and Synaptic Interactions. Handbook of Chemical Neuroanatomy, Vol. 8. Elsevier Science, Amsterdam.

69. **Veenman, C.L., A. Reiner and M.G. Honig.** 1992. Biotinylated-dextran amine as an anterograde tracer for single- and double labeling studies. J. Neurosci. Methods *41*:239-254.

70. **Wan, X-C.S., J.Q. Trojanowski and J.O. Gonatas.** 1982. Cholera toxin and wheat germ agglutinin conjugates as neuroanatomical probes: their uptake and clearance, transganglionic and retrograde transport and sensitivity. Brain Res. *243*:215-224.

71. **Wouterlood, F.G., A. Pattiselanno, B. Jorritsma-Byham, M.P.M. Arts and G.E. Meredith.** 1993. Connectional, immunocytochemical and ultrastructural characterization of neurons injected intracellularly in fixed brain tissue, p. 47-74. *In* G.E. Meredith and G.W. Arbuthnott (Eds.), Morphological Investigations of Single Neurons In Vitro. IBRO Handbook Series: Methods in the Neurosciences, Vol. 16. Wiley and Sons, Chichester.

72. **Záborszky, L. and L. Heimer.** 1989. Combinations of tracer techniques, especially HRP and PHA-L, with transmitter identification for correlated light and electron microscopic studies, p. 49-96. *In* L. Heimer and L. Záborszky (Eds.), Neuroanatomical Tract-Tracing Methods 2. Recent Progress. Plenum Press, New York.

# CHAPTER 11

# Theory and Applications of Confocal Microscopy

**Brian Matsumoto[1], Irene L. Hale[1] and Theresa R. Kramer[2]**

[1]Neuroscience Research Institute and Department of Biological Sciences, University of California, Santa Barbara, CA; [2]Department of Ophthalmology and Pathology, University of Arizona, School of Medicine, Tucson, AZ, USA

## SUMMARY

*Light microscopy is undergoing a rapid development as a research tool for the cell biologist. The recent integration of computers and optics has provided cell biologists the tools for measuring physiological processes, for studying structures that are much smaller than the wavelength of light and for reconstructing the three-dimensional architecture of tissues. An instrument type that has greatly benefited from this integration is the confocal microscope. This instrument uses an intense beam of light to raster across the specimen; it then integrates the scans to form an image. Since the effect of light scatter is minimized, the micrograph is of very high contrast. Many of these microscopes use a computer to control the illuminating light and to digitally enhance the data. These attributes enable the microscopist to study and analyze cellular processes and structures that are beyond the capabilities of traditional light and electron microscopy.*

## INTRODUCTION

Advances in light microscopy allow the biologist to resolve cellular structures that are smaller than half the wavelength of light, to quantify changes in ionic composition of living cells and to reconstruct the three-dimensional architecture of an entire cell. The increased utility of modern microscopy represents advances made in optical, video and computer technology.

An elegant example of the marriage between computer and optics is the confocal microscope. To obtain three-dimensional reconstruction of cells, these instruments take advantage of the computer's ability to coordinate image illumination and image acquisition. This joining of technologies enables the biologist to control precisely the position and pattern of illuminating light as well as quantifying changing patterns of light intensity. As with any scientific instrumentation, the technical skill of the operator determines the efficiency with which the instrument is used. A cursory knowledge of microscopy and computers can, at best, limit the information obtained during an experiment. A failure to exercise rigorous control on imaging can, at worst, introduce errors that destroy the validity of the experiment. This review is written with the intent that it will provide an introduction to the potentials of confocal microscopy for the cell biologist. In addition to summarizing publications within the field, it is the aim of the author to provide the reader an intuitive feel for the workings and operation of a confocal microscope. It is hoped that this review would stimulate new investigators to take advantage of this instrument's potential for their ongoing research.

## OPTICAL PRINCIPLES

A typical light microscope illuminates and images all points of the specimen within the field of view concurrently. Such an instrument, which is used typically for visual observation, represents an example of broad-field or wide-field imaging. In such a microscope, events affecting individual elements within the field of view can be regarded as occurring in "real time" because each point can be studied simultaneously with its neighbor. This phenomenon is familiar to many of the readers, since all optical instruments used for visual work have this characteristic (Figure 1, see color plate in Addendum, p. A-28).

However, confocal microscopes depart from this pattern because they do not illuminate broad regions of the specimen. Instead, in its most basic form, the specimen is illuminated and imaged by a single spot of light (Figure 2, see color plate in Addendum, p. A-28). To acquire structural information along the plane perpendicular to the optical axis of the microscope (x-y plane), it is necessary to move the spot across the sample and collect data from numerous points. Information from each spot can be stored temporarily so that it can be integrated with other points to build up an image that has width and height. The concept of converting a two-dimensional image into a sequence of signals was known prior to the development of the confocal microscope (50,71). Even though these earlier microscopes were not confocal, their resultant images had higher contrast because the non-illuminated areas cannot cause light to scatter and reduce the contrast in the region that is being illuminated.

A second component of the confocal microscope is an aperture within the optical train. This diaphragm is positioned and sized so that only light arising from the plane of focus passes through its opening (32). Light arising from the regions above or below the plane of focus is intercepted by the aperture's opaque borders. In summary, a confocal microscope creates high-contrast images by two actions. One, by restricting illumination to a small part of the specimen, and two, by intercepting out-of-focus light before it reaches the detector. These advantages can be achieved with transmitted, reflected or fluorescent imaging modalities (32).

Perhaps the most dramatic realization of these advantages is achieved when studying thick samples stained with fluorescent dyes. A typical epifluorescent microscope can, under ideal optical conditions, discern a fluorescent structure that is smaller than the limit of resolution. For example, Weber and coworkers detected single microtubules with a standard immunofluorescent microscope (63). Single microtubules are only 25 nm in diameter, while the limit of resolution of the light microscope is approximately 200 nm. Such results are obtainable under nearly ideal optical conditions. The sample, spread cultured cells, has a minimum height, and this physically minimizes the influence of light arising from outside the plane of focus. Specific antibody staining provides a fluorescent image that has very high contrast. Unfortunately, many samples, such as thick sections obtained by frozen sectioning, are difficult to image at such high resolution because light arising from above or below focus causes a diffuse haze that reduces contrast. A confocal microscope obviates these problems by optically restricting the detection of light to the plane of focus. Thus the optical advantages of having a thin preparation can be realized under more general conditions. This ability enables the confocal microscopist to fully exploit the sensitivity of the light microscope for studying small structures. For example, the microtubules of retinal photoreceptors (29,43) and oocytes (23,24) can be studied effectively with this instrument (Figure 3, see color plate in Addendum, p. A-29). Such a feature has advantages for morphological studies in that it provides the investigator the means to localize structures that are sub-resolution in size. Unlike electron microscopy that relies on thin sections, it is possible with the confocal microscope to follow cytoskeletal elements as they span many microns within the cell's cytoplasm (Figure 4).

Although the confocal microscope provides a demonstrable increase in resolution (67), it should be realized that the detection of submicroscopic structures is a powerful advantage and may provide more information than can be achieved by achieving higher resolution. This idea is dramatically illustrated in the field of microtubule research. For example, Allen and coworkers report that light microscopy can show microtubules from a Nomarski microscope image that is video-enhanced (1,2). These studies demonstrate that vesicular transport can occur on single microtubules. In addition, a single microtubule can support bidirectional movement. Similiar studies on microtubule dynamics may be done with a confocal microscope. Amos' laboratory shows that the bending of single microtubules can be seen with a confocal microscope (3).

## TYPES OF CONFOCAL MICROSCOPES

There are many types of confocal microscopes that are available commercially to the cell biologist. The diversity in microscope design reflects the ingenuity of the engineers seeking to develop instruments with more sensitivity as well as faster scanning rates. Unfortunately, it seems that a faster scanning rate is achieved at the expense of resolution and sensitivity.

The first reported confocal microscope is a stage or object scanning instrument (46,47). In this instrument, the specimen is moved relative to the light that is held stationary along the optical axis of the microscope. Optically, this design is most efficient for obtaining the highest resolution. The correction for chromatic and spherical aberration is more easily achieved when only the center of the lens is used for illumination and imaging (32). Such conditions allow the optical engineer to simplify lens design for optimal correction of spherical and chromatic aberration. Unfortunately, this the-

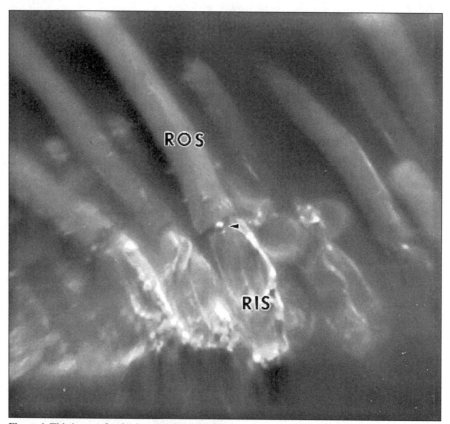

**Figure 4. This is a confocal micrograph of a photoreceptor treated with cytochalasin D and stained with rhodamine phalloidin.** The anti-actin drug disassembles f-actin cables; however, residual actin remains in the photoreceptor cell body. The concentration of f-actin in the connecting cilium (arrowhead) increases following drug treatment. This is a micrograph taken with a cooled intensified CCD camera mounted on a bilateral scanning confocal microscope (Meridian's *InSIGHT PLUS*™). A Simulated Fluorescence Process algorithm takes 8 sections (0.5 μm separation) and renders them to provide a three-dimensional reconstruction of the photoreceptor. The micrograph illustrates that the actin spot lies between the rod outer segment (ROS) and the rod inner segment (RIS). This image illustrates the increased depth of field that is possible with confocal microscopy.

234

oretical advantage in optical performance is obtained by increasing the time it is needed to acquire an image.

An example of this type of microscope is the ACAS 570™ (or Ultima™) made by Meridian Instruments (Okemos, MI, USA). In theory, a stage scanning microscope has the potential of providing the highest quality images; however, in practice, investigators seem too willing to sacrifice some optical quality for faster scans. Especially in the study of living organisms, the beam scanning microscopes are the preferred instrument.

Perhaps the most commonly available confocal microscopes are those that hold the specimen stationary and scan it with a single beam of laser light. The laser beam is targeted and scanned with galvanometer-controlled mirrors. Mirror control can be adjusted to suit experimental conditions. For example, the microscopist can increase the speed of acquiring a scan by either limiting the region to be illuminated or by limiting the resolution of the desired image (5). For example, to achieve extremely rapid scanning, the laser may be scanned as a line, and measurements of fluorescence intensity taken as a function of its position along that line. The utility of this design is exemplified by the number of manufacturers who sell instruments of this type. Such companies include Bio-Rad (Hercules, CA, USA), Leica (Deerfield, IL, USA), Zeiss (Thornwood, NY, USA), Nikon (Melville, NY, USA) and Olympus (Lake Success, NY, USA).

The stage and beam scanning confocal microscopes suffer one limitation. Their speed in acquiring an image is relatively slow. In the case of stage scanning microscopes, the acquisition time may be on the order of tens of seconds, while the beam scanning confocal microscope may require a second to acquire an image. There are designs of confocal microscopes that seek to achieve image acquisition at video rates—that is, on the order of 30 images per second. An interesting example of such an instrument is the bilateral scanning confocal microscope that was developed by Brakenoff (8). This microscope achieves a higher rate of scanning by using a slit of laser light to scan the specimen in a single sweep. The slit can be scanned at video rates, and it is possible to view the specimen through the oculars like a regular light microscope. As in the more traditional light microscope design, the system is modular. It is possible to attach a variety of image capture devices such as a 35-mm film camera to a cooled intensified CCD camera.

Another strategy for increasing the scan rate for imaging samples involves acousto-optical deflectors (AOD). This is a solid state device that has no moving parts (5,68). These two microscope designs, the bilateral scanning and the AOD driver, are examples of scanning a single laser beam at faster rates. A final strategy for faster image acquisition is scanning with multiple spots or multiple bars. This strategy is employed in the tandem-scanning confocal microscope that uses a spinning disc (53). The multiplicity of incident illumination beams and imaging scans makes the microscope capable of acquiring images at video rates. It is possible to use this instrument like a traditional microscope and view the specimen directly through the oculars.

Although the tandem-scanning confocal microscope is simple to use, its application in biological research is limited. The major problem is that it provides a dim image. This is a minor handicap in reflection light mode—one can compensate for dimness by using a light source of greater intensity. Thus, many laboratories use this microscope in this mode. This microscope is capable of generating three-dimensional images (9,10) and providing good views of Golgi-impregnated neurons (22). There are several laboratories that are taking advantage of this microscope's fast video rate to study living tissue with reflected light. The anterior portions of the eye, the cornea, is a popular tissue to study. Such work may have potential in clinical work (6,17,36). Unfortunately, image dimness limits its potential in fluorescence microscopy. This disadvantage can be overcome by using low-light video cameras to image fluorescent samples (69).

## RESOLUTION: VERTICAL AND TEMPORAL

As with any new technique, confocal microscopy is used most effectively when there is an appreciation of its limitations. A fundamental concept of any microscopy is resolution—or the ability to distinguish separate objects as being distinct. Lateral resolution, the ability to distinguish two discrete spots as separate, is defined as a function that is directly proportional to the wavelength of light and inversely proportional to the numerical aperture of the objective and condenser (31). Many biologists already have an appreciation of lateral resolution from their work with bright field microscopy. For most work, effective resolution with visible light is about 200 nm. An advantage of the point scanning confocal microscope is that it can increase resolution by 50% (67).

The scanning confocal microscope's capability of imaging delicate detail within thick structures as well as its method of illumination requires that the microscopist thinks of resolution as occurring along the vertical axis and along time. To efficiently reconstruct a cell's three-dimensional architecture requires an understanding of vertical resolution. A misunderstanding of this concept can result in either obtaining too few sections, which limits effective three-dimensional reconstruction, or obtaining too many sections, which needlessly complicates and delays three-dimensional reconstruction. Like lateral resolution, vertical resolution is dependent on the numerical aperture of the objective lens (32). Lenses with the highest numerical aperture have the finest resolution. However, it should be realized that z-axis resolution is always less than lateral resolution. Confocal microscopists report that vertical resolution with a 60 power lens and a numerical aperture of 1.4 is 0.5–0.8 μm (12). In comparison, the lateral resolution of such a lens is 0.2 μm. Thus, there is less accuracy in measuring the separation of structures along the vertical than along the lateral axis.

## Temporal Resolution

Another extension of the definition of resolution is caused by the scanning of the specimen. In other words, the studying of dynamic structures requires attention to the fourth dimension—time. At one level, confocal microscopes are scanning instruments, and objects lying in different parts of the fields may not be imaged simultaneously. This problem is obvious in the point scanning confocal microscopes; however, it is also a problem with microscopes that provide video-rate imaging. Although such microscopes, (i.e., Meridian *IN-SIGHT PLUS*™ and the Bio-Rad Thruview) provide confocal images visible through the ocular, it should be noted that the scan rates are finite. Occasionally these instruments are described as having "real time" characteristics. It would be more correct to describe these microscopes as generating images at video rates. This indicates that scan rates are finite and occur at a speed that enables either the investigator's visual systems or a video camera to integrate the individual scan into a single image.

Most point scanning confocal microscopes require at least a second to fully scan the field of view at maximum resolution. Under most conditions, the scans generate an image that is noisy, and it is desirable to average a number of scans to reduce the effects of random noise. Thus, it is not unusual to take 10 to 60 s to acquire a single micrograph. Again, such constraints will place a limit on the type of dynamic events that can be studied with the microscope. An experienced microscopist is prepared to sacrifice some spatial information to obtain higher scan rates. Thus the region to be scanned or the extent of the scan can be limited to increase the rate that data can be acquired by the microscope.

## Problems with Thick Specimens

Since confocal microscopists deal with thick specimens, it should be remembered that the sample's optical properties influence the imaging system of the microscope. An obvious effect is the interference of overlying structures on the incident beam illuminating structures of interest. In extreme cases, such as algae with a spiral arrangement of chloroplasts, the overlying structures will intercept the incident laser beam and reduce the fluorescence of underlying structures (16).

A more subtle effect of increased specimen thickness is its influence on the optical properties of the microscope. For example, the determination of stage movement (a factor for calculating vertical depth of the image within the specimen) is dependent on the index of refraction of the mounting medium (41). The higher the index of refraction, the less the apparent movement of the stage when focus changes. Failure to correctly compensate for altered degrees of stage movement can result in the reconstruction of a spherical object as an elliptical ovoid. Depending on conditions, the error in accurately measuring stage movement can be as high as 30% (41).

An additional problem when using thick specimens is the progressive loss of optical correction as one focuses into the sample. Many microscopists forget that modern objective lenses can perform optimally only under stringent conditions. For example, large numerical aperture dry lenses are optimized for specimens that lie immediately below a 0.17 mm thick coverslip (31). Unfortunately, variations in the manufacture of these items, as well as the difficulty of mounting the specimen immediately underneath the coverslip, causes an increase in spherical aberration. Frequently this can be corrected empirically by use of a correction collar that moves lens elements within the objective. For high-resolution work, confocal microscopists will use oil immersion objectives. Traditionally, such lenses were designed for specimens mounted in a medium with an index of refraction of 1.515 (i.e., Permount®, Fisher Scientific, Pittsburgh, PA, USA). Confocal microscopists use these lenses with specimens mounted under coverslips in mounting medium, such as water, glycerol or oil of wintergreen. The loss in optical correction can become significant in studies that require the highest resolution (31,41).

## APPLICATIONS OF CONFOCAL MICROSCOPY

Because of its versatility, it is impossible to do a complete review of all the applications of confocal microscopy in biological research in an article of this length. However, a few observations may be pertinent in the application of the confocal microscope to biological research. First, the majority of studies use the confocal microscope in the fluorescent light mode. Although there are studies that use the microscope with reflected light (i.e., those cited in the paragraph on the tandem-scanning confocal microscope), such studies are a minority. In part, the application of fluorochromes for studying biological processes is extremely versatile. Fluorescence microscopy is in itself a powerful technique. At the most basic level, fluorochromes are used to localize the position of a molecule within a cell, such as in indirect antibody staining. As mentioned earlier, extremely minute structures can be identified with such procedures. Additionally, because fluorescent compounds can be small and nontoxic, they can be applied to the study of living cells. Potential-sensitive dyes can both localize living mitochondria and quantify the electrical potential across their cisternea (40). The tomography of an entire cell can be evaluated when its cytoplasm is perfused with fluorescent dyes, such as Lucifer yellow (61). There are dyes that can be used for measuring pH and calcium concentration. This versatility encourages the application of confocal microscopy to all studies that use traditional epifluorescence microscopy. Such work spans structural studies on cellular organelles to physiological studies on the dynamics of living cells. The confocal microscopist's ability to precisely control and manipulate the illumination and imaging light of this microscope is advantageous in studying both fixed and living cells.

The confocal microscope is capable of viewing thick specimens that would normally require embedding, sectioning and then staining. The capability of avoiding histological preparative techniques is advantageous for studying tissues with delicate cellular structure that is susceptible to damage. This is true in developmental biology where oocytes and developing embryos can be easily damaged by histological preparative procedures. The potential for confocal microscopy in this work is illustrated by some of the investigations of David Gard. His laboratory employs the confocal microscope for studying the distribution of microtubule arrays during the development of the African clawed frog (*Xenopus laevis*). His studies suggest that the germinal vesicles, an inclusion that is rich in acetylated microtubules, might serve as a microtubule organizing center during normal development as well as a nucleating site for microtubule regrowth following drug disassembly (23). Additional studies showed that germinal vesicles have a high content of gamma tubulin (25). In later studies, Gard fluorescent-labeled tubulin, micro-injected into oocytes could be identified with the laser confocal microscope in a series of time-lapse studies (24).

These studies are examples of the sensitivity of the confocal microscope for the detection of structures that are smaller than the limit of resolution of the light microscope. Gard discusses the importance of fixatives in the preservation of the microtubules (25). It is significant that although the laser scanning confocal microscope is an instrument based on light, he employs histological preparative techniques that are similiar to those used in electron microscopy. For example, the light microscope fixatives methanol-acetic acid or Carnoy's do not preserve microtubule integrity. Cold methanol or acetone fixatives, such as those used in immunocytochemistry, also proved unsatisfactory. The optimum fixative is a modification of Karnovsky's fixative (34) and contains 3.7% formaldehyde, 0.25% glutaraldehyde and 0.5 µm taxol in a modified microtubule assembly buffer (23,26). The fixative is similiar to that employed by Matsumoto and Hale, who used 4% paraformaldehyde and 0.1% glutaraldehyde in a cacodylate buffer for preserving ocular tissue (29,43). These reports suggest that for high-resolution work, tissues should be treated with the same care that is employed for samples to be studied with the electron microscope.

As mentioned earlier, virtually all structures that have been studied by traditional light microscopy are amenable to studies with the confocal microscope. In oocytes, it is possible to localize glycoproteins (69,70), chromosomes (54,56,57,72) and endoplasmic reticulum (59).

In addition to studies on fixed cells, the confocal microscope is being used to study the dynamics of living oocytes. Calcium plays an important role in the dynamics of all living cells, and the confocal microscope can be a powerful tool for quantifying variations of this ion in response to different conditions (37). Because ultraviolet lasers are not universally available to confocal microscopists, there is a tendency to use the longer wavelength calcium dyes. Investigators who wish to use the shorter wavelength dyes have the option of using either a tandem-scanning confocal microscope with mercury illuminator or a laser capable of exciting short wavelength dyes (28,35,49).

The oocyte is easily studied with a confocal microscope. It is a large, isolated cell that can be mounted on a glass slide for direct observation with the microscope. It is more difficult to study cells within tissues. They are usually much smaller and are usually studied *in situ*. An example of a preparation that is amenable to direct observation is found in the cornea. This tissue can be studied directly because of the transparency of the corneal stroma.

However, most cells in tissue must be isolated from the organism before they can be studied. Although routine histological preparative techniques can be employed with confocal microscopy, it has been found in neuroanatomical studies that use a much thicker tissue slice (approximately 100 µm thick) can be successfully analyzed. Such thick preparations enable the microscopist to study the three-dimensional architecture of neurons without the need to mechanically section the cell. In neuroanatomy, the confocal microscope has proven to be a powerful tool for studying the three-dimensional architecture of neurons (62). The confocal microscopist can study traditional Golgi-impregnated neurons with reflectance microscopy (10,22). A recent development within this field involves injecting biotin within the neuron and then reacting it with avidin-horseradish peroxidase. Treating the sample with diaminobenzidine (DAB) creates a reaction product detectable with reflectance imaging (19,20,64). It should be noted that the capability of identifying DAB reaction products is present in the electron microscope. This enables an investigator to pursue a correlative light and electron microscope study of a neuron (21). Alternatively, fluorescent techniques, such as intracellular injection with the dye Lucifer yellow (58), can be used to identify neurons. With a double-barrel pipet, an investigator can electrically record from the cell of interest and then mark it for later morphological analysis (60). Fluorescent staining of neurons can also be accomplished by a biotin injection and by using the appropriate avidin fluorochrome or antibody staining to an appropriate protein (13,14,42,45,51,52).

Neuroanatomical studies on the physiology of living neurons is also being actively pursued. There are now many publications of calcium dynamics within neurons (30,33,44). Although this review cites many references in the developmental biology and neuroanatomy, the same studies are being carried out in a variety of tissues. For example, cardiac (15,38,39,48), kidney (4,18,35) and epithelial cells (6,7) are all being studied effectively by confocal microscopy.

## FUTURE PROSPECTS OF CONFOCAL MICROSCOPY

Confocal microscopes are still new instruments to the cell biologist. However, there are already many reports on this microscope's application for the study of living cells. Future improvements in microscope design will include increasing the speed of image acquisition and heightening the sensitivity in the photodetection system. Such developments are necessary for long-term studies of living cells. The increase in image acquisition rates will allow the investigator to increase the temporal resolution in quantifying dynamic changes within the cell. An increase in sensitivity will allow the investigator to reduce incident illumination to decrease risks in phototoxicity.

An additional development for the confocal microscope will be the development of new light sources for imaging different dyes. Presently there is an active interest in the development of ultraviolet light sources for this microscope (28,48,49). Along with the developments of extending the wavelengths available to the microscopist will be the development of light sources that possess multiple wavelengths. Such illuminators will improve the capability of the confocal microscopist to localize multiple dyes within the same tissue. The development of the krypton-argon laser is an example of a new laser developed for multiple dye localization with the confocal microscope (11,12).

All of these developments will increase the confocal microscope's utility to the cell biologist. We can expect the microscope's high resolving power and its precise control of illuminating and imaging light to extend our understanding of cell physiology and structure.

## REFERENCES

1. **Allen, R.D., J.L. Travis, N.S. Allen and H. Yilmaz.** 1981. Video-enhanced contrast, differential interference contrast (AVEC-DIC) microscopy: a new method capable of analyzing microtubule-related motility in the reticulopodial network of Allogromia laticollaris. Cell Motil. *1*:291-302.
2. **Allen, R.D., D.G. Weiss, J.H. Hayden, D.T. Brown, H. Fujiwake and M. Simpson.** 1985. Gliding movement of and bidirectional transport along single native microtubules from squid axoplasm. Evidence for an active role of microtubules in cytoplasmic transport. J. Cell Biol. *100*:1736-1752.
3. **Amos, L.A. and W.B. Amos.** 1991. The bending of sliding microtubules imaged by confocal light microscopy and negative stain electron microscopy. J. Cell Sci. Suppl. *14*:95-101.
4. **Andrews, P.M., W.M. Petroll, H.D. Cavanagh and J.V. Jester.** 1991. Tandem scanning confocal microscopy (TSCM) of normal and ischemic living kidneys. Am. J. Anat. *191*:95-102.
5. **Art, J.J. and M.B. Goodman.** 1993. Rapid scanning confocal microscopy. Methods Cell Biol. *38*:47-77.
6. **Auran, J.D., M.B. Starr, C.J. Koester and V.J. LaBombardi.** 1994. In vivo scanning slit confocal microscopy of Acanthamoeba keratitis. A case report. Cornea *13*:183-185.
7. **Auran, J.D., C.J. Koester, R. Rapaport and G.J. Florakis.** 1994. Wide field scanning slit in vivo confocal microscopy of flattening-induced corneal bands and ridges. Scanning *16*:182-186.
8. **Brakenhoff, G.J. and K. Visscher.** 1992. Confocal imaging with bilateral scanning and array detectors. J. Microsc. *165*:139-146.
9. **Boyde, A.** 1985. Stereoscopic images in confocal (tandem scanning) microscopy. Science *230*:1270-1272.
10. **Boyde, A.** 1992. Three-dimensional images of Ramon y Cajal's original preparations, as viewed by confocal microscopy. Trends Neurosci. *15*:246-248.
11. **Brelje, T.C. and R.L. Sorenson.** 1992. U.S. Patent 5,127,730.
12. **Brelje, T.C., M.W. Wessendorf and R.L. Sorenson.** 1993. Multicolor laser scanning confocal immunofluorescence microscopy: practical application and limitations. Methods Cell Biol. *38*:97-181.
13. **Brodin, L., M. Ericsson, K. Mossberg, T. Hokfelt, Y. Ohta and S. Grillner.** 1988. Three-dimensional reconstruction of transmitter-identified central neurons by "en bloc" immunofluorescence histochemistry and confocal scanning microscopy. Exp. Brain Res. *73*:441-446.
14. **Carlsson, K., P. Wallen and L. Brodin.** 1989. Three-dimensional imaging of neurons by confocal fluorescence microscopy. J. Microsc. *155*:15-26.

15. **Chacon, E., J.M. Reece, A.L. Nieminen, G. Zahrebelski, B. Herman and J.J. Lemasters.** 1994. Distribution of electrical potential, pH, free Ca2+, and volume inside cultured adult rabbit cardiac myocytes during chemical hypoxia: A multiparameter digitized confocal microscopic study. Biophys. J. *66*:942-952.

16. **Cheng, P.C. and R.G. Summers.** 1990. Image contrast in confocal microscopy, p. 179-195. *In* J.B. Pawley (Ed.), Handbook of Biological Confocal Microscopy. Plenum Press, New York.

17. **Chew, S.J., R.W. Beuerman and H.E. Kaufman.** 1992. In vivo assessment of corneal stromal toxicity by tandem scanning confocal microscopy. Lens Eye Toxic Res. *9*:275-292.

18. **Cornell-Bell, A.H., L.R. Otake, K. Sadler, P.G. Thomas, S. Lawrence, K. Olsen, F. Gumkowski, J.R. Person and J.D. Jamieson.** 1993. Membrane Glycolipid trafficking in living, polarized pancreatic acinar cells: Assessment by confocal microscopy. Methods Cell Biol. *38*:221-240.

19. **Deitch, J.S., K.L. Smith, C.L. Lee, J.W. Swann and J.N. Turner.** 1990. Confocal scanning laser microscope images of hippocampal neurons intracellularly labeled with biocytin. J. Neurosci. Methods *33*:61-76.

20. **Deitch, J.S., K.L. Smith, J.W. Swann and J.N. Turner.** 1990. Parameters affecting imaging of the horseradish-peroxidase-diaminobenzidine reaction product in the confocal scanning laser microscope. J. Microsc. *160*:265-278.

21. **Deitch, J.S., K.L. Smith, J.W. Swann and J.N. Turner.** 1991. Ultrastructural investigation of neurons identified and localized using the confocal scanning laser microscope. J. Electron Microsc. Tech. *18*:82-90.

22. **Freire, M. and A. Boyde.** 1990. Study of Golgi-impregnated material using the confocal tandem scanning reflected light microscope. J. Microsc. *158*:285-290.

23. **Gard, D.L.** 1991. Organization, nucleation, and acetylation of microtubules in Xenopus laevis oocytes: A study by confocal immunofluorescence microscopy. Dev. Biol. *143*:346-362.

24. **Gard, D.L.** 1992. Microtubule organization during maturation of Xenopus oocytes: Assembly and rotation of the meiotic spindles. Dev. Biol. *151*:516-530.

25. **Gard, D.L.** 1994. Gamma-tubulin is asymmetrically distributed in the cortex of Xenopus oocytes. Dev. Biol. *161*:131-140.

26. **Gard, D.L.** 1993. Ectopic spindle assembly during maturation of Xenopus oocytes: evidence for functional polarization of the oocyte cortex. Dev. Biol. *159*:298-310.

27. **Gard, D.L.** 1993. Confocal immunofluorescence microscopy of microtubules in amphibian oocytes and eggs. Methods Cell Biol. *38*:242-265.

28. **Girard, S. and D.E. Clapham.** 1994. Simultaneous near ultraviolet and visible excitation confocal microscopy of calcium transients in Xenopus oocytes. Methods Cell Biol. *40*:263-284.

29. **Hale, I.L. and B. Matsumoto.** 1993. Resolution of subcellular detail in thick tissue sections: immunohistochemical preparation and fluorescence confocal microscopy. Methods Cell Biol. *38*:289-324.

30. **Hernandez-Cruz, A., F. Sala and P.R. Adams.** 1990. Subcellular calcium transients visualized by confocal microscopy in a voltage-clamped vertebrate neuron. Science *247*:858-862.

31. **Inoué, S.** 1986. Video Microscopy. Plenum Press, New York.

32. **Inoué, S.** 1990. Foundation of confocal scanned imaging in light microscopy, p. 1-14. *In* J.B. Pawley (Ed.), Handbook of Biological Confocal Microscopy. Plenum Press, New York.

33. **Jaffe, D.B., S.A. Fisher and T.H. Brown.** 1994. Confocal laser scanning microscopy reveals voltage-gated calcium signals within hippocampal dendritic spines. J. Neurobiol. *25*:220-233.

34. **Karnovsky, M.J.** 1965. A formaldehyde-glutaraldehyde fixative of high osmolarity for use in electron microscopy. J. Cell Biol. *27*:131a.

35. **Kurtz, I. and C. Emmons.** 1993. Measurement of intracellular pH with a laser scanning confocal microscope. Methods Cell Biol. *38*:183-194.

36. **Lemp, M.A., P.N. Dilly and A. Boyde.** 1985. Tandem-scanning (confocal) microscopy of the full-thickness cornea. Cornea *4*:205-209.

37. **Lechleiter, J.D. and D.E. Clapham.** 1992. Spiral waves and intracellular calcium signalling. J. Physiol. (Paris) *86*:123-128.

38. **Lipp, P. and E. Niggli.** 1993. Ratiometric confocal Ca(2+)-measurements with visible wavelength indicators in isolated cardiac myocytes. Cell Calcium *14*:359-372.

39. **Lipp, P. and E. Niggli.** 1994. Modulation of Ca2+ release in cultured neonatal rat cardiac myocytes. Insight from subcellular release patterns revealed by confocal microscopy. Circ. Res. *74*:979-990.

40. **Loew, L.M.** 1993. Confocal microscopy of potentiometric fluorescent dyes. Methods Cell Biol. *38*:195-209.

41. **Majlof, L. and P. Forsgren.** 1993. Confocal microscopy: important considerations for accurate imaging. Methods Cell Biol. *38*:79-97.

42. **Mason, P., S.A. Back and H.L. Fields.** 1992. A confocal laser microscopic study of enkephalin-immunoreactive appositions onto physiologically identified neurons in the rostral ventromedial medulla. J. Neurosci. *12*:4023-4036.

43. **Matsumoto, B. and I. Hale.** 1993. Preparation of retinas for studying photoreceptors with confocal microscopy. Methods Neurosci. *15*: 54-71.

44. **Melamed, N., P.J. Helm and R. Rahamimoff.** 1993. Confocal microscopy reveals coordinated calcium fluctuations and oscillations in synaptic boutons. J. Neurosci. *13*:632-649.

45. **Mesce, K.A., K.A. Klukas and T.C. Brelje.** 1993. Improvements for the anatomical characterization of insect neurons in whole mount: the use of cyanine-derived fluorophores and laser scanning confocal microscopy. Cell Tissue Res. *271*:381-397.

46. **Minsky, M.** 1957. U.S. Patent 3013467 Microscopy Apparatus.

47. **Minsky, M.** 1988. Memoir on investing the confocal microscope. Scanning *10*:128-138.

48. **Niggli, E. and W.J. Lederer.** 1990. Real-time confocal microscopy and calcium measurements in heart muscle cells: towards the development of a fluorescence microscope with high temporal and spatial resolution. Cell Calcium *11*:121-130.

49. **Niggli, E., D.W. Piston, M.S. Kirby, H. Cheng, D.R. Sandison, W.W. Webb and W.J. Lederer.** 1994. A confocal laser scanning microscope designed for indicators with ultraviolet excitation wavelengths. Am. J. Physiol. *266*:303-310.

50. **Nipkow, P.** 1884. German patent 30,105.

51. **Onstott, D., B. Mayer and A.J. Beitz.** 1993. Nitric oxide synthase immunoreactive neurons anatomically define a longitudinal dorsolateral column within the midbrain periaqueductal gray of the rat: Analysis using laser confocal microscopy. Brain Res. *610*:317-324.

52. **Ornatowska, M. and J.A. Glasel.** 1992. Two- and three-dimensional distributions of opioid receptors on NG108-15 cells visualized with the aid of fluorescence confocal microscopy and anti-idiotypic antibodies. J. Chem. Neuroanat. *5*:95-106.

53. **Petran, M., M. Hadravsky, M.D. Egger and R. Galambos.** 1968. Tandem-scanning reflected light microscope. J. Opt. Soc. Am. *58*:660-664.

54. **Robinson, J.M. and B.E. Batten.** 1989. Detection of diaminobenzidine reactions using scanning laser confocal reflectance microscopy. J. Histochem. Cytochem. *37*:1761-1765.

55. **Segal, M. and D. Manor.** 1992. Confocal microscopic imaging of [Ca2+]i in cultured rat hippocampal neurons following exposure to N-methyl-D-aspartate. J. Physiol. (Lond) *448*:655-676.

56. **Summers, R.G., S.A. Stricker and R.A. Cameron.** 1993. Applications of confocal microscopy to studies of sea urchin embryogenesis. Methods Cell Biol. *38*:265-287.

57. **Summers, R.G., C.E. Musial, P.C. Cheng, A. Leith and M. Marko.** 1991. The use of confocal microscopy and STERECON reconstructions in the analysis of sea urchin embryonic cell division. J. Electron Microsc. Tech. *18*:24-30.

58. **Stewart, W.W.** 1981. Lucifer dyes—highly fluorescent dyes for biological tracing. Nature *292*:17-21.

59. **Terasaki, M. and J.A. Laurinda.** 1993. Imaging endoplasmic reticulum in living sea urchin eggs. Methods Cell Biol. *38*:211-220.

60. **Turner, J.N., D.H. Szarowski, K.L. Smith, M. Marko, A. Leith and J.W. Swann.** 1991. Confocal microscopy and three-dimensional reconstruction of electrophysiologically identified neurons in thick brain slices. J. Electron Microsc. Tech. *18*:11-23.

61. **Turner, J.N., J.W. Swann, D.H. Szarowski, K.L. Smith, D.O. Carpenter and M. Fejtl.** 1993. Three dimensional confocal light microscopy of neurons: fluorescent and reflection stains. Methods Cell Biol. *38*:345-366.

62. **Wallen, P., K. Carlsson, A. Liljeborg and S. Grillner.** 1988. Three-dimensional reconstruction of neurons in the lamprey spinal cord in whole-mount, using a confocal laser scanning microscope. J. Neurosci. Methods *24*:91-100.

63. **Weber, K., P.C. Rathke and M. Osborne.** 1978. Cytoplasmic microtubular images in glutaraldehyde-fixed tissue culture cells by electron microscopy and by immunofluorescence microscopy. Proc. Natl. Acad. Sci. USA *75*:1820-1824.

64. **Welsh, M.G., J.M. Ding, J. Buggy and L. Terracio.** 1991. Application of confocal laser scanning microscopy to the deep pineal gland and other neural tissues. Anat. Rec. *231*:473-478.

65. **Williams, D.A.** 1990. Quantitative intracellular calcium imaging with laser-scanning confocal microscopy. Cell Calcium *11*:589-597.

66. **Williams, D.A.** 1993. Mechanisms of calcium release and propagation in cardiac cells. Do studies with confocal microscopy add to our understanding? Cell Calcium *14*:724-735.

67. **Wilson, T. and C. Shepard.** 1984. Theory and Practice of Scanning Optical Microscopy. Academic Press, New York.

68. **Wright, S.J., V.E. Centonze, S.A. Stricker, P.J. DeVries, S.W. Paddock and G. Schatten.** 1993. In-

troduction to confocal microscopy and three-dimensional reconstruction. Methods Cell Biol. *38*:1-45.

69. **Wright, S.J., J.S. Walker, H. Schatten, C. Simerly, J.J. McCarthy and G. Schatten.** 1989. Confocal fluorescence microscopy with the tandem scanning light microscope. J. Cell Sci. *94*:617-624.

70. **Yoshida, M., D.G. Cran and V.G. Pursel.** 1993. Confocal and fluorescence microscopic study using lectins of the distribution of cortical granules during the maturation and fertilization of pig oocytes. Mol. Reprod. Dev. *36*:462-468.

71. **Young, J.Z. and F. Roberts.** 1951. A flying-spot microscope. Nature *167*:231.

72. **Zernicka-Goetz, M., J.Z. Kubiak, C. Antony and B. Maro.** 1993. Cytoskeletal organization of rat oocytes during metaphase II arrest and following abortive activation: a study by confocal laser scanning microscopy. Mol. Reprod. Dev. *35*:165-175.

# CHAPTER 12

# Standardization and Practical Guidelines of Image DNA Cytometry in Clinical Oncology

## Ursula G. Falkmer[1] and Gerhard W. Hacker[2]

[1]Department of Oncology and Pathology, Karolinska Institute and Hospital, Stockholm, Sweden; [2]Immunohistochemistry and Biochemistry Unit, Institute of Pathology, County Council Hospital, Salzburg, Austria

## SUMMARY

*DNA cytometry—both flow cytometry (FCM) and image cytometry (ICM)—has become an established technique in the field of analytical cellular pathology. These techniques are mainly used to obtain information about the nuclear DNA content of neoplastic parenchymal cells in human malignant tumors. The DNA ploidy pattern and the S-phase fraction offer tumor cell biological data that are supplementary to the often purely descriptive cytodiagnostic and histopathologic techniques. Diverging results among different laboratories have limited the value of DNA cytometry as a supplementary prognosticating tool, and this is the main reason why only a few clinical oncologic trials have the DNA data included in their protocols. The differences of the results seem to depend mainly on the methodological procedures used. In an effort to evaluate and to minimize sources of errors, particularly those originating from technical steps, the results of concomitant FCM and ICM DNA assessments on a large series of normal human cells and of cells of a broad spectrum of malignant tumors have been used to analyze each of the methodological steps in detail. Based on these observations, standardized procedures for sampling, fixation, staining and measurement were developed, as well as for data storage and evaluation of the ICM DNA histograms, together with guidelines for quality assurance.*

## INTRODUCTION

Cytometric assessments of the nuclear DNA content of neoplastic parenchymal cells have been extensively performed in clinical oncology for more than 20 years (2,9,34,36). The DNA ploidy (stemline position given as the DNA Index [DI]) and the S-phase fraction (SPF) (2,24,34) of tumor cells can be used to assess the degree of the neoplastic potential (genomic deviation/instability and proliferation) of the cells investigated. A highly malignant tumor is usually associated with an aggressive neoplastic disease and thus with a

245

poor prognosis for the patient (2,10,11,18, 32,40,42).

Two techniques of DNA cytometry to measure the DNA ploidy and SPF are image cytometry (ICM) (2,9,36) and flow cytometry (FCM) (34,38). With the recent advances in computer technology, a number of analysis systems have been developed for ICM and FCM. As a result, many clinico-pathological laboratories now have the capability to use DNA cytometry. The DNA cytometric data provide additional information to the fundamental cytodiagnostic and histopathologic assessments of the patient's tumor (2,5,12, 28,30,33,43,46).

Using ICM and FCM, comprehensive studies have been performed on most kinds of normal and neoplastic cell types. The studies have demonstrated that crude changes in nuclear DNA content are a common event in many human neoplastic lesions (2,35,38). The nuclei of the neoplastic cells of benign tumors generally display euploid (diploid and tetraploid) DNA histograms, and those of malignant tumors frequently express DNA aneuploid ones (2,3,10,11,18,30,32,33,36,40,42,43,46). However, there are considerable numbers of exceptions to this rule, and they make the use of DNA cytometric results as a diagnostic tool somewhat controversial (1,27,29,50).

In most entities of solid malignant tumors, the clinical treatment decisions for the individual patient are based on the conventional tumor grading (TNM grading) (41) as well as fundamental histopathological and clinical facts. However, for a more precise prediction of the course of the disease, additional information is required. Thus, several large groups of solid malignant tumors, such as carcinomas of the breast, the prostate, the gastrointestinal tract, the urinary bladder, the ovary and the uterus, have been investigated by DNA cytometry in numerous studies with the aim of obtaining reliable prognostic information (2,8,10,11,18,30–32,36,40,42,45,49). Nevertheless, the results of these studies are not always unequivocal and the prognostic impact of cytometric DNA assessments even with practical clinical implications for the individual patient is still rather limited.

Methodological problems of DNA cytometry among different laboratories have created controversies about the value of cytometric DNA assessments in various malignant neoplastic diseases (13–15,20,26,48). Several methodological sources of error are inherent in each of the cytometric techniques and are notorious in influencing the DNA cytometric results. Today, the methodological problems among different FCM laboratories are still not resolved (4,6,12,22,23,37,39,49). In clinical ICM DNA cytometry, the methodological problems are, at least, of the same magnitude as in FCM. ICM DNA histogram evaluations and interpretations are even more confusing than in FCM (2,7,12,19,44). The aim of the present investigation is to scrutinize every single step in the ICM procedure in order to reveal possible sources of errors and to give practical advice to minimize them.

ICM was the first technique introduced for DNA cytometry (2,9,36). By means of common ICM equipment, the light transmission per "pixel" (picture element) in the cell nuclei stained with absorption dyes is measured. These values are transformed into those representing the local density, integrated over the whole field measured, and given as the integrated optical density (IOD). The IOD is assumed to be equivalent to the amount of DNA present in the nuclei. For ICM DNA assessments, it is necessary to prepare the cells on glass slides, preferably as single-cell preparations.

The most commonly used technique for the assessments of the nuclear DNA content in neoplastic cells is FCM (34). In FCM, the fluctuation in light intensity of a fluorochrome dye binding to the nuclear DNA is measured. The amount of fluorescence from the stain is assumed to be equivalent to the amount of DNA present. For FCM DNA assessments, it is necessary to prepare the cells in a monocellular suspension.

In our laboratories, we have done comprehensive studies of >10000 DNA assessments by using both cytometric methods concomitantly on the same tumor material. This has allowed us to analyze the difficulties in each one of the methodological steps (1,13–15,17,27,29,31,50). The investigations were made on carcinomas of the breast, prostate, gastrointestinal tract and uterus, as well as on tumors or tumorlike lesions of the endocrine system, soft tissue and bone sarcomas (1,2,5,13,14,17,18,27,29,30,31,33,43). Based on these observations, standardized procedures for sampling, fixation, staining and measurement were developed, as well as for data storage and evaluation of the ICM DNA histograms, together with guidelines for quality assurance.

## STEP-BY-STEP CONSIDERATION OF THE ICM PROCEDURE

### Sampling and Fixation

The best cellular material for ICM DNA assessments are single-cell preparations that are obtained by several techniques and optimized for each tumor entity. From tumor nodules in the patient, the fine needle aspiration (FNA) method or the imprint technique (from cut biopsy specimens) can be applied. Both techniques yield representative cellular material. They can also be used on freshly excised or fresh-frozen tumor specimens.

The aspirated cellular material is gently smeared on a clean glass slide, air-dried and fixed overnight in 10% neutral buffered formalin. Imprints from freshly cut biopsy specimens, e.g., from true-cut needle biopsy specimens, are fixed in the same manner, as well as imprints obtained on freshly excised and fresh-frozen tumor specimens after rapid thawing. The amount of cells depends on the histopathological structure of the neoplastic lesion. Samples from tumor specimens with a stroma containing hyaline sclerotic structures, as in mammary carcinomas and pancreatic adenocarcinomas, often show a smaller number of parenchymal tumor cells than samples from tumors where the stroma is scanty. In lesions containing only a few parenchymal cells, the FNA biopsy technique seems to be superior to the imprint technique for enriching the samples with parenchymal cells.

**Feulgen-stained threads.** When samples with Feulgen threads appear in the nuclei, the sampling procedures have been made with too much mechanical force. It occurs especially when highly vulnerable tumor cells are prepared. The nuclear membrane is opened and the DNA leaks out. Such a selective destruction of a highly vulnerable tumor cell population during the first step of the method must be avoided (Figure 1, see color plate in Addendum, p. A-30).

**High coefficient of variation (CV).** The main peaks of normal and neoplastic (benign or malignant) cells have a certain broadness, which is calculated via the CV (standard deviation divided by the modal value). When the CV is exceeding the mean value with >2× the standard deviation (SD), the reason can be that an artifact was evoked during the air-drying procedure. When the cellular material obtained is placed in a fluid, e.g., in a FNA biopsy specimen of the prostate in its secretion fluid, the drying process takes too much time. Then, the nuclei swell and give rise to falsely high CV values of their peaks (Figure 2, see color plate in Addendum, p. A-31). The air-drying step should be carried out as fast as possible, and a hair dryer or a fan at room temperature can be used to accelerate this process.

High CV values can also reflect an inappropriate fixation. Specimens fixed in Bouin's fluid cannot be used for DNA cytometry. The reasons are essentially unknown but the picric acid in the Bouin's fluid may have some influence. Samples in which there is a suspicion of inappropriate fixation, should —for safety's sake—be refixed in neutral buffered 10% formalin over-night. Also, after refixation in 10% formalin, the specimens cannot be appropriately stained by the Feulgen procedure and they display a peculiar, faint density.

**Cell losses.** When using the imprint technique, the cells of certain types of specimens fail to stick to the surface of glass slides. This is a problem for those with a high fat content and those from a benign, encapsulated mesenchymal tumor. In order to avoid this difficulty, we use poly-L-lysine-coated glass slides, a technique that helps considerably.

## Staining

A number of staining procedures are used for ICM DNA assessments, and these procedures differ from one another considerably. Feulgen staining is the most commonly applied method and serves as the standard procedure. It includes acid (5 $N$ HCl) hydrolysis of the DNA and the highly specific staining reaction of the DNA molecule with Schiff's reagent. The reagent binds covalently to DNA (12,21) and does not react with RNA or proteins. Thus, RNA and proteins do not have to be removed. The efficiency of acid hydrolysis varies among different cell types, which is related to the compactness of their chromatin structure (13).

**Pale nuclei, high CV.** The hydrolysis time and the temperature are the critical steps of the Feulgen procedure. Therefore, the rinsing procedure that terminates the hydrolysis has to be made with great care. If there is any HCl left, over-denaturation and destruction of the DNA are common. The nuclei become pale and the density values measured become low. The peaks of such DNA histograms are characteristically broad with high CVs (Figure 2). Another common cause for the appearance of pale nuclei and broad CVs in the DNA histograms is that the highly light-sensitive Schiff reagent has not been protected against light. The reagent has to be kept in the dark before *and* during the staining. Also, the stained specimens have to be protected against light as much as possible.

## Measurement and Data Storage

Modern, microscope-based and CCD-camera-equipped image analysis systems allow a visual inspection and identification of the cell to be analyzed. The DNA assessments are performed by measuring the light transmission per "pixel", adequately transforming these values into those representing the local density, and followed by integration over the whole field of the nuclei. At this stage, background determination for each object and automatic particle detection by thresholding procedures are made. The measured values are stored in a "list-mode file" on a "cell-by-cell" basis. For multilaboratory studies the measurements can be exported as data files. Data acquisition, analysis and storage all become accessible "on-line" during measurement. DNA histograms can be displayed as one-dimensional frequency distributions of cell count vs. DNA content, or preferably as two-parameter "dot-plots" of DNA content vs. nuclear area (Figure 3).

The actual measurement of the cell nuclei of interest is made according to the "at random principle". It is of great importance that ICM measurements register the randomly selected cell nuclei and all events on the glass slide that are not selected by the investigator. These non-randomly selected cells can automatically be "detected"

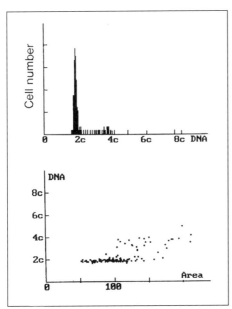

Figure 3. A DNA histogram displayed as one-dimensional frequency distribution, (the upper diagram). The x-axis contains 250 channels (0.04c per channel). In most assessments, the 2c (DI 1.0) peak of the cell population is placed in channel 50. In the lower diagram the same assessment is shown as a two-parameter "dot-plot" of DNA content (y-axis) vs. the nuclear area (x-axis). The main G0/G1 peak shows quite a narrow peak in the DNA distribution (CV 4.1%) despite a nuclear area ratio of 1:3.

and registered during the measuring process without influence of the investigator. These values turn up as a second histogram at the end of the measurement (Figure 4).

**Deficient linearity stability.** Linearity of the measuring equipment should be checked frequently because the equipment and their inherent parts are not always stable in linearity. Deviations can influence the calculation of the DI (see below) and the DNA ploidy classification.

**Low x-axis resolution.** In general, low CV values are important for the correct determination of the DI, especially so for the so-called near-diploid cell populations. The CV values of the main peaks in the DNA histograms are influenced by the staining and by the technical solutions used in the software of the ICM instrument. To judge the quality of a CV value of a certain cell population measured, the standard value referred to is the CV value found by the FCM technique. This method offers DNA histograms with narrow peaks (low CVs). The CV for lymphocytes measured by FCM is <2.0% and a similar value is obtained for the human cerebellum cortex cell nuclei. The ICM densitometric method has a lower resolution than the microfluorescent method FCM measurements. However, with optimal ICM systems, the CV values obtained by both methods should be similar. The procedures for an adequate accuracy control of the resolution of an ICM system are described below. When the CV values offered by the instrument are too high, each component of the equipment (microscope, camera, electronic compounds and software) has to be checked.

**Poor reproducibility.** It is of the utmost importance to check whether the results of two subsequent measurements of specimen give the same DNA ploidy and SPF DNA histograms. When the cellular solution prepared for FCM is measured several times, no great differences among the DNA histograms exist. This is due to the principle of FCM that each particle passing the measuring office is registered in the FCM histogram, i.e., all particles in the solution are

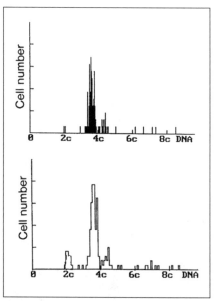

Figure 4. The DNA histogram (upper diagram) shows only the "at random" selected tumor cell nuclei with one main peak on 3.8c and a second small peak on 4.2c positions of the x-axis. During this selected measurement, the "detecting mode" is automatically registering all other cells (lower diagram) without influence of the observer. It turns up on the screen when the measurement is finished. When the selected measurement is done at random, corresponding to the whole cell population, both histograms define the same peaks. Stroma cell nuclei and leukocytes are present as one peak only in the detected histogram.

detected. This is called the zero-resolution of FCM. In contrast, the ICM technique already allows the observer to have a selective measurement of the objects. This selection is subjective and can influence the reproducibility of the DNA histograms. Therefore, the intra- and inter-observer histogram reproducibility has to be established. Of course, minor differences occur when the cell nuclei have been measured after a rather careful selection according to certain morphological criteria. Raising the total number of cell nuclei assessed, as well as the correlation among the subjectively "selected" and the objectively "detected" histogram (Figure 4) can give adequate insights into the magnitude of this source of error.

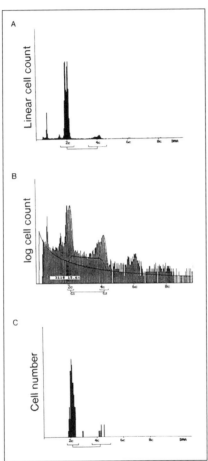

**Figure 5. FCM DNA histograms obtained from a ductal carcinoma of the breast.** In (A) the number of cells represented on y-axis is linear, whereas in (B) it is logarithmic. A near-diploid cell population can easily be recognized. The logarithmic y-axis exposes the non-Gaussian distribution of both the G0/G1 and the corresponding G2/M peaks more obviously than the linear one. The corresponding ICM DNA histogram (C) did not enclose the near-diploid population at all despite a fairly good resolution.

## Histogram Evaluation

In analogy with the evaluation procedure used in DNA assessments by FCM, the ICM DNA histogram should contain data allowing an evaluation of a complete description of the "DNA-ploidy-establishing population(s)". The rightmost peak on the abscissa will then be defined as the "DNA-ploidy-establishing population". It should be possible, as it is in the FCM DNA histogram evaluations, to calculate the "functional" cell cycle compartments of this population, the G0/G1-fraction, the SPF and the G2/M-fractions. The position of the main population should be given as DI, as well as the CV for the main peak of the population. To avoid subjective evaluation procedures, automatic programs have been developed. They are of the same type as those already available in the existing, well-established software used for evaluation of FCM DNA histograms.

**Inadequate number of nuclei assessed.** It is wise to let the number of cells assessed be decided upon by the pattern of the DNA histogram obtained. If the cytometric DNA data of

251

all cells measured are found to be distributed so that they seem to form a Gaussian peak, it is statistically acceptable to restrict the number of cells analyzed to approximately 100. The reason for such a decision is that it is plausible to assume that these cells actually correspond to a random sample out of the whole cell population. An obviously non-Gaussian distribution of a peak in a DNA histogram can actually represent a "double peak", comprising two different cell populations. If such a suspicion arises, the total number of cells assessed has to be raised until an acceptable resolution among the two peaks can be obtained. However, the limit of the resolution of the ICM method has to be considered. Such samples should be measured concomitantly by the FCM method (Figure 5).

If a subpopulation of the cells is found to be distributed outside the main peak of a DNA histogram—as is always the case for the cells representing the SPF—the total number of cells measured has to be adequately raised according to conventional statistical rules. The number of cells measured does not significantly influence the time of analyzing one sample. The really critical point in ICM DNA cytometry is the number of measurable cells available on the slide. However, the results of the SPF found by ICM should be confirmable by FCM or by other methods, e.g., immunohistochemical staining with proliferation specific markers, such as Ki-67/MIB-1.

**Definition problems as to the "2c" position.** In FCM, the position of the peaks on the x-axis (DI) is often defined by internal reference cells (chicken erythrocytes, trout erythrocytes or human lymphocytes). However, the use of different types of reference cells such as lymphocytes or granulocytes can, especially in ICM, cause problems in the definition of the real 2c position and require a series of correcting factors. The best standard of DNA content is the normal tissue component that represents the normal counterpart of the neoplastic cells (39). This recommendation can also be followed for the ICM method. The normal 2c component is easily recorded by means of the detecting mode during the actual selective measurement as mentioned above. During the step of histogram evaluation, the normal counterpart is recognized in the detected histogram and used to define the position of the population measured (Figure 6). As a rule, in solid tumors such as breast and prostate carcinomas, as well as in soft tissue and bone sarcomas, the detected histograms contain at least >10% of all nuclei assessed as a 2c peak. This peak corresponds to the normal counterpart, e.g., stromal cells and leukocytes admixed with the parenchymal tumor cell population. Thus, another important practical application of using a detecting mode (see above), is a more accurate and objective calculation of the position of the non-diploid peaks.

### Ploidy Definitions

The interpretations of the DNA histograms obtained by ICM are performed in analogy with those recommended by FCM (39).

**Euploidy vs. aneuploidy.** The ranges for DNA diploidy, tetraploidy and aneuploidy are almost the same for ICM and FCM when the ICM assessments have been made on cytodiagnostic specimens and performed on high-resolution equipment. Thus, a DNA diploid histogram should contain one single major peak with a DI ranging from 0.90 to 1.10. An ICM histogram of tetraploid type should display a major peak with a DI among 1.80 and 2.20. Another prerequisite is that this peak should contain at least 15% of all cells measured. When the ICM histogram shows a peak with a DI outside the diploid and tetraploid ranges, it is classified as DNA aneuploid. In order to fulfill the criteria of an aneuploid peak in a DNA histogram, the aggregated measured values should be based on at least 5%–10% of all cells measured. If more than one aneuploid peak/population is found, the DNA histogram is interpreted as multiploid.

**SPF calculation.** Provided that statistically enough nuclei have been assessed, the percentage of SPF is calculated according to the cell-cycle-model. Due to the selective measurement of tumor cells, the ICM DNA histograms of solid tumors show to a great extent only one population with or without an SPF. Thus, the SPF belonging to an aneuploid population is more distinctly visible than in FCM, where the histogram is superimposed by the DNA content of non-tumor cells or of nuclear aggregates. Also, the SPF in DNA diploid histograms is more reliable than in FCM due to the principle of the selective DNA assessment on tumor cells.

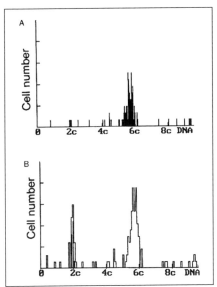

Figure 6. An ICM DNA histogram (A) showing the results of measurements, made on an imprint preparation of a malignant fibrous histiocytoma in which the tumor cell nuclei were selected at random. The G0/G1 peak of the tumor cell population has a DI of 2.95 or 5.9c. In the concomitantly registered "detected" DNA histogram (B), the main peak is confirmed and an additional peak in the 2c region is automatically recorded without influence of the observer. This normal counterpart confirms the 2c position in the DNA histogram. It is the most suitable internal standard and is recognized in more than 90% of the DNA histograms of all solid tumor specimens.

**Biological significance.** As a crude rule of thumb, it can be said that neoplastic diseases in which the cytometric nuclear DNA pattern of the tumor cells found to be of aneuploid type with a high percentage of SPF run a clinical course of a rapidly progressive type with extensive metastatic spread and early death of the patient. Those with an euploid, diploid or tetraploid DNA pattern in combination with a low percentage of SPF of the neoplastic cell nuclei are less aggressive (3,5,8,10,11,18, 25,28,30–33,40,42,43,45,49). However, there are many well-known exceptions (1,27,46,50) and even

253

opposites (29). A malignancy grading of a certain neoplastic disease by DNA cytometric results requires the correlation among the DNA ploidy/SPF and all other clinical and biological tumor variables, e.g., histopathological features, tumor size, clinical stage of the disease. Thus, the interpretation of a certain DNA ploidy and a certain SPF can only be made tumor entity-specific (6,12,22, 37,39,47).

## QUALITY ASSURANCE

For the clinical use of ICM DNA cytometry, constant and comparative intra- and inter-laboratory DNA results are expected. For multicenter oncologic studies, including patients with tumors of a certain DNA ploidy and SPF, the results from different laboratories have to be pooled. As a prerequisite for quality assurance of the DNA procedure, it is important to avoid or at least to be aware of the problems inherent in the different methodological steps.

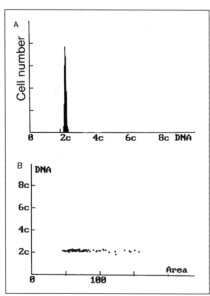

### Basic Technical Tests (Object: One Nucleus)

**Regular resolution and stability controls.** This step should be done initially and regularly when any components of the DNA equipment is changed (microscope, camera, frame grabber and IOD assessment). One and the same cell nucleus is assessed 100 times in one and the same position of the measure field. The CV of such a peak has to be <0.8 for high-resolution and stable equipment.

**Regular light stability controls.** This step has to be carried out when the DNA equipment is installed and when parts of the equipment are changed (microscope, light setting and optical devices, camera). One and the same nucleus is assessed 100 times at different positions of the measure field. The CV of the peak obtained should still be <3%.

**Figure 7. The ICM DNA histogram (A) shows the assessment of the nuclei of an imprint made from human cerebellum cortex.** These cells are ideal as biological test objects. All nuclei are in the G0 phase without proliferating nuclei. Thus, the DNA histograms always show sharp peaks. The CV of the main peak of 100 nuclei measured at random here reflects mainly their nuclear/chromatin structure and their staining conditions. The numeric value of the CV depends on the resolution capacity of the ICM equipment and its variation depends on the stability in handling the equipment and the Feulgen staining procedure. The area of these nuclei shows great variability. The dot plot (B), DNA against nuclear area, shows a linear correlation within an area ratio of the nuclei of 1:3.

**Initial and individual focus controls.** This step has to be done initially for each microscope and continuously for each user. One and the same nucleus is assessed 100 times, constantly kept at the same position, but focused upon at different levels. The CV of this peak has to be <1%.

## Basic Biological Tests (Object: One Cell Population)

**Resolution and stability controls.** This step shows the resolution of the equipment on one cell nuclei and on a whole cell population. The CV depends mainly on a certain nuclear/chromatin structure and their staining conditions. A glass slide with a standard cell preparation (e.g., imprint of human cerebellum cortex cell nuclei) can be used. At least 100 cell nuclei are assessed at random. The CV of the main peak should be <4%. The dot plot of DNA against nuclear area has to be linear within an area variation of the nuclei measured of 1:3 (Figure 7).

**Inter-laboratory staining control.** Glass slides containing a standard cell preparation (e.g., human cerebellum glia cell nuclei) prepared and fixed at the reference laboratory according to the guidelines are stained in each collaborating laboratory following the staining procedure. The CV of the randomly measured 100 nuclei should be similar to the CV of the cell population on the glass slide stained at the reference laboratory.

## Stability Control of the Feulgen Staining

The Feulgen staining procedure is a critical step for ICM DNA cytometry. Its reproducibility is of the utmost importance. The cerebellum imprints, which are prepared, fixed and stained according to the guidelines, are used for the continuous registration of the "absolute" integrated optical density (IOD) of 100 measured cerebellum cell nuclei by otherwise unchanged technical conditions. The standard deviation of the absolute IOD values on the continuous staining occasions should be <4% (Figure 8).

**Figure 8. The reproducibility/stability of the Feulgen staining at the laboratory is shown as the continuous registration of an absolute IOD value.** Here, the weakly stained human cerebellum cortex cell nuclei (at least 100 nuclei are measured as described in Figure 7) were used as reference cells. The standard deviation of the absolute IOD values during 79 staining occasions was 3.6%.

## DISCUSSION

Cytometric DNA assessments, either by means of FCM or ICM, have been performed on a wide variety of human neoplasms for more than two decades. During these years, considerable controversies existed over the results published from different laboratories, even in the same kind of tumors (6,22, 37,47). In fact, the clinical use of

DNA assessments for diagnostic, prognostic or predictive purposes in malignant neoplastic diseases has become restricted due to the growing amount of contradictory results in the literature. In the DNA cytometry consensus conference, a series of important methodological sources of errors has been focused upon, and guidelines for clinical FCM DNA cytometry are strongly recommended (23,39). In general, the authors agreed that DNA cytometric data should be included in clinical studies on a series of different solid tumors in order to obtain additional information about the prognostic and predictive power of the results of cytometric DNA assessments. Today, the methodological problems among different laboratories are still not solved (4). It remains to be seen whether the guidelines and standardization protocols recommended above will be used in daily practice, and whether the results then obtained in the different laboratories will show better concordance.

The methodological problems of ICM DNA cytometry are at least of the same magnitude as in FCM. In our earlier studies we found that a series of methodological sources of errors played an important role when controversies arise among different laboratories about the value of cytometric DNA assessments (13–15,17). To elucidate these methodological problems, we have performed comprehensive DNA studies on various kinds of solid tumors using both ICM and FCM concomitantly. Although ICM and FCM are fundamentally different techniques for DNA assessments of tumor cell nuclei, they give essentially the same results. During these studies we found that a high reproducibility of the different technical steps of the ICM procedure is of decisive importance. Under standardized conditions the methodological discrepancies among the ICM and FCM results can be minimized. Thus, the standardization procedure as well as the steps of quality assurance for ICM proposed above are all based on the practical work on DNA histograms from a broad series of solid human tumors.

To solve the problem of the ICM DNA histogram evaluation is an extremely challenging task. A large number of different histogram evaluation algorithms make it difficult to compare the ICM DNA results with those found by FCM (2,7,19,44). The evaluation program proposed above is the result of the practical evaluation of concomitant ICM and FCM DNA histograms obtained from thousands of DNA assessments made on various kinds of human solid tumors. The evaluation program is made under consideration of the statistical rules and in a rather straightforward way. There is no artifact recognition and elimination as is necessary in FCM histogram evaluation programs. From a methodological point of view, the results obtained by this ICM DNA evaluation program are in good agreement with the data from assessments of both DNA ploidy and SPF (high, intermediate or low) obtained by FCM.

When new laboratory guidelines for a method like ICM are established with the purpose of minimizing technical errors during a sequence of different steps, it is of the utmost importance to be aware of their practical applicability. In different laboratories, both the guidelines for standardization and for quality assurance are possible to integrate in the daily DNA assessments of clinical tumor material (16). They are of great value to trace technical faults in equipment or in the handling of it. Mistakes during the steps of preparation, fixation and staining can easily be detected. The influence of inherent fluctuations depending on structural/ chromatin pecularities in nuclear DNA assessments can be monitored separately from the "real" methodological errors.

## ACKNOWLEDGMENTS

This work was supported by grants from the Swedish Medical Research Council (Project 102), the Swedish Cancer Society (Project 930364), the Cancer Society of Stockholm and the King Gustaf V Jubilee Fund, Stockholm, Sweden.

## REFERENCES

1. **Alanen K.A., U.G. Falkmer, P.J. Klemi, H. Joensuu and S. Falkmer.** 1992. Flow and image cytometric study of pancreatic neuroendocrine tumours: frequent DNA aneuploidy and an association with the clinical outcome. Virchows Arch. A Pathol. Anat. *421*:121-125.
2. **Auer, G., U. Askensten and O. Ahrens.** 1989. Cytophotometry. Hum. Pathol. *20*:518-526.
3. **Auer, G., H. Kato, M. Nasiell, V. Roger, A. Zetterberg and L. Karlén.** 1980. Cytophotometric DNA-analysis of atypical squamous metaplastic cells, carcinoma in situ, and bronchogenic carcinoma, p. 1467-1476. In H.E. Nieburgs (Ed.), Prevention and Detection of Cancer, Part 2. Proceedings of the Third International Symposium on Detection and Prevention of Cancer, April 26–May 1, 1976. Marcel Dekker, New York.
4. **Baldetorp, B., P.O. Bendahl, M. Fernö, K. Alanen, U. Delle, U.G. Falkmer, B. Hansson-Aggesjö, T. Höckenström, A. Lindgren, L. Mossberg, S. Nordling, H. Sigurdsson, O. Stål and T. Visakorpi.** 1995. Reproducibility in DNA flow cytometry analysis of breast cancer; comparison of 12 laboratories' results for 67 sample homogenates. Cytometry *22*:115-127.
5. **Bauer, H.C.F., A. Kreicbergs, C. Silfverswärd and B. Tribukait.** 1988. DNA analysis in the differential diagnosis of osteosarcoma. Cancer *61*:1430-1436.
6. **Bauer, K.D., B. Bagwell, W. Giaretti, M. Melamed, R.J. Zarbo, T.E. Witzig and P.S. Rabinowitch.** 1993. Consensus review of the clinical utility of DNA flow cytometry in colorectal cancer. Cytometry *14*:486-491.
7. **Böcking, A. and W. Auffermann.** 1985. Algorithm for DNA cytophotometric diagnosis and grading of malignancy in squamous epithelial lesions of the larynx with DNA cytophotometry. Cancer *56*:1600-1604.
8. **Bose, K.K., D.C. Allison, R.H. Hruban, S. Piantadosi, M. Zahurak, W. Dooley, P. Lin and J.L. Cameron.** 1993. A comparison of flow cytometric and absorption cytometric DNA values as prognostic indicators for pancreatic carcinoma. Cancer *71*: 691-700.
9. **Caspersson, T.** 1950. Cell Growth and Cell Function. A Cytochemical Study. Norton, New York.
10. **Clark, G.M., L.G. Dressler, M.A. Owens, G. Pounds, T. Oldakar and W.L. McGuire.** 1989. Prediction of relapse or survival in patients with node-negative breast cancer by DNA flow cytometry. N. Engl. J. Med. *320*:627-633.
11. **Desphande, N., I. Michell, D. Allen and R. Millis.** 1983. Deoxyribonucleic acid (DNA) content of carcinomas and prognosis in human breast cancer. Int. J. Cancer *32*:693-696.
12. **Falkmer, U.G.** 1994. Image DNA cytometry: diagnostic and prognostic applications in clinical tumor pathology. Cell Vision *1*:79-83.
13. **Falkmer, U.G.** 1989. Methodological aspects on DNA cytometry. Image and flow cytometry on malignant tumors of the breast, the endometrium, the brain, and the salivary glands. Thesis. Karolinska Institute, Stockholm.

# Image DNA Cytometry

14. **Falkmer, U.G.** 1991. Methodological aspects on the value of flow and image cytometric nuclear DNA assessments of the neoplastic cells of prostatic adenocarcinoma. Acta Oncol. *30*:201-203.
15. **Falkmer, U.G.** 1992. Methodological sources of errors in image and flow cytometric DNA assessments of the malignant potential of prostatic carcinoma. Hum. Pathol. *23*:360-367.
16. **Falkmer, U.G.** 1994. Methodological guidelines for interlaboratory studies by means of image DNA cytometry. Anal. Cell. Pathol. *6*: 282.
17. **Falkmer, U.G., T. Hagmar and G. Auer.** 1990. Efficacy of combined image and flow cytometric DNA assessments in human breast cancer. Anal. Cell Pathol. 2:297-312 .
18. **Fallenius, A., S. Franzén and G. Auer.** 1988. Predictive value of nuclear DNA content in breast cancer in relation to clinical and morphological factors. A retrospective study of 409 consecutive cases. Cancer *62*:521-530.
19. **Forsslund, G. and A. Zetterberg.** 1990. Ploidy level determinations in high-grade and low-grade malignant variants of prostatic carcinoma. Cancer Res. *50*:4281-4285.
20. **Frierson, H.F.** 1991. The need for improvement in flow cytometric anlysis of ploidy and S-phase fraction. Am. J. Clin. Pathol. *96*:439-441.
21. **Gaub, J.J., G. Auer and A. Zetterberg.** 1975. Quantitative cytochemical aspects of a combined Feulgen naphtol-yellow-S staining procedure for the simultaneous determination of nuclear and cytoplasmatic proteins and DNA in mammalian cells. Exp. Cell Res. *92*:323-332.
22. **Hedley, D.W., G.M. Clark, C.J. Cornelisse, D. Killander, T. Kute and D. Merkel.** 1993. Consensus review of the clinical utility of DNA cytometry in carcinoma of the breast. Cytometry *14*:482-485.
23. **Hedley, D.W., V. Shankey and L.L. Wheeless.** 1993. DNA cytometry consensus conference. Cytometry *14*:471.
24. **Hiddemann, W., J. Schumann, M. Andreff, B. Barlogie, C.J. Herman, R.C. Leif, B.H. Mayall, R.F. Murphy and A.A. Sandber.** 1984. Convention on nomenclature for DNA cytometry. Cytometry *5*:445-446.
25. **Hirano, T., B. Franzén, H. Kato, Y. Ebihara and G. Auer.** 1994. Genesis of squamous cell lung carcinoma. Sequential changes of proliferation, DNA ploidy and p53 expression. Am. J. Pathol. *144*:296-302.
26. **Homburger, H.A., R. McCarthy and S. Deodhar.** 1989. Assessment of interlaboratory variability in analytical cytology. Results of the College of the American Pathologists Flow Cytometry Study. Arch. Path. Lab. Med. *113*:667-672.
27. **Hulting, A.L., U. Askensten, B. Tribukait, J. Wersäll, G. Auer, L. Grimelius, S. Falkmer and S. Werner.** 1989. DNA evaluation in growth hormone producing pituitary adenomas: flow cytometry versus single cell analysis. Acta Endocrinol. (Copenh.) *121*:317-321.
28. **Kimura, T., T. Sato and K. Onodera.** 1993. Clinical significance of DNA measurements in small cell lung cancer. Cancer *72*:3216-3222.
29. **Kogner, P., G. Barbany, O. Björk, M.A. Castello, A. Donfrancesco, U.G. Falkmer, F. Hedberg, H. Kouvidou, H. Persson, G. Raschella and C. Dominici.** 1994. TRK mRNA and low affinity nerve growth factor receptor mRNA expression and triploid DNA content in favorable neuroblastoma tumors. Adv. Neuroblastoma Res. *4*:137-145.
30. **Matsura, H., H. Kuwano, M. Morita, S. Tsutui, Y. Kido and M. Mori.** 1991. Predicting recurrence time of esophageal carcinoma through assessment of histologic factors and DNA ploidy. Cancer *67*:1406-1411.
31. **Moberger, B., G. Auer and G. Forsslund.** 1984. The prognostic significance of DNA measurements in endometrial carcinoma. Cytometry *5*:430-437.
32. **Montgomery, B.T., O. Nativ, M.L. Blute, G.M. Farrow, R.P. Myers, H. Zincke, T.M. Therneau and M.M. Lieber.** 1990. Stage B prostate adenocarcinoma. Flow cytometric nuclear DNA ploidy analysis. Arch. Surg. *125*:327-331.
33. **Nasiell, K., I. Näslund and G. Auer.** 1984. Cytomorphologic and cytochemical analysis in the differential diagnosis of cervical epithelial lesions. Anal. Quant. Cytol. *3*:196-201.
34. **Riley, R.S., E.J. Mahin and W. Ross.** 1993. Clinical Applications of Flow Cytometry. Igaku-Shoin Medical Publisher, New York.
35. **Sandberg, A.A.** 1990. The Chromosomes in Human Cancer and Leukemia, 2nd ed. Elsevier, Amsterdam.
36. **Sandritter, W. (Ed.)** 1964. 100 Years of Histochemistry in Germany. Schattauer, Stuttgart.
37. **Shankey, T.V., O.-P. Kallioniemi, J.M. Koslowski, M.L. Lieber, B.H. Mayall, G. Miller and G.J. Smith.** 1993. Consensus review of the clinical utility of DNA content cytometry in prostate cancer. Cytometry *14*:497-500.
38. **Shapiro, H.M.** 1988. Practical Flow Cytometry, 2nd ed. Alan R. Liss, New York.
39. **Shankey, T.V., P.S. Rabinovitch, B. Bagwell, K.D. Bauer, R.E. Duque, D.W. Hedley, B.H. Mayall**

and L. Wheless. 1993. Guidelines for implementation of clinical DNA cytometry. Cytometry *14*:472-477.

40. Sigurdsson, H., B. Baldetorp, A. Borg, M. Dalberg, M. Fernö, D. Killander and H. Olsson. 1990. Indicators of prognosis in node-negative breast cancer. N. Engl. J. Med. *322*:1045-1053.

41. Sobin, L.H. and D.E. Henson. 1994. TNM Supplement 1993. A commentary on uniform use. Cancer *74*:2385.

42. Stål, O., J.M. Carstensen, S. Wingren, L.E. Rutqvist, L. Skoog, C. Klintenberg and B. Nordenskjöld. 1994. Relationship of DNA ploidy and S-phase fraction to survival after first recurrence of breast cancer. Acta Oncol. *33*:423-429.

43. Steinbeck, R.G., J. Moege, K.M. Heselmeyer, W. Klebe, W. Neugebauer, B. Borg and G. Auer. 1993. DNA content and PCNA immunoreactivity in oral precancerous and cancerous lesions. Eur. J. Cancer B Oral Oncol. *29B*:279-284.

44. Stenkvist, B. and G. Strande. 1990. Entropy as an algorithm for the statistical description of DNA cytometric data obtained by image analysis microscopy. Anal. Cell Pathol. *2*:159-165.

45. Tribukait, B. 1987. Flow cytometry in assessing the clinical aggressiveness of genito-urinary neoplasms. World J. Urol. *5*:108-122.

46. Wallin, G., U. Askensten, M. Bäckdahl, L. Grimelius, G. Lundell and G. Auer. 1989. Cytochemical assessment of the nuclear DNA distribution pattern by means of image and flow cytometry in thyroid neoplasms and in non-neoplastic thyroid lesions. Acta Chir. Scand. *155*:251-258.

47. Wheeless, L.L., R.A. Badalament, R.W. deVere White, Y. Fradet and B. Tribukait. 1993. Consensus review of the clinical utility of DNA cytometry in bladder cancer. Cytometry *14*:478-481.

48. Wheeless, L.L., J.S. Coon, C. Cox, A.D. Deitch, R.W. deVereWhite, Y. Fradet, L.G. Koss, M.R. Melamed, M.J. O'Connell, J.E. Reeder, R.S. Weinstein and R.P. Wersto. 1991. Precision of DNA flow cytometry in inter-institutional analyses. Cytometry *12*: 405-412.

49. Witzig, T., C. Loprinzi, N. Gonchoroff, H. Reiman, S. Cha, H. Weiand, J. Katzman, J. Paulsen and C. Moertel. 1991. DNA ploidy and cell kinetic measurements as predictors of recurrence and survival in stages B$_2$ and C colorectal adenocarcinomas. Cancer *68*:879-888.

50. Zedenius, J., A. Auer, M. Bäckdahl, U. Falkmer, L. Grimelius, G. Lundell and G. Wallin. 1992. Follicular tumors of the thyroid gland: diagnosis, clinical aspects and nuclear DNA analysis. World J. Surg. *16*:589-594.

259

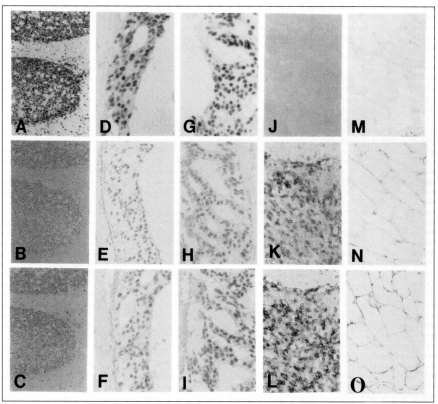

Figure 1. Comparison of the intensity of MW heating AR-IHC staining on routinely formalin-fixed paraffin-embedded sections between pH 1.0 (A,D,G,J), pH 6.0 (B,E,H,K) and pH 10.0 (C,F,I,L) by using monoclonal antibodies to MIB1 (A–C), ER (D–I), and MT1 (J–L). The strongest intensity of MIB1 and ER was obtained by using AR solution at pH 1.0, while the strongest intensity of MT1 was achieved by using AR solution at pH 10.0. Tris-HCl buffer was used for A–F, acetate buffer was used for G–I, sodium diethylbarbiturate-HCl buffer was used for J–L. An increased intensity of immunostaining with monoclonal antibody to dystrophin by using a non-heating AR-IHC method developed at BioGenex (N), in contrast to the section without pretreatment showing negative staining result (M), and a stronger positive staining result obtained by using MW heating AR-IHC (O). Magnification, ×100.

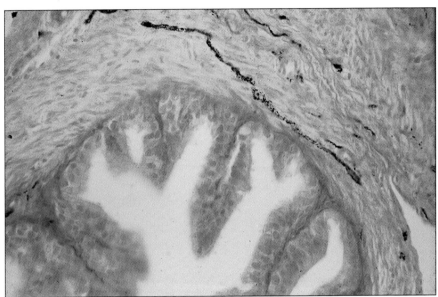

**Figure 1. Calcitonin gene-related peptide (CGRP) containing nerve fiber (black) in rat Fallopian tube.** IGSS with silver lactate autometallography (10,79), polyclonal rabbit primary antibodies, goat-anti-rabbit IgG adsorbed to 5-nm colloidal-gold particles, Bouin's-fixed 5-μm thick paraffin section, counterstained with eosin. From joint collaboration with J.M. Polak and D.R. Springall, London, UK. Magnification = 360×.

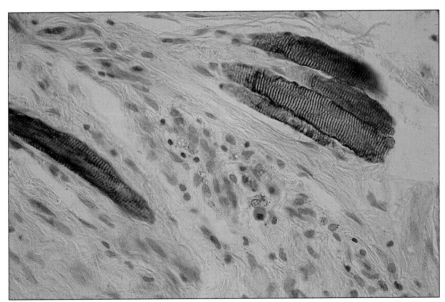

**Figure 2. Striation of muscle fibers in human skin detected by monoclonal mouse antibodies to desmin.** IGSS with silver acetate autometallography (31), goat anti-mouse IgG immunogold reagent with 5-nm gold particles, formalin-fixed 4-μm paraffin section. A very high resolution is obtained, showing an intense black signal distinctly standing out against the H&E-counterstained background morphology. Magnification = 400×.

A-2

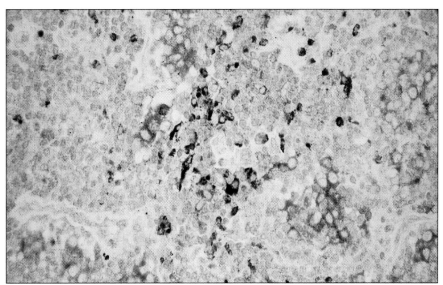

**Figure 3. Double immunostaining with IGSS (black) and alkaline phosphatase anti-alkaline phosphatase (APAAP, red) methods.** Metastatic carcinoma cells in lymph node stained with monoclonal antibodies to cytokeratins (KL-1, red) and polyclonal antibodies to carcinoembryonic antigen (CEA, black). A few cells contain both antigens, which is better recognizable by setting different planes of focussing in the light microscope. Formalin-fixed 4-mm thick paraffin section, counterstained with hematoxylin. Magnification = 360×.

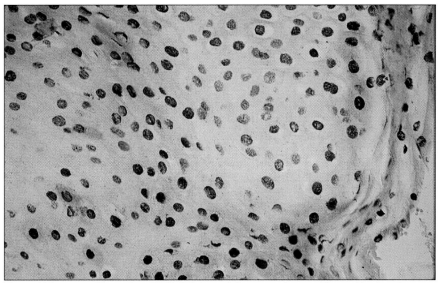

**Figure 4. Human papillomavirus (HPV 6/11) in condyloma acuminatum.** The combination of conventional ISH with Streptavidin-Nanogold allows detection limits far below those of other techniques. Additional CARD-amplification with biotinylated tyramide (Zehbe, Hacker, et al., unpublished observations) allows single copy sensitivity and can potentially replace *in situ* PCR and *in situ* 3SR techniques, which are most often associated with reproducibility and specificity problems. Formalin-fixed 4-μm-thick paraffin section, counterstained with Nuclear Fast Red.

A-3

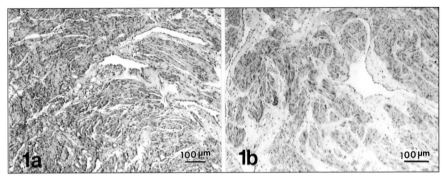

**Figure 1.** a) Immunostaining using polyclonal rabbit antibody to ANP on rabbit right atrium with conventional indirect immunoperoxidase procedure. Tremendous background staining is present in the connective tissue in addition to the atrial myocardiocytes. b) When using the same antibodies on the same tissue following the PSC protocol, the specific immunostaining in the cytoplasm of the atrial myocardiocytes was clearly visible and the intense background staining was completely removed.

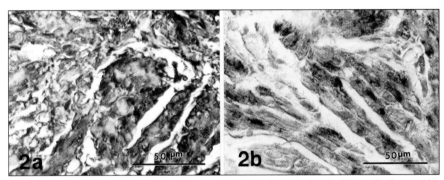

**Figure 2.** a) Higher magnification of ANP immunostaining using polyclonal rabbit antibody on rabbit atrium. Immunoreactivity was present in the connective tissue region in addition to the myocardiocytes. The specific immunostaining was obscured by the extensive background reaction. b) On the same tissue using the same antibodies, PSC procedure removed all the background and the specific reaction sites become clearly visible.

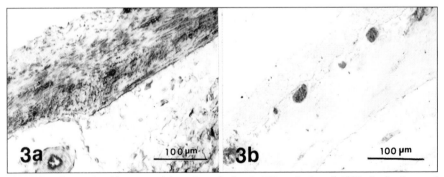

**Figure 3.** a) Conventional immunostaining using polyclonal rabbit antibody to S-100 in the duodenum of rabbit. Extensive background staining was present in the muscle layer, the connective tissue and the vasculature, in addition to the Schwann cells in the nerve bundles and ganglion cells. b) Using the same antiserum on the same tissue samples following the PSC procedure, the background staining was completely removed and S-100 immunoreactivity was clearly demonstrated in the Schwann cells.

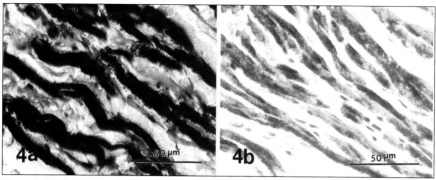

**Figure 4.** a) ISS using polyclonal rabbit ANP antibody on rabbit tissue following a well-established IGSS procedure. Extensive background staining was present in the connective tissue in addition to the specific staining sites. b) Using the same antiserum on the same tissue but following the PSC protocol, the unspecific reaction was completely removed and ANP immunoreactivity was demonstrated in the cytoplasm of the atrial myocardiocytes.

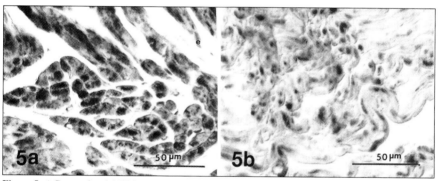

**Figure 5.** a) Conventional ANP immunostaining in the atrial myocardiocytes of rat. They were abundantly demonstrated in the cytoplasm with little background staining. b) When using the same antibodies on the same tissue following the PSC protocol, the specific staining was clearly demonstrated in the cytoplasm of the atrial myocardiocytes. In comparison with 5a, the immunoreactivity was not changed or slightly decreased and the background was even cleaner, indicating that the PSC protocol does not significantly weaken the detecting sensitivity of the antibodies and may be used routinely on tissue types other than rabbit to achieve a clean background staining.

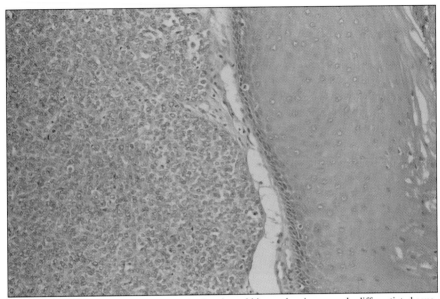

**Figure 1. MW-fixed, 1.5-h processing cycle**. Esophageal biopsy showing a poorly differentiated squamous cell carcinoma beneath the squamous mucosa.

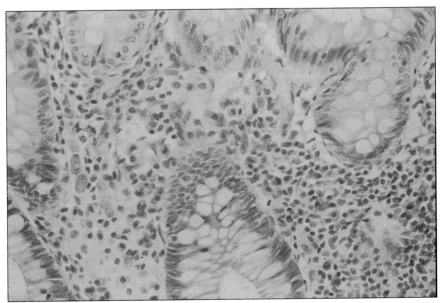

**Figure 2. MW-fixed, 3.5-h processing cycle.** Segmental resection of colon for diverticular disease. Note the good cytomorphologic preservation.

A-6

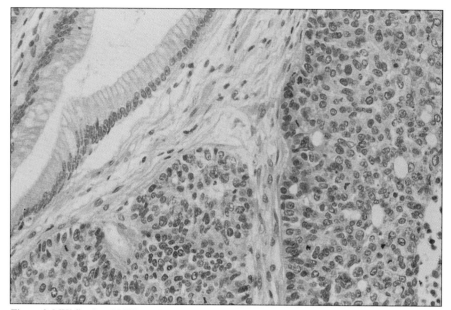

**Figure 3. MW-fixed and MW-accelerated processing; total time 30 min.** Carcinoid tumor showing excellent preservation of tumor and overlying gastric mucosa.

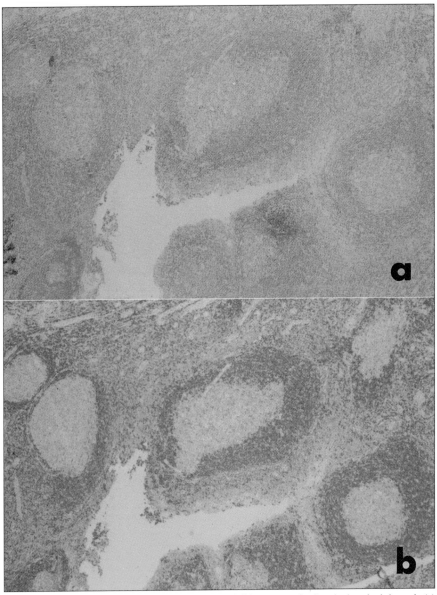

**Figure 4. Streptavidin-biotin peroxidase staining for *bcl*-2 oncoprotein.** Standard method shown in (a) and immunostaining following MW retrieval in 10 mM citrate buffer, pH 6.0, in (b).

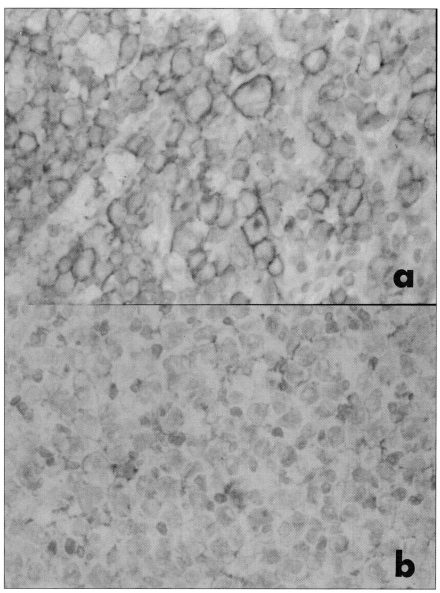

**Figure 5. Malignant lymphoma, follicular, mixed large and small cell type.** Stained for light chain immunoglobulin following MW antigen retrieval in 4 M urea. Definite cytoplasmic and membranous staining for Igλ in (a) (note dot-like cytoplasmic staining of Golgi in large cell in center of field) and negative staining for IgK.

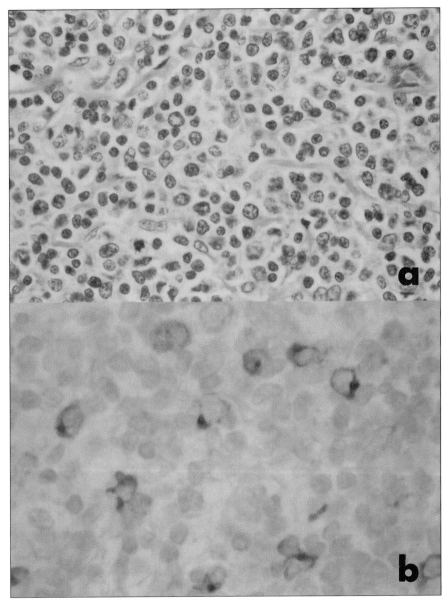

**Figure 6. Malignant lymphoma, diffuse, large and small cell type.** Note the good morphologic preservation of the hematoxylin and eosin (H&E)-stained section in (a) after MW fixation and 3.5-h cycle processing through absolute alcohol and chloroform and the definite cytoplasmic staining for IgK in (b) following MW antigen retrieval in 4 M urea. Besides cyoplasmic staining of large cells, there is also dot-like accentuation in the Golgi areas.

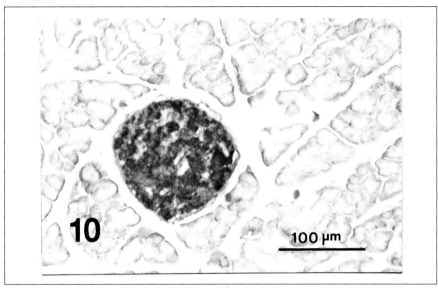

**Figure 10. Conventional ABC immunostaining of β cells of rat pancreas using monoclonal anti-insulin antibodies.** A very strong specific immunostaining and no background staining were obtained when the following conditions were used: PA overnight incubation at 4°C, biotinylated secondary antibody (BSA) incubation for 60 min at room temperature (RT), streptavidin-horseradish peroxidase conjugate (EC) for 30 min at RT, substrate-chromogen (AEC) for 5 min at RT and counterstaining with hematoxylin.

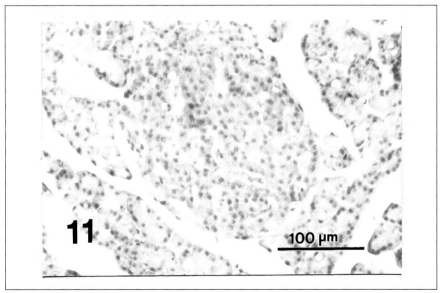

**Figure 11. Suboptimal condition of MW-stimulated immunostaining of β cells of rat pancreas using monoclonal anti-insulin antibodies.** A total time of 60 s MW irradiation at 75% power level and Tsp at 37°C, followed by a 5-min incubation at RT for each of the three stages of ABC method, resulted in weak specific staining and no background staining.

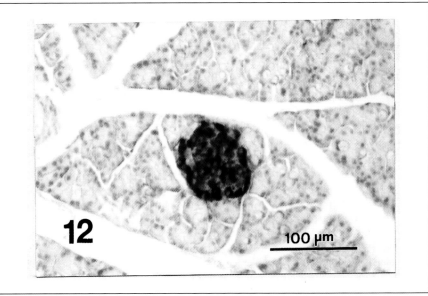

**Figure 12. Optimal MW-stimulated immunostaining of β cells of rat pancreas using monoclonal anti-insulin antibodies.** A total time of 3 min MW irradiation at 75% power level and Tsp at 37°C followed by a 2-min post-MW incubation inside the oven for each of the three stages of ABC method, resulted in very strong specific staining and no background staining.

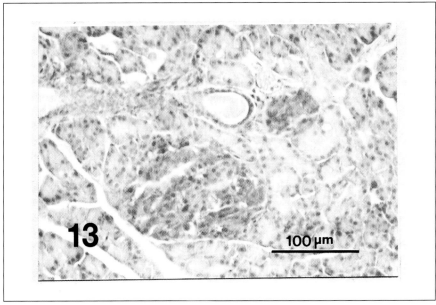

**Figure 13. Stimulating effect of MWs in immunostaining.** Compared with Figure 12, the same duration of 5 min incubation at room temperature for each of the three stages of ABC method resulted in weaker staining of the β cells of rat pancreas using monoclonal anti-insulin antibodies.

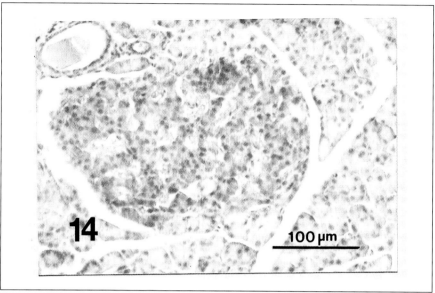

**Figure 14. Enhancing effect of post-MW incubation in MW-stimulated immunostaining.** Compared with Figure 12, a total time of only 3 min MW irradiation at 75% power level and Tsp at 37°C for each of the three stages of ABC method resulted in weaker staining of the β cells of rat pancreas using monoclonal anti-insulin antibodies.

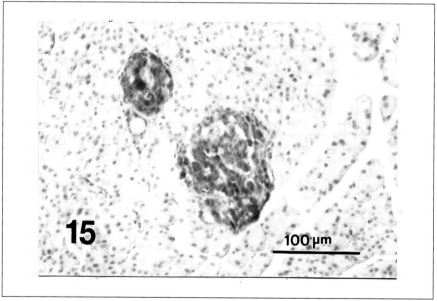

**Figure 15. Additive effect of each stage of ABC method in MW-stimulated immunostaining.** Compared with Figure 14, an additional 3 min post-MW incubation at RT for the BSA and EC steps resulted in stronger staining of the β cells of rat pancreas using monoclonal anti-insulin antibodies. However, when compared with the optimal staining as shown in Figure 12, omission of the post-MW incubation of the PA step resulted in a weaker specific immunostaining.

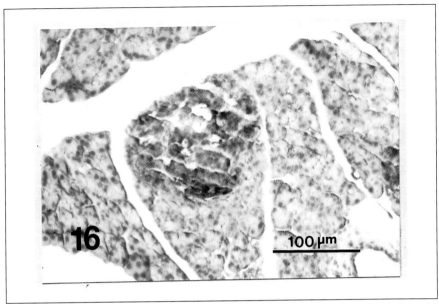

**Figure 16. Reduction of MW irradiation time resulting in weaker staining.** Compared with Figure 12, a total time of 2 min MW irradiation at 75% power level and Tsp at 37°C, followed by 2 min post-MW incubation inside the oven for each of the three stages of ABC method, resulted in weaker staining of the β cells of rat pancreas using monoclonal anti-insulin antibodies.

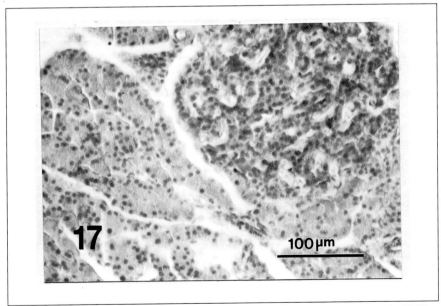

**Figure 17. The maximal allowable temperature of antibody-antigen immunoreactions in ABC immunostaining is 55°C.** With Tsp at 55°C for PA and BSA steps, and Tsp at 37°C for EC step, the same duration of MW irradiation and post-MW incubation for each of the three stages resulted in weak specific immunostaining and increased background staining, compared with Figure 12.

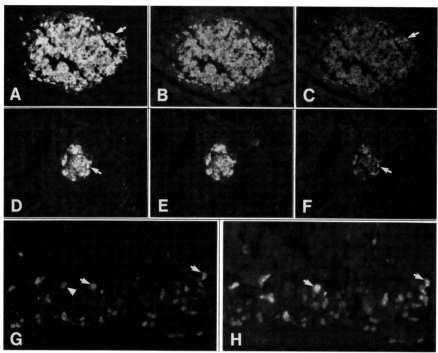

**Figure 1. Triple immunofluorescence with antibodies directed at IAPP (A and D; FITC), proinsulin (B and E; TRITC) and somatostatin (C and F; AMCA) in pancreatic islets from rats treated with dexamethasone (A–C) and streptozotocin (D–F).** In A–C, the islet is enlarged due to the dexamethasone-treatment. The B cells display co-localization of IAPP- (A) and proinsulin-like immunoreactivity (B); in addition, a subpopulation of peripherally located somatostatin-immunoreactive cells contains IAPP-like immunoreactivity (A and C; arrows). In C, note that the intense fluorescence from the FITC-coupled antibodies in A has escaped the AMCA filter and is visible as a dull green fluorescence. In D–F, streptozotocin treatment has caused B cell destruction, resulting in a disruption of the normal islet architecture with islet cells scattered. IAPP-like immunoreactivity predominantly occurs in insulin cells (D and E) but also to a lesser extent in somatostatin cells (D and F; arrows; islets in A–F are shown at the same magnification; ×200). In rat antrum (G and H), IAPP-like immunoreactivity demonstrated by a monoclonal antibody (G; FITC) occurs in a major population of somatostatin-immunoreactive cells (H; TRITC) as examplified by arrows. However, some IAPP immunoreactive cells (arrowheads) lack somatostatin (1, G and H; ×300).

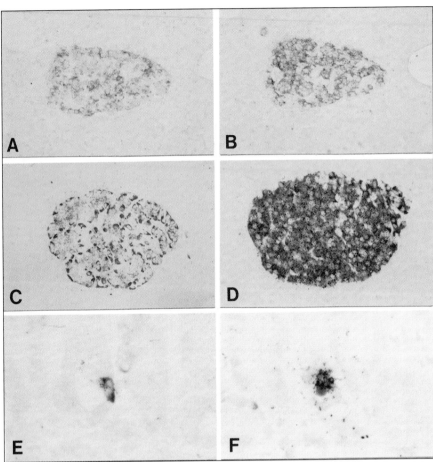

**Figure 3. Combined *in situ* hybridization with IAPP probes and immunoperoxidase.** In human islets (A–C), the IAPP probe labels centrally located clusters of cells; these cells are also IAPP- (A) and proinsulin-immunoreactive (B; A and B are sections of the same islet), although they lack glucagon-like immunoreactivity (C). In a rat islet (D), the IAPP probe-labeling and IAPP-like immunoreactivity is perfectly congruent and localized to a central core of islet cells (B cells; islets are shown at the same magnification; ×200). In the antral mucosa of the mouse stomach, an IAPP-immunoreactive endocrine cell labeled by the IAPP probe is shown (E); IAPP mRNA is also seen in a somatostatin-immunoreactive cell (F; E and F; ×400).

A-16

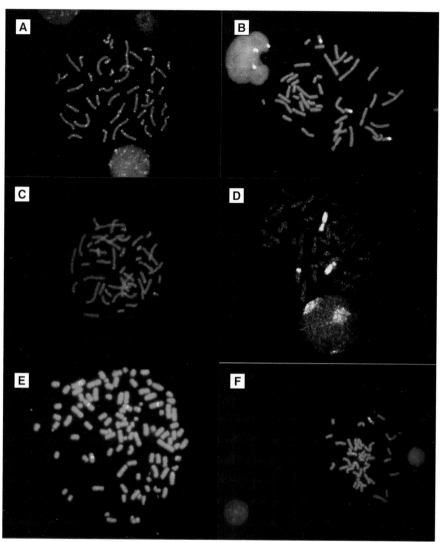

**Figure 2. Metaphase cytogenetics.** FISH with different types of DNA probes on normal and abnormal human metaphase chromosomes. (A) Double hybridization using alphoid and telomeric repeat sequence probes for the simultaneous visualization of both centromeres and telomeres in a cytogenetically normal individual. (B) Biotinylated DNA specific for the alpha satellite subfamily of chromosome 18 highlighting three copies of chromosome 18 in a newborn suspected with trisomy 18. (C) CISS hybridization with a digoxigenin-labeled cosmid probe (D21S65) showing hybridization signals on the Down's syndrome critical region of chromosome 21 at band q22.3. (D) WCPP on hybridization shows uniformly "painted" green fluorescence over the length of chromosome 7. The distal part of chromosome 13 is also highlighted, indicating a translocation of a part of chromosome 7 to 13. (E) Biotin-labeled c-*erb* B2 cosmid gene probe demonstrating the site of integration of amplified genes on different chromosomes from a breast cancer patient. (F) FISH analysis of a patient suspected of having Prader-Willi syndrome. Hybridization of 15q11-q13 band specific probes indicates one normal copy of chromosome 15. However, the other homologue does not exhibit a hybridization signal at this critical band, which indicates deletion of band q 11q13 in this patient and confirms Prader-Willi syndrome.

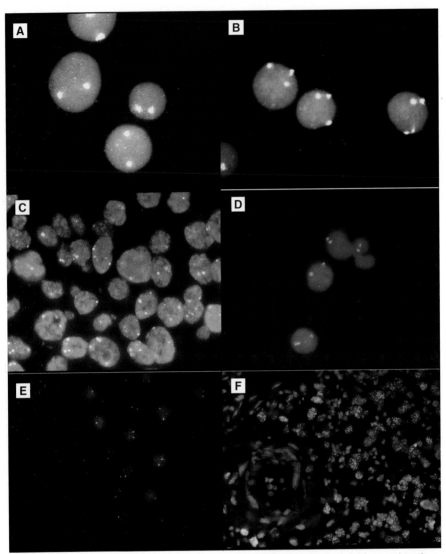

**Figure 3. Interphase cytogenetics.** Single- and dual-color FISH on interphase nuclei from various tissue samples using centromeric, classical satellite and ongogene probes. (A) Dissociated nuclear suspension from normal breast epithelium showing two hybridization domains for chromosome 7 using biotinylated alphoid probe D7Z1. (B) Interphase analysis for trisomy 16 in a dissociated nuclear suspension from an aborted fetus. Centromeric probe D16Z2 from Oncor was used in this assay. (C) A 4-μm-thick paraffin specimen from an intraductal comedo carcinoma showing monosomic, disomic, and trisomic clones when hybridized with alphoid probe D1Z5 specific for the centromeric region of chromosome 1. (D) Dual-color FISH for the direct identification of X and Y chromosomes on a male bone marrow interphase nuclei using two different fluorescent labels. X alpha satellite BODIPY TR-labeled and Y classical-FITC labeled. (E) Interphase bone marrow cells from a patient suspected of having CML. The double color FISH using *bcr/abl* translocation probe (Oncor) demonstrates the presence of the translocation where the green *abl* oncogene and the red *bcr* region fused together at interphase confirming the presence on CML. (F) Detection of digoxigenin labeled N-*myc* oncogene amplified several times in interphase nuclei (intense yellow signal) on a 4-μm-thick paraffin-embedded neuroblastoma specimen.

A-18

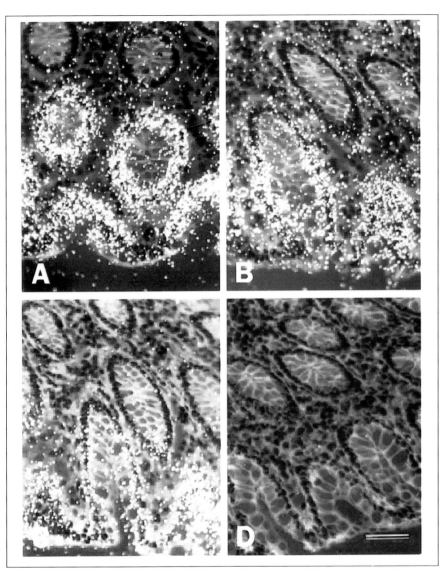

**Figure 2. *In situ* hybridization results from riboprobes generated by ribovector vs. PCR showing that comparable results were obtained.** Photomicrograph of sections of colon subjected to *in situ* hybridization with α-$^{33}$P-UTP labeled cRNA probes directed against guanylin, a endogenous ligand for the STa receptor that is expressed in the luminal epithelial cells of the colon (6). **A.** Section of colon probed with guanylin antisense cRNA probe corresponding to nucleotides 6–544 of the cDNA generated from a linearized riboclone by *in vitro* transcription. **B.** Section of colon probed with guanylin antisense cRNA probe generated from the same riboclone as A, except that PCR was used to generate the template that was subsequently used for *in vitro* transcription. Primers directed against the T3 and T7 promoters were used for the PCR (Table 2). **C.** Section of colon probed with guanylin antisense cRNA probe generated from the same gene fragment within a non-riboclone plasmid using PCR. Specific primers having guanylin sequence and RNA promoter sequence were used (Table 2). **D.** Section of colon probed with guanylin sense cRNA probe (negative control) generated from the same gene template used in "C", using the RNA polymerase corresponding to the opposite promoter to produce the sense probe. Bar = 50 μm.

A-19

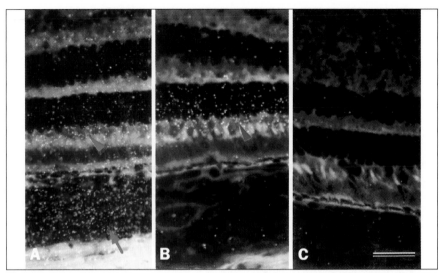

**Figure 3.** *In situ* **hybridization results from riboprobes labeled with either** $^{33}$**P-UTP or** $^{35}$**S-UTP, showing decreased background and increased signal-to-noise ratio obtained with the** $^{33}$**P-labeled probe.** Photomicrograph of sections of retina subjected to *in situ* hybridization with cRNA probes directed against retina-specific guanylyl cyclase (18). Radiolabeled cRNA probes were transcribed from the PCR fragment of the retGC kinase domain corresponding to nucleotides 1541 to 2374 of the full-length cDNA clone. The radiolabeled probes were labeled using either $\alpha$-$^{35}$S-UTP or $\alpha$-$^{33}$P-UTP. **A.** $^{35}$S labeled antisense probe. Specific hybridization is present over the outer nuclear layer (arrowhead). Nonspecific hybridization is present over the choroid (arrow), which was also present in the sections probed with the sense negative control (not shown). **B.** $^{33}$P-labeled antisense probe generated from the same template as used for A. Nonspecific hybridization is markedly reduced, and the specific signal over the outer nuclear layer is more clear (arrow head). **C.** Section of retina probed with $^{33}$P-labeled sense probe, showing lack of hybridization. Bar = 50 μm.

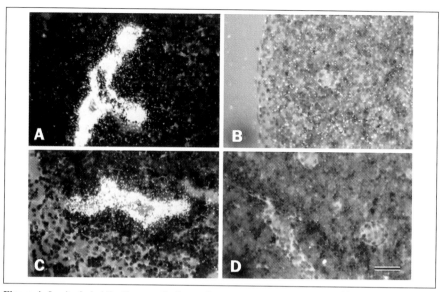

**Figure 4.** *In situ* **hybridization results from riboprobes labeled with either** [33]P-UTP **or** [35]S-UTP, **showing comparable results when a gene having a high level of mRNA expression is evaluated.** Photomicrographs of sections of lymph node subjected to *in situ* hybridization with cRNA probes directed against GlyCAM 1, an endothelial glycoprotein that acts as a ligand for L-selectin (10). Radiolabeled cRNA probes were transcribed from the PCR fragment corresponding to nucleotides 92–458 of the Sgp50 cDNA. **A.** α-[35]S-UTP labeled cRNA antisense probe showing specific hybridization to the high endothelial venule (HEV). **B.** α-[35]S-UTP labeled cRNA sense probe showing lack of hybridization. **C.** α-[33]P-UTP labeled cRNA antisense probe showing a similar pattern of hybridization to the HEV as noted in A. **D.** α-[33]-P-UTP labeled cRNA sense probe showing lack of hybridization. Bar = 50 μm.

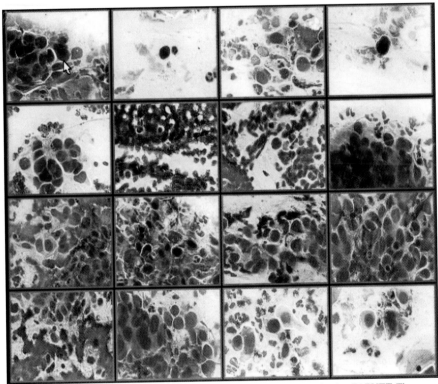

**Figure 1. Summary screen of the first case of carcinoma *in situ* diagnosed using PAPNET.** The smear was screened in September 1993.

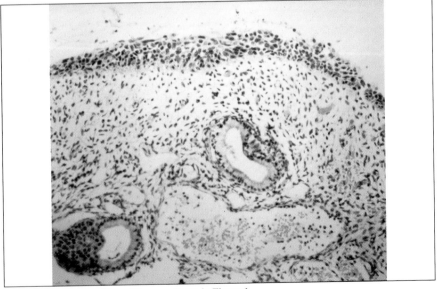

**Figure 2. Histologic section of the case shown in Figure 1.**

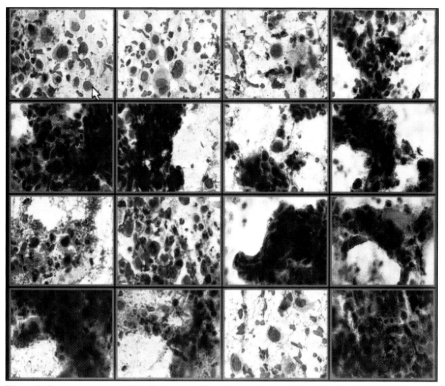

**Figure 3. Summary screen of the first cases of invasive carcinoma detected using PAPNET.** Note that some tiles contain "soft signs," such as necrosis and old blood.

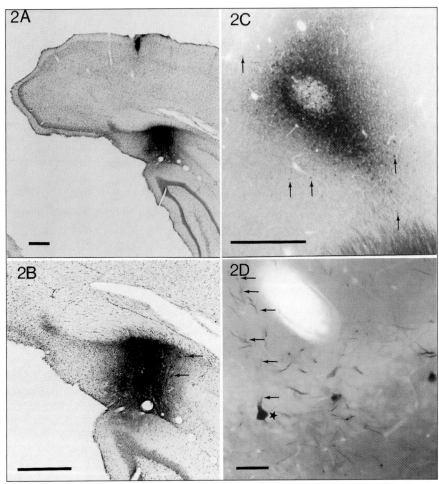

**Figure 2. Color micrographs at low (A) and high (B) magnification, of a typical PHA-L injection site into the dorsal subiculum (hippocampus) of the rat.** Arrows in B point to neurons with well-filled processes at the injection site. A typical biocytin injection site (C) in the piriform cortex of the rat prepared for electron microscopy. Arrows point to neurons labeled at the injection site and (D) shows cells retrogradely labeled in the vicinity of the injection site. This material has also been prepared for electron microscopy. Note the well-filled processes (arrows) of labeled cell (star). Scale bars in C = 500 μm and D = 50 μm.

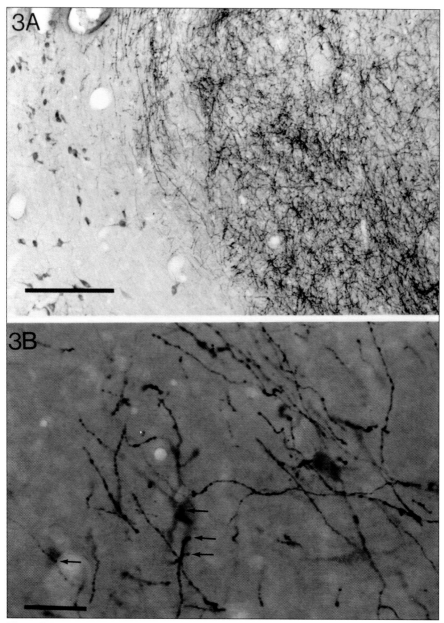

**Figure 3. Material prepared with dual-label immunocytochemistry.** The micrograph in (A) shows PHA-L-labeled subicular fibers (DAB-nickel) and choline acetyltransferase-immunoreactive cells (DAB) in the nucleus accumbens and septum of the rat. The photomicrograph (B) shows close appositions between PHA-L containing varicosities (arrows) and choline acetyltransferase-immunoreactive neurons in nucleus accumbens. Scale bar in A = 250 μm and in B = 50 μm.

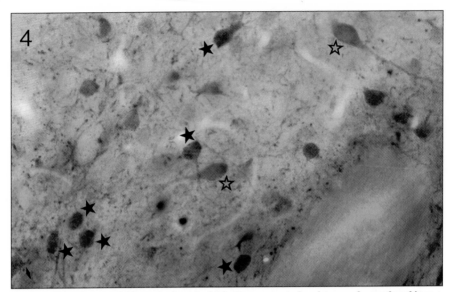

**Figure 4. This micrograph illustrates how dual-label immunocytochemistry can be employed in correlative light and electron microscopy.** This figure illustrates neurons (open stars) immunoreacted for choline acetyltransferase (DAB) and others (filled stars) immunoreactive for leucine[5]enkephalin (SIG). Note the homogeneous staining of the DAB precipitate (open stars) and the coarse grainy appearance of the SIG cells (filled stars). The puncta are presumptive terminals of the two cell types. However, due to their small size, it is impossible to distinguish the DAB- from the SIG-reacted varicosities.

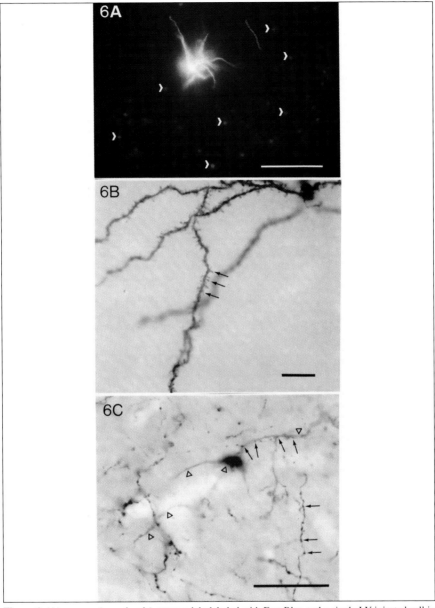

**Figure 6.** (A) neurons (arrowheads) retrogradely labeled with Fast Blue and a single LY-injected cell in nucleus accumbens of the rat. (B) Following immunocytochemistry for LY, dendritic segments complete with spines (arrows) contain the amorphous brown DAB deposit. Note the great detail visible. (C) A neuron and its processes (open arrow heads) that were labeled by Fast Blue from the mesencephalon, intracellularly injected with LY and further immunoreacted, is seen in a sea of blue-black fibers and varicosities labeled with PHA-L (DAB-nickel chromogen). The tracer, PHA-L, was placed in the ventral subiculum of the hippocampus *in vivo*, and transported anterogradely to nucleus accumbens. Note that several PHA-L-labeled varicositites appear juxtaposed to the processes of the intracellularly injected cell. Scale bar in A = 100 μm, in B = 10 μm and in C = 50 μm.

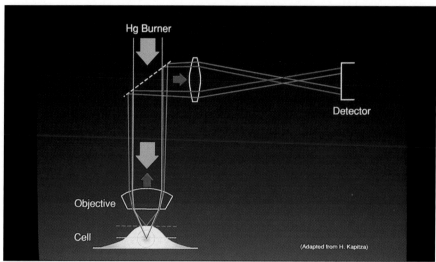

(Adapted from H. Kapitza)

**Figure 1. This ray diagram of an epifluorescent microscope illustrates the cause of reduced signal-to-noise ratio in thick samples.** The incident light (green arrow) irradiates the full thickness of the cell (pink). Fluorochromes that lie outside the plane of focus, such as those in a plane above focus (dashed blue line) emit light (solid blue line), which reaches the Detector. This light, which arises from a plane outside focus, creates a fog that reduces the contrast between fluorochrome and background. Unless the cell is very thin, it is difficult to obtain the highest signal-to-noise ratio, i.e., glowing structure against a black background.

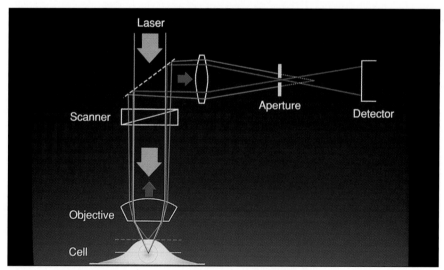

**Figure 2. This is a ray diagram of a laser scanning confocal microscope.** Unlike an epifluorescent microscope, this microscope only irradiates a tiny spot of the specimen. It is necessary to use a galvanometer driven mirror (scanner), to raster the light across the specimen's surface. The irradiation of the specimen by a spot of light increases contrast because the unilluminated regions cannot generate light scatter. Within the optical train there is an aperture whose opening is positioned so that light from the plane of focus (red line) can pass through diaphragm. In contrast, light arising from above the plane of focus (blue line) is intercepted by the opaque borders of the aperture. The application of a point source of light and an appropriately sized and positioned aperture provides a high-contrast image.

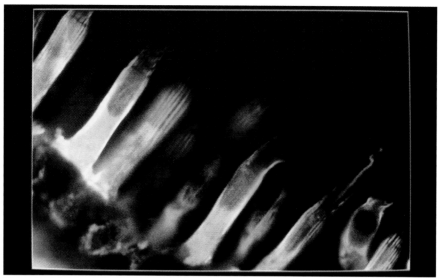

**Figure 3. This is a confocal micrograph of a photoreceptor stained with fluorescein phalloidin and an indirect antibody stain to tubulin (rhodamine).** Actin is pseudocolored green and tubulin pseudocolored red. The cells are in a 100-μm-thick slice of retinal tissue, and the high contrast of the micrograph is typical for a confocal microscope. Fluorescent structures lying above and below the plane of focus do not degrade the focused image. In a standard epifluorsecent microscope, the spots of actin cannot be seen because of light scatter. The concentration of actin in the connecting cilium is extremely minute. The cilium is only 200 nm in diameter. This is a single plane image taken with a Bio-Rad 600.

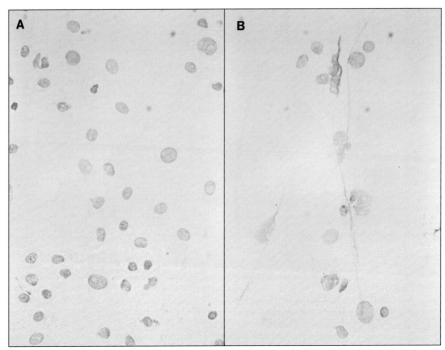

**Figure 1. Medium-power photomicrograph of the nuclei of a poorly differentiated ductal carcinoma of the breast (Feulgen-stained imprint preparation made according to the guidelines).** Most of the neoplastic cell nuclei are well preserved with an intact nuclear membrane (A). There are, however, in the same imprint preparation more vulnerable nuclei (B) showing extensive leakage of the DNA outside. This artifact is visible as Feulgen-stained threads in among the nuclei.

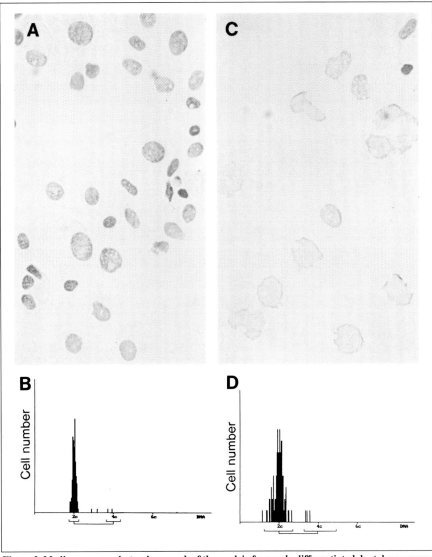

**Figure 2. Medium-power photomicrograph of the nuclei of a poorly differentiated ductal mammary carcinoma (Feulgen-stained imprint preparation and according to the guidelines).** As in Figure 1, the neoplastic nuclei are well preserved, keeping their size and Feulgen stainability during the fast air-drying process (A). The main peak (G1/G0) of the DNA histogram of the population has a certain broadness, calculated as the CV to be 4.1% (B). When the cellular material is placed in fluid, the nuclei swell (C), and the main peak of the DNA histogram of such nuclei shows falsely high CV values (in this very case 12.5%) (D).